Air Pollution Chemistry

Photo: Courtesy Birmingham Post and Mail

Air Pollution Chemistry

J. D. BUTLER

Department of Chemistry
The University of Aston
Birmingham
England

ACADEMIC PRESS · 1979
LONDON · NEW YORK · SAN FRANCISCO
A Subsidiary of Harcourt Brace Jovanovich, Publishers

ACADEMIC PRESS INC. (LONDON) LTD
24/28 Oval Road
London NW1 7DX

United States Edition published by
ACADEMIC PRESS INC.
111 Fifth Avenue
New York, New York 10003

Library of Congress Catalog Number: 78-18677
ISBN: 0-12-147950-1

PRINTED BY J. W. ARROWSMITH LTD., BRISTOL BS3 2NT

Preface

Definition of Air Pollution

The presence in breathable air of chemical elements or compounds in sufficient quantity to constitute injury to health or life over short or long time periods.

Academic scientists are often attracted to working strictly within their own disciplines, indeed this is the way in which fundamental discoveries are made. These discoveries then lead to other developments in pure science or can be applied beneficially to other fields of endeavour. This book is concerned with the latter – it deals with the application of chemistry in all its branches to air pollution.

In this context the topic inevitably becomes multi-disciplinary. The presence of chemicals in air poses quesions about their toxicity, concentration, distribution, stability and influence on climate. Current ideas on these subjects are discussed to introduce the reader to the methods employed which seek to answer these questions. The sequence of topics considered commences with some established relationships between air pollutants and chemical toxicity. This is followed by a consideration of emission sources, the circulation of chemicals on a global scale, air sampling techniques, chemical analysis, chemical reactions in the troposphere and stratosphere, dispersion theory and finally urban atmospheres.

The sections on the global circulation of chemicals, along with chemicals in the troposphere and stratosphere, include an account of the role of the hydroxyl radical in atmospheric reactions. An appreciation of the participation of hydroxyl radicals in atmospheric reactions represents one of the most significant and exciting discoveries that has been made in chemistry in the last decade. Other advances which are important in air pollution are the improvements in analytical techniques which have enabled smaller and smaller concentrations of compounds to be detected and quantitatively assessed. These procedures are mainly dependent on physico-chemical instrumentation and are discussed in some detail in Chapter 4.

The coverage and treatment presented in this book are based on second and third year undergraduate lecture courses to students. However, although designed primarily for chemists, a brief introductory account of the relevant theory has been given where considered necessary. It is hoped that the adoption of this approach will make the text intelligible to other scientists and engineers who are interested in environmental issues. The references given in the text are not intended to be exhaustive. They have been selected for their fundamental contribution to the subject under discussion. Those intending to undertake research topics should find them a valuable introduction to the literature.

During the writing of this book I have received much help from my colleagues in the Department of Chemistry and my students Drs D. R. Middleton, S. D. MacMurdo and D. R. Davis, Mr. P. D. E. Biggins and Mr. P. Crossley. I am especially indebted to Mr. D. M. Colwill of the Transport and Road Research Laboratory, and to Dr. A. C. Chamberlain of AERE; Dr. L. E. Reed of the Central Unit on Environmental Pollution of the DoE; Dr. J. S. S. Rey of Warren Spring Laboratory; Dr. J. McK. Ellison of DHSS and Dr. B. T. Commins, also to Mr. A. Archer of the Environmental Department of the City of Birmingham for discussions on various aspects of pollution problems considered in the text. My thanks to Dr. A. G. Clarke of the Department of Fuel and Combustion Science, University of Leeds, for supplying information on asbestos given in Chapter 4, and my gratitude to the authors and publishers of papers that have been quoted, for permission to reproduce their work. Finally, my thanks to my wife, Margaret, for her patience and encouragement during the writing of this book.

February 1979 J. D. BUTLER

Contents

Preface v

1 Health Factors 1

1.1 Introduction 1
1.2 Symptoms of chronic exposure 2
1.3 Causes 2
1.4 Particulate pollutants 4
1.5 Gaseous pollutants 33
1.6 Radioactivity 39
References 44

2 Sources, Sinks and Removal Mechanisms of Emissions to the Atmosphere 50

2.1 Introduction 50
2.2 Global aspects of pollution–emissions to atmosphere 50
2.3 Details of anthropogenic emissions 63
2.4 Elimination of atmospheric pollutants 102
References 120

3 Air Sampling and Collection 125

3.1 Sampling time 125
3.2 Air sampling principles 132
3.3 Collection of gaseous pollutants 133
3.4 Collection of particulate pollutants 145
References 165

4 Analysis of Pollutants by Instrumental Methods 168

4.1 Trace gas analysis 168
4.2 Trace metal analysis by instrumental methods 193
4.3 Chemical analysis of polynuclear hydrocarbons (PNHs) 206

4.4 Detection and determination of asbestos 249
4.5 Analysis of airborne radioactivity 251
4.6 Statistical methods for the presentation of results 256
References 259

5 Atmospheric Reactions 267

5.1 Introduction 267
5.2 Natural ozone occurrence 270
5.3 Tropospheric chemical reactions 273
5.4 Stratospheric chemical reactions 305
References 315

6 Meteorological Aspects of Pollutant Dispersions 320

6.1 Thermodynamic properties of the atmosphere 320
6.2 Plume dispersion theory 322
6.3 Estimation of plume parameters: standard deviations σ_y and σ_z as a function of downwind distance from source 325
6.4 Applications to specific dispersion problems 329
References 343

7 Urban Atmospheres 345

7.1 Introduction 345
7.2 Particulate aerosol characteristics in urban atmospheres 345
7.3 Gases–urban diffusion models 372
References 377

Author Index 381

Subject Index 395

1

Health Factors

1.1 Introduction

The planet Earth upon which we live is surrounded by an envelope of gas about half of which is concentrated into a lower layer no more than 5 km thick called the troposphere. This layer of gas comprising mainly nitrogen, 78·0%, and oxygen, 20·9% by volume, supports life and sustains man's various activities and civilizations. In the last 150 years or so increasing industrialization with accompanying concentration of population into conurbations has resulted in often severe cases of localized atmospheric pollution.

The fact that this vast envelope of gas would not tolerate for ever man's abuses without retribution was brought home to many people with the advent of the atomic bomb. The extensive publicity following atmospheric contamination by radioactive fall-out produced an international awareness which recognized the inherent dangers of such activities. Subsequently, we have come to realize that other processes previously considered acceptable, such as combustion or chemical synthesis, if carried out on a large enough scale without safeguards may upset the balance of nature. The disciplines of occupational medicine, epidemiology, pharmacology and toxicology are designed to investigate and assess the cause of disease. Much of our knowledge is fragmentary. Although we may know quite a lot about the symptoms and effects of high concentrations of pollutants, we are frequently ignorant of the consequences of long-term low-level exposures.

If we are to pass on to succeeding generations a planet which is inhabitable, we must acknowledge now the need for good husbandry. Science and technology when directed towards this goal can supply the answers to many of our problems. This can be done provided we are prepared to face the consequences of this policy. The more advanced nations of the world must adopt a responsible attitude so that international co-operation is achieved. Air pollution knows no boundaries: it is global in extent.

1.2 Symptoms of chronic exposure

Nausea, vomiting, irritation of eyes, nose and throat, pains and constriction in the chest with coughing, laboured breathing and severe headaches are typical symptoms of people exposed to high air-pollution episodes. Under these circumstances the incidence of asthma and bronchitis can reach epidemic proportions. Amongst older persons, those more than 65 years of age, there is an increase in mortality, particularly from those who already suffer from chest or heart complaints.

1.3 Causes

The worst examples of air pollution disasters are well documented (see Table 1.1). In winter, effluent from industrial and domestic chimneys, and the exhaust from engines have combined under atmospheric conditions of near-freezing temperatures and low windspeeds to produce a low lying stagnant layer of polluted air. In contrast, the photochemical smogs of Los Angeles and other urban areas with Mediterranean-type climates, arise from the interaction of sunlight with motor vehicle exhaust gases.

TABLE 1.1 *Major air pollution disasters in winter*

Location	Date	Duration	Deaths attributed to episode	Reference
Meuse Valley, Liege, Belgium	December 1930	3 days	60	1
Donora, Pa, USA	October 1948	5 days	20	2
London, UK	December 1952	5 days	4000	3
	January 1956	3 days	1000	
	December 1957	3 days	700–800	
	December 1962	5 days	700	
New York, USA	January/ February 1963	15 days	200–400	4
	November 1966	3 days	168	5

TABLE 1.2 *Concentrations of various metals in human blood, μg(100 ml)⁻¹*

Metal	Concentration: Range	Concentration: Mean	Population location	Reference
Hg	0·5–2·05	0·95	Ann Arbor	6
	0·3–5·7	1·43	South American Indians	6
Pb	<1–81	14·6	Ann Arbor	6
	0·0–3·87	0·83	South American Indians	6
		17·7	Pasadena	7
	<1–109·7	13·17	19 different places USA	8[a]
	10–53	29	Japan	10
Cd	<0·1–9·6	1·71	Ann Arbor	6
	0·07–3·72	0·57	South American Indians	6
	0·5–14·16	1·77	19 different places USA	8
	0·34–5·35	0·85		9
	0·5–5·8	1·7	Japan	10
Zn		466	Pasadena	7
	60–1987	530	19 different places USA	8
	780–1600	1200	Japan	10
Cu	31–281	100	Ann Arbor	6
	43–259	99	South American Indians	6
		71	Pasadena	7
	16–348	89	19 different places USA	8
	79–170	110	Japan	10
Cr		10·7	Pasadena	7
	1·28–5·54	2·76		9
	1·6–8·0	4·5	Japan	10
Co		23·8	Pasadena	7
Ni		32·7	Pasadena	7
	0·9–45·5	4·2		9
	4·0–12·0	6·9	Japan	10
Fe		38 800	Pasadena	7
Mn		4·0	Pasadena	7
	1·2–27·0	6·4	Japan	10
Mo		7·0	Pasadena	7
		2·0	Japan	10
V		12·6	Pasadena	7
Al		20·7	Pasadena	7
Ba		6·9	Pasadena	7
Ca		6310	Pasadena	7
Sr		3·9	Pasadena	7
Ag		2·4	Pasadena	7
B	3·9–36·5	11·4		9
As	5·0–6·0	5·6	Japan	10
Sb	1·0–35·0	9·6	Japan	10

Statistically significant correlation coefficients between pairs of elements reported in reference 10.

Total Hg–Cd	0·96	Ni–Sb	0·53
Methyl Hg–Cu	0·88	Cd–Zn	0·62
Total Hg–Zn	0·62	Mn–Ni	0·52
Total Hg-MethylHg	0·56	Cu–Mn	0·51

[a] The 19 different places mentioned comprise 15 companies in the Cincinnati area together with 4 other areas in New York, Denver, Miami and Portland.

1.4 Particulate pollutants

1.4.1 Metals

A whole range of metals and chemical compounds found in the environment are known to be harmful. The elements Cu, Zn, Fe, Ca, Sr, Mn, Mo and Cr are essential to biological processes. Their concentration within the body may vary in normal compared with diseased conditions. Ni, Cd, Ba, Al, Pb and Hg are always present in human tissue, but they are generally

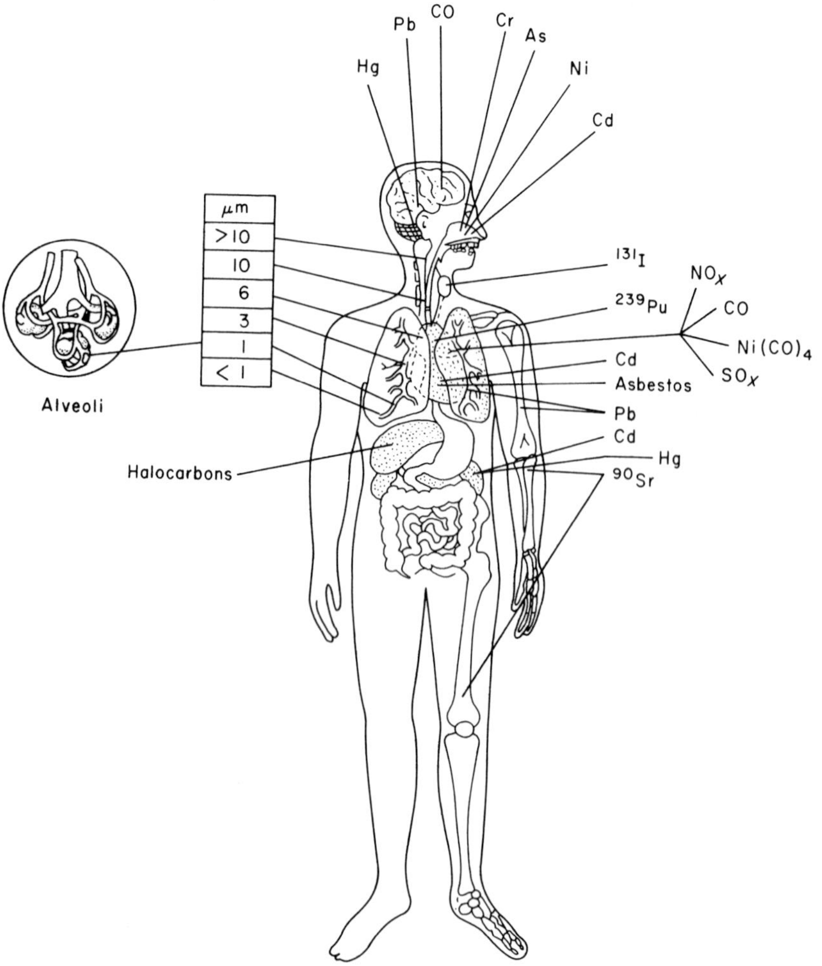

FIG. 1.1 Target areas and organs of the human body for some chemical elements and compounds commonly present in polluted air

regarded as non-essential in the biological sense. Table 1.2 gives values of some elements found in human blood of populations in various parts of the world.

Experience has shown that we are susceptible to particular pollutants which attack at specific sites or organs of the body, and Fig. 1.1 illustrates some of these areas where disease can develop as a consequence of such exposure. Penetration of the respiratory tract by dust and finely divided particulate matter depends upon particle size. Only particles less than 10 μm in diameter reach beyond the trachea and of these only those less than 1 μm get as far as the alveoli (Fig. 1.1). The vital function of the thin-walled alveoli sacs is to allow exchange of gas, namely oxygen, to take place.

Some guide of permissible exposure to metals and chemical compounds is available through the Threshold Limit Value (TLV) concept. These values refer to the maximum levels that should be tolerated assuming an 8 h exposure day^{-1} for a 40 h working week. Table 1.3 indicates the TLVs for some transitional metals known to be present in air along with some reference to their toxicity. The TLVs shown for metals in Table 1.3 and for gases in Table 1.15 are those applicable to the USA, UK, Argentina, Norway and Peru. Other nations set different standards and these have recently been compared by Winell [11] (Table 1.4), for a wide range of chemicals encountered in working environments. Some of the differences between the USSR and USA hygenic standards, and the reasons for them, have been reviewed by Roschin and Timofeevskaya [12].

1.4.1.1 Lead

Of all the heavy metals featured in Table 1.3 increasing concern has been expressed about the quantities of lead present in the environment. This is because, especially in city populations, the blood lead levels of around 20–30 μg (100 ml)$^{-1}$ of whole blood are approaching more closely than any other pollutant, concentrations that are regarded as medically unacceptable (Table 1.5). According to Patterson [29], for instance, the degree of exposure to lead of US citizens constitutes a "chronic lead insult" and are nearing values at which symptoms of chronic poisoning might be expected. Patterson goes on to suggest that these symptoms may manifest themselves in subtle ways, such as an increased irritability on the part of the population. Bryce-Smith [30] and Bryce-Smith and Waldron [31] have highlighted the neurotoxic nature of lead as a poison and the latter have attempted to show a correlation between hyperactivity and criminality. This relationship, however, was based on observations reported by Lob and Desbaume [32] on prisoners in a Lausanne gaol and these authors have now issued a correction to their original paper which admits that their

TABLE 1.3 *Threshold limit values (TVLs) and toxic symptoms of som… heavy metals* [28]

Metal	TLV ($mg\ m^{-3}$)	Toxicity	References
Cd	0·1 fume 0·2 dust	Wilson's disease, extremely hazardous, accumulation in kidneys and lungs, 40–50 $mg\ m^{-3}$ over 1 h fatal; 9 $mg\ m^{-3}$ over 5 h fatal.	See section 1.4.I.2
V	0·05 fume 0·5 dust	Acute bronchospasms, emphysema.	13
Mn	5·0	Aching limbs and back, nervousness, drowsiness and lack of control of bladder action. Can induce pneumonia, nasal congestion and bleeding.	14, 15
As	0·5 0·05 parts 10^{-6} as AsH_3	Can cause cancer of the skin, lungs and liver. Has teratogenic properties, may cause abnormalities in offspring.	16, 17
Hg	0·05 inorganic vapour and salts 0·01 alkyl compounds	Gastrointestinal symptoms, attacks central nervous sytem, tremor and neuro-psychiatric disturbances.	18, 19, 20
Fe	10·0 0·01 parts 10^{-6} as $Fe(CO)_5$	Normally regarded as non-toxic but inhalation may cause siderosis a dust disease which imparts a red discoloration to the lungs.	21, 22
Zn	5·0	Inhalation of ZnO fume causes fever, muscular pain, nausea and vomiting, normal recovery time 24–48 h.	23
Pb[a]	0·15	Performs no biological useful function, can cause anaemia, paralysis of limbs, brain damage and death.	See section 1.4.1.1
Ni	1·0 0·001 parts 10^{-6} as $Ni(CO)_4$	Can cause dermatitis; inhalation can produce cancer of the sinus and lung. 12–36 h after exposure to nickel carbonyl dizziness, nausea and vomiting develop, followed by rapid respiration and possibly death between 4th–12th day after exposure.	22, 24, 25
Cr	1·0	Can cause dermatitis; inhalation can produce lung cancer.	26, 27
Cu	0·1 fume 1·0 dust	A balance between absorption and excretion is essential. When this is disturbed excessive retention can cause disease of the liver and central nervous system.	

[a] See also Waldron, H. A. (1974). The blood lead threshold. *Archives of Environmental Health* **29**, 271–27…

chemical analytical procedures in the first publication for the estimate of lead were in error [33].

Symptoms of lead poisoning are abdominal cramp, constipation, loss of appetite, anaemia, insomnia, irritability, motor-nerve paralysis and encephalopathy. When diagnosed in children it may result in subsequent kidney disease, mental retardation, recurrent seizures, cerebral palsy or optic nerve atrophy, most of which can be attributed to inhibition of growth of the central nervous system [34, 35]. In 1964 Moncrieff *et al.* [36] showed that mentally subnormal children had elevated blood lead concentrations. Two papers, one from the USA by Pueschel *et al.* [37] and the other from the UK by Betts *et al.* [38] demonstrated that there was a relationship between blood lead concentration and haemoglobin concentration in the blood of children. At blood lead concentrations of greater than $40 \mu g$ $(100\ ml)^{-1}$ there seems to be a fall in blood haemoglobin in infants and young children. As Hammond [39] points out, however, there is contradictory evidence showing that children exposed to high lead concentrations do not always react in this manner. This does not mean that the reported results are wrong. It could be that additional factors need to be taken into account. For example, it is known from animal experiments on the rat that an iron deficiency results in a higher concentration of lead in the body. It has been suggested, therefore, that nutritional factors may play some part in the mechanism of lead uptake or retention in the body. It is known that nutritionally deprived children, perhaps in some cases suffering from anaemia, are more susceptible than others, but currently the problem is not resolved.

It has been known for many years that lead compounds are abortifacients and that women working with lead have a high miscarriage rate. Studies by Barltrop [40] have demonstrated that placental transfer of lead begins as early as the twelfth week of gestation and that the total lead content of the fetus increases through pregnancy. Lancranjan *et al.* [41] have recently shown that workmen displaying moderately increased absorption of lead have an alteration in their spermatogenesis which causes a substantial decrease of fertile ability.

The intake of lead by the body comes from lead in food, water and air. Unless special circumstances are involved, such as occupational exposure at work or living in the proximity of a lead manufacturing or processing plant, the lead present in urban air comes from motor vehicles. The majority of lead exhausted from motor vehicles is less than $1\ \mu m$ in diameter and therefore readily penetrates the lung as far as the alveoli (Fig. 1.1), from which transfer to the bloodstream takes place. The distribution of lead retained in the body, together with the biological half-life times for various organs, according to Grundy [42] is given in

TABLE 1.4 *Work environment hygienic standards in different countries* [11]

	USA—Osha (1974) parts 10^{-6}	mg/m^{-3}	BRD 1974 mg/m^{-3}	DDR 1973 mg/m^{-3}	Sweden 1973 mg/m^{-3}	CSSR 1969 mg/m^{-3}	USSR 1972 mg/m^{-3} (c)
Acetaldehyde	200	360	360	100	90	—	5
Acetic acid	10	25	25	20	25	—	5
Acetone	1000	2400	2400	1000	1200	800	200
Acetonitrile	40	70	70				10
Acrolein	0·1	0·25	0·25	0·25	0·25	0·5	0·7
Aldrin		0·25	0·25				0·01
Allyl alcohol	2	5	5	5	5	3	2
Ammonia	50	35	35	25	18	40	20
Ammonium sulphamate	—	15	15	—	—	—	10
Amyl acetate	100	525	525	200	525	200	100
Aniline	5	19	19	10	19	5	0·1
p-Anisidine	0·1	0·5	0·5	—	—	—	1
Antimony and compounds (as Sb)	—	0·5	0·5	0·5	0·5	—	0·3–2
Arsenic and compounds (as As)	—	0·5	0	0·3	0·05	0·3	0·3
Arsine	0·05	0·2	0·2	0·2	0·05	0·2	0·3
Benzene	10	30	0	50	30	50	5
Benzoyl peroxide	—	5	5	—	—	—	5
Benzyl chloride	1	5	5	5	—	—	0·5
Beryllium	—	0·002	0	0·002	0·002	—	0·001
Boron oxide	—	15	15	—	—	—	10
Boron trifluoride	1 (c)	3 (c)	3	—	—	—	1
Bromoform	0·5	5	—	—	—	—	5
1,3-Butadiene	1000	2200	2200	500	—	500	100
2-Butanone	200	590	590	300	440	—	200
Butyl acetate	150	710	950	400	710	400	200
Butyl alcohol	100	300	300	200	150	100	10
Butylamine	5	15	15	—	—	—	10
Cadmium (metal dust and soluble salts)	—	0·2	—	0·1 (a)	0·05	—	0·1
Cadmium oxide fume (as Cd)	—	0·1	0·1	0·1 (a)	0·02	0·1	0·1
Camphor	2	12	2	—	—	—	3
Carbaryl (Sevin)	—	5	5	—	—	—	1
Carbon disulphide	20	60	60	50	30	30	10
Carbon monoxide	50	55	55	55	40	30	20
Carbon tetrachloride	10	65	65	50	65	50	20

Chlorine	1	3	1·5	1	3 (c)	3	1
Chlorine dioxide	0.1	0.3	0.3	—	0.3	—	0.1
Chlorobenzene	75	350	230	50	—	200	50
Chlorodiphenyl (42% chlorine)	—	1	1	1	0·5	1	1
Chlorodiphenyl (54% chlorine)	—	0·5	0·5	1	0·5	0·5	1
Chloroprene	25	90	90	10	90	50	2
Chromic acid and chromates (as Cr)	—	0·1 (c)	0·1	0·1	0·05	0·05	0·01
Cobalt, metal fume and dust	—	0·1	0·5	0·1	0·1	0·1	0·5
Copper, fume	—	0·1	0·1	0·2 (b)	—	—	1
Copper, dusts and mists	—	1	1	—	—	—	1
Crotonaldehyde	2	6	6	—	—	—	0·5
Cumene	50	245	245	50	—	—	50
Cyclohexane	300	1050	1050	—	—	—	80
Cyclohexanone	50	200	200	—	—	—	10
Cyclopentadiene	75	200	200	—	—	—	5
2,4-D	—	10	10	—	—	—	1
DDT	—	1	1	1	—	—	0·1
Dibutylphthalate	—	5	—	—	—	—	0·5
o-Dichlorobenzene	50 (c)	300 (c)	300	150	—	—	20
p-Dichlorobenzene	75	450	450	200	—	—	20
Dichlorvos (DDVP)	0·1	1	1	—	—	—	0·2
Dieldrin	—	0·25	0·25	—	—	—	0·01
Diethylamine	25	75	75	50	—	—	30
Diethylamino ethanol	10	50	50	—	—	—	5
Diisopropylamine	5	20	—	10	—	—	5
Dimethylamine	10	18	18	—	—	—	1
Dimethylaniline (*N*-dimethylaniline)	5	25	25	—	—	—	0·2
Dimethylformamide	10	30	60	30	30	30	10
Dinitrobenzene	0·15	1	1	1	—	1	1
Dinitro-*o*-cresol	—	0·2	0·2	0·2	—	—	0·05
Dinitrotoluene	—	1·5	1·5	1	—	—	1
Dioxane	100	360	360	200	90	—	10
Epichlorhydrin	5	19	18	5	—	—	1
Ethyl acetate	400	1400	1400	500	1100	400	200
Ethyl alcohol	1000	1900	1900	1000	1900	1000	1000
Ethyl amine	10	18	18	20	—	—	1
Ethyl bromide	200	890	890	500	—	—	5
Ethyl chloride	1000	2600	2600	2000	—	—	50
Ethyl ether	400	1200	1200	500	1200	300	300
Ethyl mercaptan	10 (c)	25 (c)	1	—	—	—	1
Ethylene chlorohydrin	5	16	16	—	—	—	0·5

TABLE 1.4—*continued*

	USA—Osha (1974) parts 10^{-6}	mg/m^{-3}	BRD 1974 mg/m^{-3}	DDR 1973 mg/m^{-3}	Sweden 1973 mg/m^{-3}	CSSR 1969 mg/m^{-3}	USSR 1972 mg/m^{-3} (c)
Ethylene diamine	10	25	25	—	—	—	2
Ethylene imine	0·5	1	1	1	0	—	0·02
Ethylene oxide	50	90	90	20	36	—	1
Fluoride (as F)	—	2·5	2·5	—	2·5	1	1
Formaldehyde	2	3	1·2	2	3 (c)	2	0·5
Furfural	5	20	20	10	—	—	10
Heptachlor	—	0·5	0·5	—	—	—	0·01
Hydrazine	1	1·3	0·13	—	0·13	0·1	0·1
Hydrogen chloride	5 (c)	7 (c)	7	5	7 (c)	8	5
Hydrogen cyanide	10	11	11	5	11	3	0·3
Hydrogen fluoride	3	2	2	1	2 (c)	1	0·5
Hydrogen sulphide	20 (c)	30 (c)	15	15	15	10	10
Iodine	0·1 (c)	1 (c)	1	—	—	—	1
Isopropylamine	5	12	12	—	—	—	1
Lead, inorganic fumes and dusts	—	0·2	0·2	0·15	0·1	0·05	0·01
Lindane	—	0·5	0·5	0·2	—	—	0·05
Maleic anhydride	0·25	1	0·8	—	1	1	1
Manganese and compounds (as Mn)	—	5 (c)	5	5	2·5	2	0·3
Mercury, metal	—	0·1 (c)	0·1	0·1	0·05	0·05	0·01
Mercury, alkyl	—	0·01	0·01	0·01	0·01 (c)	—	0·005
Methyl acetate	200	610	610	200	—	200	100
Methyl acrylate	10	35	35	20	—	—	20
Methyl alcohol	200	260	260	100	260	100	5
Methyl amine	10	12	12	—	—	—	1
Methyl bromide	20 (c)	80 (c)	80	50	—	—	1
Methyl chloride	100	210	105	100	—	100	5
Methyl chloroform	350	1900	1080	500	540	500	20
Methyl cyclohexane	500	2000	2000	—	—	—	50
Methyl isocyanate	0·02	0·05	0·05	—	—	—	0·05
α-Methyl styrene	100 (c)	480 (c)	480	—	—	—	5
Methylene chloride	500	1740	1750	500	350	500	50
Molybdenum, soluble compounds	—	5	5	—	—	—	4
Molybdenum, insoluble compounds	—	15	15	10	—	—	6

Morpholine	20	70	70	—	—	—	0·5
Naphtha (coal tar)	100	400	—	—	—	200	100
Naphthalene	10	50	50	20	—	—	20
Nickel carbonyl	0·001	0·007	0·7	—	0·007	—	0·0005
Nickel, metal	—	1	0	0·5	0·01	—	0·5
p-Nitroaniline	1	6	6	—	—	—	0·1
Nitrobenzene	1	5	5	5	5	5	3
p-Nitrochlorobenzene	—	1	1	1	—	1	1
Nitroethane	100	310	310	—	—	—	30
Nitrogen dioxide	5	9	9	10	9 (c)	10	5
Nitromethane	100	250	250	—	—	—	30
1-Nitropropane	25	90	90	50	—	—	30
2-Nitropropane	25	90	90	50	—	—	30
Ozone	0·1	0·2	0·2	0·2	0·2	0·1	0·1
Pentachlorophenol	—	0·5	0·5	0·5	0·5	—	0·1
2-Pentanone	200	700	700	—	—	—	200
Perchloroethylene	100	670	670	300	200	250	10
Phenol	5	19	19	20	19	20	5
Phosgene	0·1	0·4	0·4	0·5	0·2 (c)	0·4	0·5
Phosphine	0·3	0·4	0·15	0·1	0·4	0·1	0·1
Phosphorus (yellow)	—	0·1	0·1	—	—	0·03	0·03
Phthalic anhydride	2	12	5	10	12	5	1
Propargyl alcohol	1	2	2	—	—	—	5
n-Propyl acetate	200	840	840	400	—	400	200
Propyl alcohol	200	500	—	—	—	500	10
Propylene dichloride (1,2-Dichloropropane)	75	350	350	50	—	—	10
Propylene oxide	100	240	240	—	—	—	1
Pyridine	5	15	15	10	15	5	5
Quinone	0·1	0·4	0·4	—	—	—	0·05
Selenium compounds	—	0·2	0·1	0·1	0·1	—	0·1
Sodium hydroxide	—	2	2	2	2 (c)	—	0·5
Stoddard solvent	500	2950	—	—	600	—	300
Styrene	100	420	420	200	210	200	5
Sulphur dioxide	5	13	13	10	5	10	10
Sulphuric acid	—	1	1	1	1	1	1
Tellurium	—	0·1	0·1	—	—	—	0·01
1, 1,2, 2-Tetrachloroethane	5	35	7	10	—	—	5
Tetraethyl lead (as Pb)	—	0·075	0·075	0·05	0·075	—	0·005
Tetrahydrofuran	200	590	590	200	—	—	100
Tetranitromethane	1	8	8	—	—	—	0·3
Thallium	—	0·1	0·1	—	—	—	0·01

TABLE 1.4—*continued*

	USA—Osha (1974) parts 10^{-6}	mg/m^{-3}	BRD 1974 mg/m^{-3}	DDR 1973 mg/m^{-3}	Sweden 1973 mg/m^{-3}	CSSR 1969 mg/m^{-3}	USSR 1972 mg/m^{-3} (c)
Thiram (tetramethylthiuramdisulphide)	—	5	5	1	—	—	0·5
Toluene	200	750	750	200	375	200	50
Toluene-2,4-diisocyanate	0.02 (c)	0.14 (c)	0·14	0·1	0·07 (c)	0·07	0·5
α-Toluidine	5	22	22	10	—	5	3
Trichloroethylene	100	535	260	250	160	250	10
1, 2, 3-Trichloropropane	50	300	300	—	—	—	2
Triethylamine	25	100	100	20	—	—	10
Trinitrotoluene	0·2	1·5	1·5	1·5	—	1	1
Triorthocresylphosphate	—	0·1	—	0·1	—	—	0·1
Turpentine	100	560	560	300	560	—	300
Uranium, soluble compounds (as U)	—	0·05	0·05	—	—	—	0·015
Uranium, Insoluble compounds (as U)	—	0·25	0·25	—	—	—	0·075
Vanadium, V_1O_5 dust (as V)	—	0·5 (c)	0·5	0·5	0·5	—	0·5
Vanadium, V_1O_5 fume (as V)	—	0·1 (c)	0·1	0·1	0·05 (c)	—	0·1
Vinyl chloride	1	3	—	500	3	—	30
Vinyl toluene	100	480	480	—	—	—	50
Xylene	100	435	870	200	435	200	50
Xylidine	5	25	25	10	—	5	3
Zinc oxide fume	—	5	5	5	5	5	6
Zirconium compounds (as Zt)	—	5	5	—	—	—	4–6

(a) as CdO (b) as CuO (c) ceiling value
Abbreviations: Federal Republic of Germany, BRD; German Democratic Republic, DDR; Czechoslovakia CSSR; US standards also apply to Argentina, Norway, Peru and UK.
Osha: Occupational Safety and Health Administration.

Table 1.6. Other estimates, however, by Rabinowitz *et al.* [43], indicate that only about 28% of the total lead assimilated by the body can be derived from the atmosphere. Although most cases of lead poisoning reported in children arise from pica (the habit of chewing articles unfit for food, especially when they are coated with old lead paint), the importance of inhalation of the metal is undeniable and cannot be overlooked.

For obvious reasons, among them the possible impairment of mental performance, there is a reluctance on the part of volunteers to participate

TABLE 1.5 *A guide to categories of lead absorption taken from the* British Medical Journal **4**, 501 (1968)

Test	A normal	B acceptable	C excessive	D dangerous
Blood lead $\mu g\ (100\ ml)^{-1}$	40	40–80	80–120	120
Urinary lead $\mu g\ l^{-1}$	80	80–150	150–250	250
Urinary coproporphyrin $\mu g\ l^{-1}$	150	150–500	500–1500	1500
δ-aminolaevulinic acid $mg(100\ ml)^{-1}$	0·6	0·6–2·0	2–4	4

(a) Workmen affected by plumbism are not permitted to return to work until their blood lead is $<50\ \mu g\ (100\ ml)^{-1}$

(b) Clinical lead poisoning has not been observed below $80\ \mu g\ (100\ ml)^{-1}$ of blood in adults or $50\ \mu g\ (100\ ml)^{-1}$ in children.

TABLE 1.6 *Lead parameters in the human body* [42]

	Total body	Bone	Liver	Kidney
Intake $g\ day^{-1}$	4×10^{-4}			
Average concentration $g\ (g\ wet)^{-1}$	$1{\cdot}1\times10^{-6}$	10^{-6}	2×10^{-6}	$1{\cdot}4\times10^{-7}$
Biological half-life days	1460	3650	1947	531
Fraction in organ of reference of total in body	1·0	0·7	0·1	0·05
Fraction from blood to organ of reference	1·0	0·28	0·08	0·14
Fraction reaching organ by ingestion	0·08	0·02	0·0064	0·01
Fraction reaching organ by inhalation	0·29	0·08	0·23	0·04

in experiments on humans. Many of our suspicions regarding the toxic effects of lead in air originate from animal experiments. There is a paucity of data relating to humans, particularly the effects of long-term low-level exposure.

Further complicating features of lead toxicity are that blood lead concentrations show seasonal variations [44]. On average blood lead concentrations in the spring are some 3–4 μg $(100\ \text{ml})^{-1}$ higher than in the autumn. This is believed to be related to solar radiation through the synthesis of vitamin D and the deposition of calcium in bone. Also, lead present in bone can be mobilized to reappear in the blood under conditions of stress or illness.

Information based on epidemiologial evidence has been given by Goldsmith and Hexter [45] that the relationship between blood lead concentration and atmospheric lead is given by the empirical equation,

$$\log_{10} \text{BLL} = 1{\cdot}265 + 0{\cdot}2433 \log_{10} \text{PBL} \tag{1.1}$$

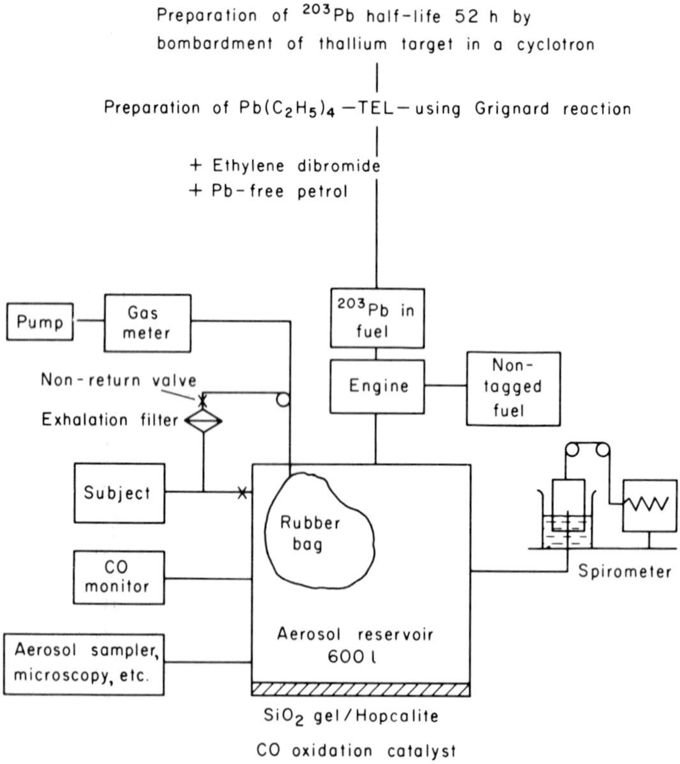

FIG. 1.2 Schematic representation of Chamberlain's experiment with inhaled lead-203 to discover the fate of lead taken up by the human body [47]

where BLL is the blood lead concentration in μg $(100\ \mathrm{ml})^{-1}$ of whole blood and PBL is the atmospheric lead concentration in $\mu\mathrm{g\,m^{-3}}$. This equation gives reasonable predictions in agreement with the blood lead concentrations and respiratory exposure to lead found by Kehoe [46].

More recently an investigation by Chamberlain *et al.* [47] was undertaken which employed radioactive lead-203. Lead labelled with lead-203 was converted into lead tetraethyl which was then added to petrol and used to fuel an internal combustion engine. Exhaust gas from this engine was inhaled and the percentage deposition in human lung, uptake by the blood and excretion of labelled lead was measured. Figure 1.2 shows the arrangement of equipment employed. Exhaust gas from a 4-stroke engine running on petrol having between 0·5 and $20\ \mathrm{g\,l^{-1}}$ additions of tagged lead-203 was fed into an aerosol reservoir of 600 l capacity. Since this gas contained carbon monoxide it could not be inhaled immediately. An oxidation catalyst comprising a mixture of silica gel and "Hopcalite" was spread over the bottom of the reservoir to convert carbon monoxide to carbon dioxide. Subjects inhaled gas from the chamber and exhaled through a filter before the exhaled gas was returned to an inflatable bag contained in the chamber. This procedure enabled a mass balance to be kept and checked on the spirometer.

Analysis of samples for tagged lead was accomplished by γ-ray counting to detect the 279 keV radiation from lead-203. Figure 1.3 shows the γ-ray spectrum recorded with a whole body counter. The full curve is the spectrum before inhalation of lead-203 and includes contributions from ambient background radiation and fall-out activity of the body. The broken curve shows the increased response in the 0–500 keV region measured one day after inhalation. Samples of blood, urine and faeces were examined by this technique for periods of up to 14 days after intake.

From a knowledge of the amounts of lead-203 inhaled and exhaled the experiments revealed that the average percentage deposition in the lung was $35 \pm 2\%$ for a 4 s cycle (i.e. 15 breaths $\mathrm{min^{-1}}$). Most of the lead-203 was removed from the lung with a half-life of about 6 h and roughly half of the amount deposited in the lung was in the blood some 50 h after inhalation. Peak concentration in samples of venous blood was observed between 25 and 85 h after inhalation. Once present in the blood, lead became distributed throughout the body, and consequently, since inhalation was not continuous, the blood lead level eventually decreased. This decrease had a biological life-time of 16 days.

The main conclusion from this work relating uptake of lead by respiration with blood lead was as follows. If particulate lead in urban areas is around $1\ \mu\mathrm{g\,m^{-3}}$, then for a 24 h exposure in a day for a subject breathing $15\ \mathrm{m^3}$ of air of which 35% of the lead is retained by the lung and 50% of this lung burden is transferred to the erythrocytes, the contribution to the blood lead assuming a blood volume of 5.4 l is given by:

$$\frac{16 \times 0{\cdot}35 \times 0{\cdot}5 \times 15}{54 \times 0{\cdot}693} = 1{\cdot}1\ \mu g\ (100\ ml)^{-1}$$

The factor, 0·693 in the denominator converts the radioactive half-life time to the mean life in a biological system.

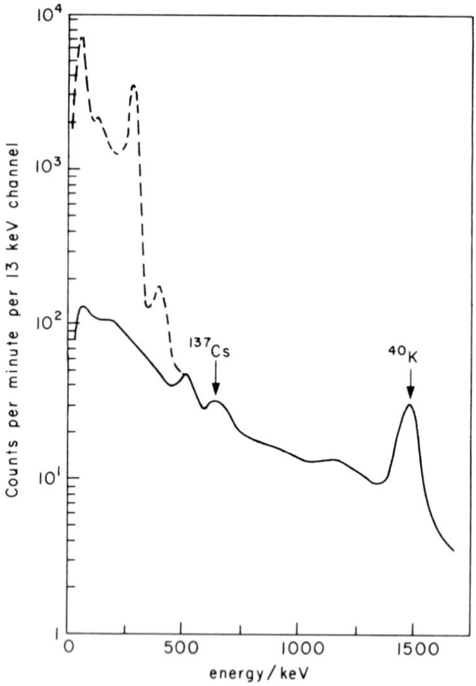

FIG. 1.3 Gamma-ray spectra recorded by whole-body counter with subject M.J.H. present. The full curve is the spectrum before inhalation of ^{203}Pb and includes contributions from ambient background radiation as well as from natural and fall-out activity in the body. The broken curve shows the increased response in the 0–500 keV region measured one day after inhalation of ^{203}Pb [47]

A rather different correspondence between airborne lead and blood lead was obtained by Knelson *et al.* [48]. In their penitentiary studies subjects were exposed to around 3 $\mu g\ m^{-3}$ and 10 $\mu g\ m^{-3}$ of lead for 23 h in every day for a period of 18 weeks. On the basis of this prolonged exposure equation (1.2) was derived:

$$\mathrm{IBL} = 0{\cdot}327L + 3.236 + (0{\cdot}914L + 0{\cdot}85)\log_{10}(L\ V R D) \times 10^{-3} \qquad (1.2)$$

where IBL is the increase in blood lead μg $(100\,\text{ml})^{-1}$, L is the air lead concentration in $\mu\text{g m}^{-3}$, V is the pulmonary ventilation in $\text{m}^3\,\text{day}^{-1}$, R is the fraction of inhaled lead that is retained and D is the duration of exposure in days. This equation predicts that quite substantial increases in blood lead may occur for respiratory doses well below those commonly encountered in ambient air.

Despite numerous investigations there is still some disagreement about how much lead present in the atmosphere ultimately enters the body or how much should be tolerated in food and drink. The WHO study of 1968 of blood lead concentrations of populations of different countries shown in Table 1.7 gives an aggregate value of 17 μg $(100\,\text{ml})^{-1}$ with a standard deviation of 11 from 801 samples. Furthermore, although the sample size was conspicuously small, there is no evidence from these data that airborne lead contributed significantly to the results. The mean value given in Table 1.7 of 23 μg $(100\,\text{ml})^{-1}$ for the UK or 21 μg $(100\,\text{ml})^{-1}$ for residents of

TABLE 1.7 *Blood lead concentrations of populations in different countries, μg $(100\,ml)^{-1}$*

Country	Sample Size	Range	Mean	Standard Deviation
Argentina	49	5–50	14	8
Chile	35	5–30	17	6
Czechoslovakia	20	0–50	20	8
Egypt	28	0–50	20	13
Finland	46	10–60	26	12
Israel	67	5–40	15	5
Italy	26	5–40	13	10
Japan	40	5–60	20	8
Holland	60	0–45	15	7
Peru	32	0–35	7	5
Poland	76	0–35	12	6
Sweden	30	5–25	10	3
Yugoslavia	46	5–60	24	16
UK	30	0–45	23	13
USA:				
California	33	10–27	17	17
New York	105	0–10	21	18
Ohio	40	10–53	16	8
New Guinea	38	10–40	22	5
TOTAL	801			
AGGREGATE			17	11

Source: Barry, P. S. I. (1975) *Post-graduate Medical Journal* **51**, 783–787.

New York City compare very favourably with those of 22 μg $(100\ \text{ml})^{-1}$ found in New Guinea. In the UK, the highest blood lead levels of around 40 μg $(100\ \text{ml})^{-1}$ found for half the residents of Aberystwyth [49], were originally attributed to airborne lead from old mine workings. However, this conclusion was shown to be invalid [50]: in this instance the culprit appears to be lead in the water supply.

On the basis of all the evidence available to the EEC regarding lead*, draft legislation currently proposes the following air quality standards:

1. An annual mean concentration of not more than 2 $\mu g\ m^{-3}$ of Pb in urban residential areas and areas exposed to sources of atmospheric lead other than that derived from motor vehicle traffic.
2. A monthly median concentration of not more than 8 $\mu g\ m^{-3}$ in areas particularly exposed to motor vehicle traffic.

This schedule was adopted by the Council of Ministers on 22 November 1973.

For additional information the reader is referred to references 51, 52 and 53.

1.4.1.2 Cadmium

Cadmium has no known biological function; it inhibits the performance of enzymes containing SH groups [54], and produces hypertension in humans [55, 56]. Cadmium poisoning was responsible for the outbreak of "itai-itai byo", literally "ouch-ouch disease", in Japan. The victims suffered multiple bone fractures and were often found to walk with a painful characteristic waddling gait. Death was generally attributed to kidney failure. The cause was traced to a lead–zinc mining operation that emitted cadmium fumes, the fall-out from which contaminated rice paddy fields.

Autopsies performed in the USA show that about 30 mg of cadmium are present in the body, some 33% in the kidneys, 14% in the liver, 2% in the lungs and 0·3% in the pancreas [57]. Since cigarettes contain cadmium—about 30 μg per packet of 20, of which approximately 70% is vaporized in the smoke—smokers carry a greater body burden of the metal than do non-smokers [58, 59].

In 1966 Carroll [60] published evidence of a relationship between cadmium in air and cardiovascular disease. Air pollution data were taken from 28 cities compiled by the National Air Sampling Network of the US Public Health Department and the standardized mortality ratios (SMR) were calculated from the county death records in which the cities were located.

* The US Environmental Protection Agency proposals for ambient air quality for lead, dated December 1977, stipulate that concentrations should not exceed 1·5 $\mu g\ m^{-3}$ on a monthly basis. Individual states must develop plans for EPA approval which demonstrate how they will attain this standard by 1982 and maintain it thereafter.

The ratio was compared to the total US death rate for the particular cause of disease. Thus, a ratio of SMR = 100 indicates that the county's mortality experience is the same as the average US rate. The results shown in Table 1.8 give a correlation coefficient $R = 0{\cdot}76$ between cadmium in air and SMR.

TABLE 1.8 *Atmospheric cadmium concentrations and age-, sex-, and race-standardized mortality ratios for diseases of the heart except rheumatic diseases, 1959–1961* [60]

City	Cd in air (ng m^{-3})	County-standardized mortality ratio
		SMR
Las Vegas	0	66·6
Eugene, Ore.	0	83·5
Medford, Ore.	1	96·7
Chattanooga, Tenn.	1	87·6
Albuquerque	1	71·7
Omaha	2	101·0
Gary, Ind.	3	103·0
Los Angeles	4	95·4
Oklahoma City	6	75·8
Phoenix, Ariz.	6	82·5
Akron, Ohio	6	95·2
Racine, Wis.	6	109·0
Wilmington, Dl.	6	115·4
Tucson, Ariz.	7	81·0
Youngstown, Ohio	7	99·0
Cincinnati	7	111·2
Canton, Ohio	9	103·9
Scranton, Pa.	11	135·2
New York	13	115·3
Columbus, Ohio	14	103·2
Charleston, W. Va.	18	115·8
Newark, N.J.	18	119·3
Indianapolis	19	102·5
Waterbury, Conn.	20	105·3
Bethlehem, Pa.	21	118·5
Philadelphia	23	127·6
Allen Town, Pa.	37	116·4
Chicago	62	123·8

The correlation between cadmium in air and disease, together with data on other atmospheric pollutants also obtained by Carroll, are given in Table 1.9. Besides heart disease except those of rheumatic origin, arteriosclerotic heart disease and hypertensive disease also show significant correlations. Zinc is the only other pollutant which correlates significantly with heart disease but since cadmium and zinc are closely related chemically and correlate ($R = 0.63$) together in air such a relationship is only to be expected. There is perhaps some comfort in noting that where cadmium-induced cancers of the testis have been diagnosed zinc salts have been prescribed to inhibit or prevent such disease conditions [61].

TABLE 1.9 *Rank correlation coefficients of age-, sex-, and race-standardized mortality ratios with atmospheric pollutants* [60]

Correlation of Cd in air and:	
	R
Heart disease except rheumatic	0·76
Arteriosclerotic heart disease	0·67
Hypertensive disease	0·61
Vascular lesions of central nervous system	−0·21
Chronic nephritis	0·28
Respiratory system cancer	0·38
Correlation of heart diseases except rheumatic and:	
Zinc in air	0·56
Lead in air	0·39
Chromium in air	0·31
Suspended particulate matter in air	0·08
Benzene soluble organic compounds in air	0·03
Correlation of Cd in air and:	
Zinc in air	0·63
Lead in air	0·29
Chromium in air	0·32
Suspended particulate matter in air	0·0
Benzene soluble organic matter in air	0·13

A relationship between air pollution by cadmium and diseases of the heart was also found in a statistical study by Hickey *et al.* [62] and in this investigation airborne vanadium was found to be implicated as well.

1.4.2 Asbestos

The name asbestos covers a class of naturally occurring fibrous silicates that can be classified into chrysotile $Mg_3Si_2O_5(OH)_4$, which is by far the most important since it accounts for 95% of all asbestos used, and amphiboles. In the amphibole classification only two types are of commercial interest: crocidolite (blue asbestos) $3Na_2O{\cdot}6FeO{\cdot}2Fe_2O_3{\cdot}16SiO_2{\cdot}H_2O$, and amosite which is an iron-rich form of anthophyllite $(Mg, Fe^{2+})_7Si_8O_{22}(OH)_2$. Other metals, chromium, cobalt and nickel, can be present in trace amounts.

Structurally chrysotile is built up from an infinite silica sheet of $(Si_2O_5)_n$ in which all the silica tetrahedra point in the same direction [62–67]. Magnesium hydroxide units attached to one side of the silica sheet produce strain which is relieved by curvature. This gives rise to a hollow cylindrical morphology (Fig. 1.4) that may be likened to a coiled roll of carpet.

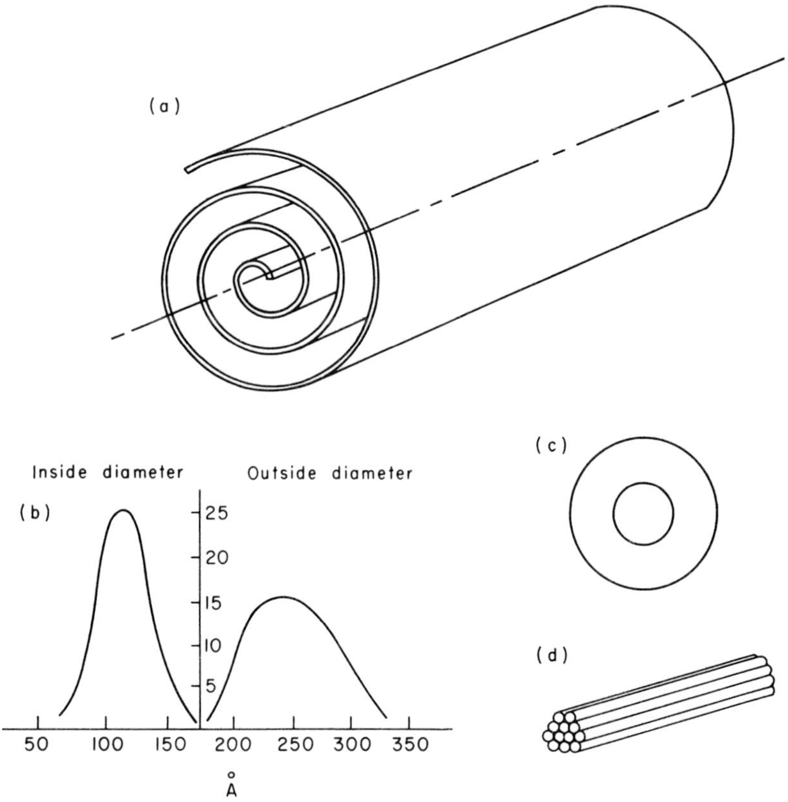

FIG. 1.4 Morphology of chrysotile. (a) Strain-induced curvature of silica sheet; (b) Frequency distribution of the inside and outside diameters of fibrils; (c) Cross-section through fibril; (d) Fibril cluster [68]

Asbestos finds wide application at the present time. It is extensively used in the building industry to strengthen cement and plasters as well as being used for heat insulation and sound absorption purposes. Clutch and brake linings of motor vehicles contain asbestos, as does fire-proof protective clothing. In addition, it is used on a large scale as a filter aid to assist in the removal of finely divided particulates or colloidal matter during filtration of liquids [69].

Reports first appeared of an association between asbestosis and lung cancer in the UK [70] and in the USA [71] in 1935. Epidemiological evidence was presented later [72, 73] that asbestos textile workers in the UK who had worked in the industry before 1930 showed a 10-fold excess risk of contracting lung cancer. Mesothelioma, a tumour which seldom occurs except in cases of asbestosis, has been shown by electron microscopy to be associated with amphiboles imbedded in lung tissue. In an investigation by Pooley [74] 65 cases were examined: 33 consisted of pleural and peritoneal mesothelioma, and the remaining 32 were control cases. Asbestos fibres were more abundant in the mesothelioma cases than in the controls. The most abundant fibres detected in the mesothelioma group were amphibole asbestos while chrysotile was more predominant in the controls. Amphibole asbestos fibres greater than 20 μm in length were easily discernible; chrysotile fibres tend to be shorter.

Asbestos fibres are known to exist in city atmospheres [75] but according to Gross [76] there is no direct evidence, at the present time, that general air pollution within urban districts contributes to an increase in mesotheliomas. Current available information on the health hazard aspect arises from an association with mining [77], manufacture or residence in the vicinity of a manufacturing plant or some other occupational exposure [78]. One of the confounding features of the disease is the time lapse between exposure and clinical symptoms of the disease. Usually a time span of between 30 and 40 years is involved. Furthermore, some comparatively recent evidence suggests that besides lung cancers, asbestos may also be responsible for cancers within the abdomen [79, 80].

1.4.3 Polynuclear hydrocarbons

There is no information as such on the sole effect on disease of low concentrations of polynuclear hydrocarbons present in polluted air. However, circumstantial evidence obtained over many years incriminates polynuclear hydrocarbons as cancer-producing agents in man (see Table 1.10). This is further supported by evidence from tumour production by these compounds in experimental animals. Because of this association and the fact that the incidence of lung cancer and smoking are related, especially for heavy smokers and smokers living in urban environments, efforts have been directed towards establishing the amount and the type of polynuclear hydrocarbons present in cigarette, pipe and cigar smoke, and community air. The possibility of a relationship between death rate with rural and urban regions was recognized as long ago as 1936 by Stocks [81]. Data presented in 1955 indicated that differences in lung cancer death rates in various age groups residing in rural, mixed and urban districts could not be attributed to smoking habits alone (Table 1.11). It should be noted that the heavier the smoker and the older the smoker the less marked the differences between rural and urban habitat. Very similar results relating the incidence of persistent coughing, production of phlegm,

TABLE 1.10 *The incidence of skin and scrotal cancers as an occupational hazard associated with polynuclear hydrocarbons*

Occupational Hazard	Reference
Wax processmen after prolonged exposure to slack or crude wax	85, 86
High temperature distillation of coal and contact with pitch and tar products	87
Spinners in textile industry in contact with lubricating oils	88,. 89
Mineral oils used as lubricating and cutting oils used by machine tool operators	90, 91

TABLE 1.11 *Comparison of lung cancer death rates per 10^5 men per year in rural, mixed and urban areas classified according to smoking habit*

Smoking category	Ages 45–54			Ages 55–64			Ages 65–74		
	Rural	Mixed	Urban	Rural	Mixed	Urban	Rural	Mixed	Urban
Non-smokers	0	0	31	0	0	147	71	0	335
Pipe smokers	0	0	104	34	59	143	145	26	232
Cigarette:									
Light	69	57	112	70	224	378	154	259	592
Moderate	90	83	138	205	285	386	362	435	473
Heavy	117	214	205	626	362	543	506	412	583

Source: Stocks, P. and Campbell, J. M. (1955). Lung cancer death rates among non-smokers and pipe and cigarette smokers. An evaluation in relation to air pollution by benzpyrene and other substances. *British Medical Journal* **2**, 923–929.

and chronic bronchitis with smoking and age have been obtained by Lambert and Reid [84]. A number of other surveys have subsequently suggested that environmental factors associated with air pollution may contribute to the disease [82, 83] Standardized mortality ratios from death due to lung cancer increase with increasing population density. This general conclusion is illustrated in Table 1.12 with reference to urban and rural areas in the USA.

Another disturbing feature arises from the higher incidence of the disease among British-born immigrants in New Zealand [92], South Africa [93], Australia [94], and the USA [95] compared with the indigenous population of these countries; and the UK has the highest lung cancer death rate in western Europe (Fig. 1.5) [96]. It seems unlikely that this increased susceptibility can be ascribed to smoking habits alone. The suggestion has been made, therefore, that exposure to environmental factors peculiar to Britain before emigration could be responsible.

TABLE 1.12 *Comparison of lung cancer death rates in Ohio with urbanization. Adjusted rates, white men, 25–64 years of age, 1947–1951*

Population area	Standardized mortality ratio
Eight metropolitan counties (cities of 10^5 population)	1·23
Seven other urbanized counties (50 000+population)	0·82
Rural counties	0·69

Source: Mancuso, T. F., MacFarlane, E. M. and Porterfield, J. D. (1955). Distribution of cancer mortality in Ohio. *American Journal of Public Health* **45**, 58–70.

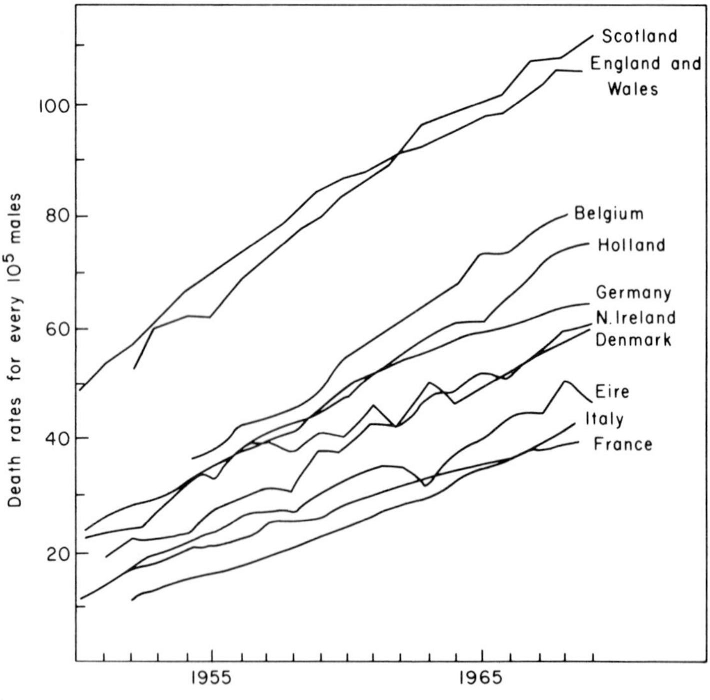

FIG. 1.5 Comparison of the number of male deaths from lung cancer in Western European countries.

It appears from these surveys that death from lung cancer can be influenced by smoking and air pollution. The next stage must be to establish the relative contributions that smoking and air pollution make to the disease. Carnow and Meier [97] have employed multiple regression analysis using information available on cigarette smoking, smoke pollution, and mortality due to lung cancer to obtain an estimate of the relative importance of these factors. In adopting such procedures one must acknowlege limitations in the model due to an assumption that the equations used are realistic in their representation of the system. Also, the terms used in the equations are subject to wide fluctuations owing to the difficulty of obtaining accurate data. Regression studies can always be criticized on these grounds; nevertheless, they can offer an empirical understanding and a useful guide.

On the assumption that death rate is related to both cigarette smoking and solid fuel consumption and that the effects are additive, the equation

$$Y = C_0 + C_1X_1 + C_2X_2 \tag{1.3}$$

may be written, where C_1 measures the increment of death rate per unit increase in cigarette consumption and C_2 measures the increment per unit increase in solid fuel consumption, Y equals the age-sex-specific lung cancer death rate, X_1 represents cigarette consumption in thousands of cigarettes per person over 15 years of age and X_2 represents the solid fuel consumption in tonnes per person per year. The age adjusted value for Y per 10^6 men, C_0, C_1 and C_2 were 749·3, 343·6, 109·1 and 142·1, respectively. These results have been interpreted by noting that the cigarette smoking coefficient C_1 is approximately 15% of the average death rate. In other words, this indicates an increment in the male deaths of 15% per 10^3 cigarettes per year. Converting the rate per 10^3 cigarettes per year to a rate per cigarette per day (i.e. $15 \times 365/10^3$) gives 5·5% increase in lung cancer death rate per cigarette per day. Thus, the smoking of one packet of 20 cigarettes per day is roughly equivalent to doubling the lung cancer death rate. For solid-fuel consumption the regression coefficient C_2 is about 19% per tonne of coal consumed per person.

Since combustion of coal produces benzo(a)pyrene, a potent carcinogen (see section 2.3), a relationship between atmospheric benzo(a)pyrene and coal consumption would be helpful for establishing a relationship between lung cancer death rate with the benzo(a)pyrene content of the air. This has been done and is expressed as an index of 1 ng m^{-3} benzo(a)pyrene unit. As a rough guide, evidence from epidemiological studies suggests that an increase of 1 benzo(a)pyrene unit corresponds to an increase of 5% in the lung cancer death rate. So far the discussion in this section has centred on polynuclear hydrocarbons, but once the influence of air pollution on disease is admitted, this immediately complicates the issues because the atmospheric aerosol contains many other ingredients. For example, experiments with animals have verified that exposure to low concentrations of benzo(a)pyrene administered with iron III oxide dust [98], or sulphur dioxide [99], produces a carcinogenic system more potent than that of the polynuclear hydrocarbon alone. If substantiated in humans this adds another dimension to the problem. The recognition that atmospheric particulates previously regarded as harmless can act as co-carcinogens to enhance the activity of known carcinogens present in urban air, immediately questions air quality standards currently thought to be acceptable.

From what has been said, it is clear that exogenic factors from air pollution sources, besides contributing to disease patterns associated with high pollution short period episodes already discussed in section 1.3, may also have an influence on disease from long-term low-level exposure spanning many years. Furthermore, it is likely that air pollution contributes synergistically to the development of diseases associated with smoking.

1.4.4 Chlorinated hydrocarbons and dioxin

Chlorinated hydrocarbons are used as insecticides, fungicides and herbicides for agricultural and horticultural purposes. Inevitably, therefore, they enter the human food chain and this is the principal route by which these compounds enter the body. An inhaled intake from typical atmospheric concentrations of 10–30 ng m^{-3} is at least two orders of magnitude below the intake from total diet [100]. Nevertheless, the fact that they are sprayed on crops in aerosol form, and that they are globally distributed in the troposphere as well as being returned to the Earth's surface during rainfall, clearly justify their inclusion as an atmospheric contaminant. Some of the more widely used chlorinated hydrocarbons are listed in Table 1.13.

Most of the information on the toxicity of DDT comes from animal experiments. It collects in fatty tissue in insects and animals, causing convulsions, degeneration of the heart muscles and disruption of the kidney function. Another chlorinated hydrocarbon called hexachlorophene used in fungicides for vegetables and citrus fruits is also used as an additive in hair shampoos and cosmetic preparations. It was used medically for the treatment of burns and for washing newborn babies as a precaution against staphylococcal disease, but both these applications are now discouraged [101].

In the late 1960s documented accounts of the harmful effects of human exposure to chlorinated hydrocarbons began to appear. In southern Japan in 1968 more than 1000 persons who had eaten rice oil contaminated by

TABLE 1.13 *The chemical name, the abbreviated name and chemical formula of some chlorinated hydrocarbons in common use*

Chemical name	Abbreviated name	Chemical formula
1,1,1-trichloro-2,2-bis(*p*-chlorophenyl)-ethane	*p,p'*-DDT	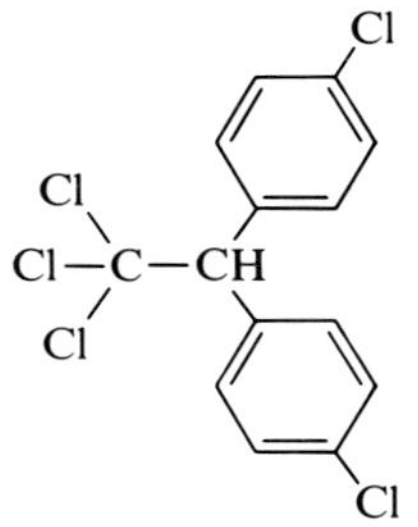
1,1,1-trichloro-2-(*o*-chlorophenyl)-2-(*p*-chlorophenyl)-ethane	*p'*-DDT	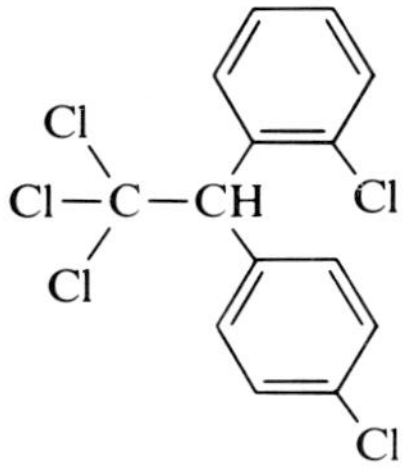
2,2-bis(*p*-chlorophenyl)-1,1-dichloroethane	TDE	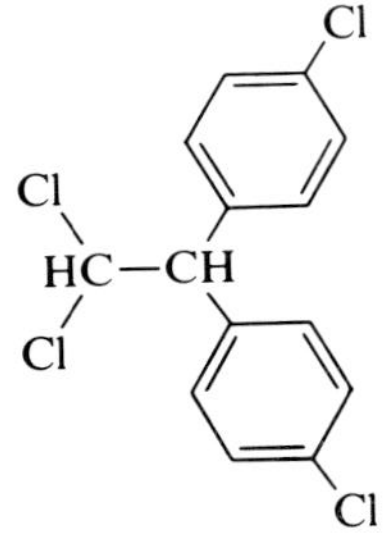

TABLE 1.13—*continued*

Chemical name	Abbreviated name	Chemical formula
1,1-dichloro-2,2-bis-(*p*-chlorophenyl)-ethylene	DDE	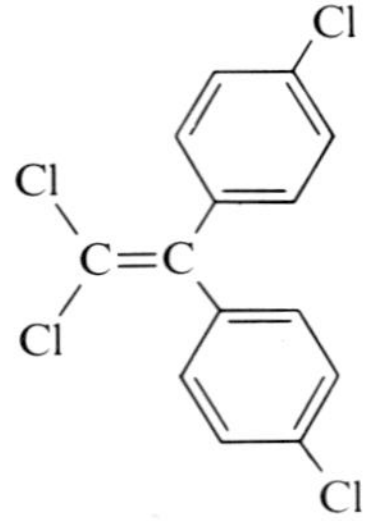
1,1-dichloro-2-(*o*-chlorophenyl)-2-(*p*-chlorophenyl)-ethylene	DDE	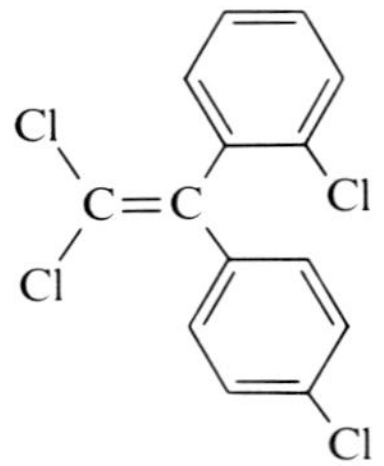
1,2,3,4,10,10-hexachloro-1,4,4a,5,8,8a-hexahydro-1,4 endo-exo-5,8-dimethanonaphthalene	Aldrin	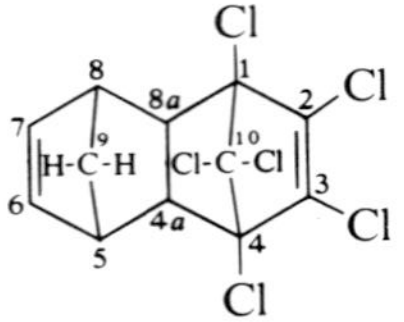

1,2,3,4,5,6-hexachlorocyclohexane (benzene hexachloride)	BHC	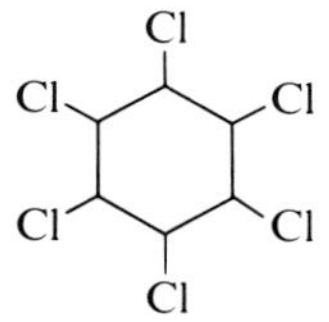
1,2,4,5,6,7,8,8-octachloro-2,3,3a,4,7,7a-hexahydro-4,7-methanoindane	Chlordane	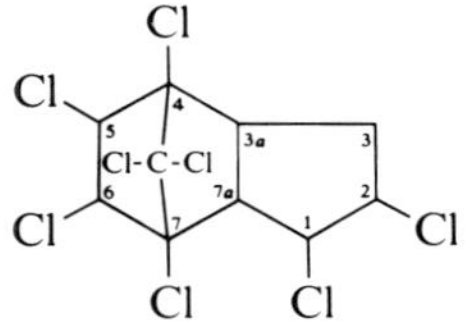
2,4-dichlorophenoxyacetic acid	2,4-D	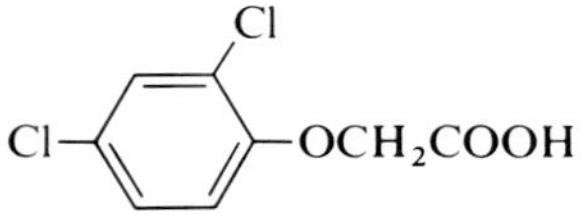
1,2,3,4,10,10-hexachloro-6,7-epoxy-1,4,4a,5,6,7,8,8a-octahydro-1,4-endo-exo-5,8-dimethanonaphthalene	Dieldrin	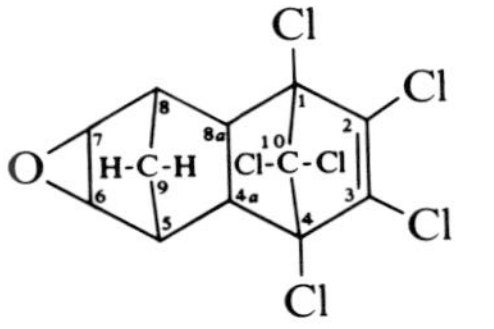
1,2,3,4,10,10-hexachloro-6,7-epoxy-1,4,4a,5,6,7,8,8a-octahydro-1,4-endo-exo-5,8-dimethanonaphthalene	Endrin	Configuration of methano-bridges different from that in Dieldrin

TABLE 1.13—*continued*

Chemical name	Abbreviated name	Chemical formula
1,4,5,6,7,8,8-heptachloro-3a,4,7,7a-tetrahydro-4,7-methanoindene	Heptachlor	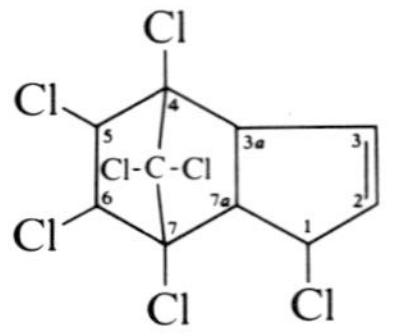
1,4,5,6,7,8,8-heptachloro-2,3-epoxy-3a,4,7,7a-tetrahydro-4,7-methanoindene	Heptachor epoxide	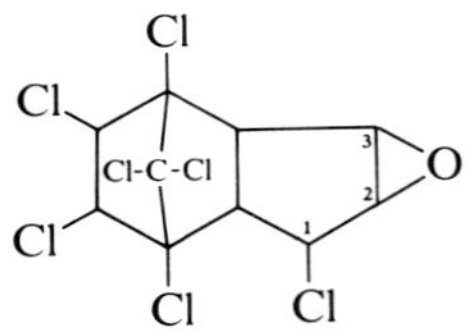
3,5,3′,5′,-tetrachlorobiphenyl	PCB	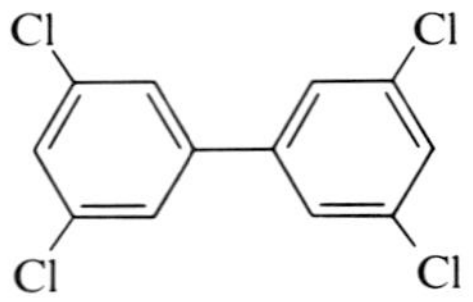

Bis(2-hydroxy-3,5,6-trichlorophenyl)methane	Hexachlorophene	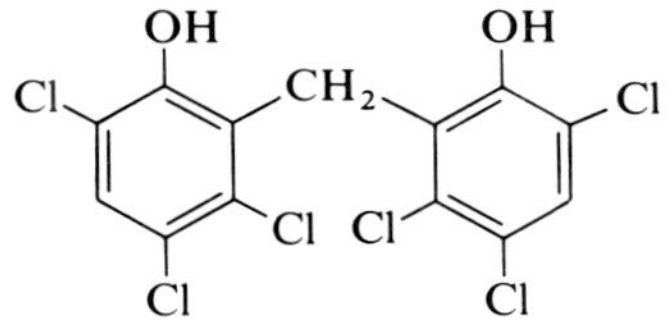
2,3,7,8-tetrachloro-dibenzo-*p*-dioxin	TCDD	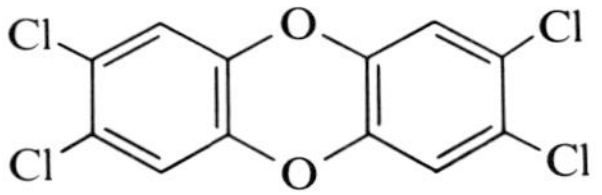

polychlorbiphenyls (PCBs) that had leaked from a manufacturing plant developed darkened skins, eye irritation and an acute skin complaint called chloracne. The outbreak was called "yusho oil disease". The symptoms became noticeable after consumption of about 0·5 g of PCB. It is not known whether any of the subsequent deaths of the affected population could be attributed to acute PCB poisoning or whether the commercial PCB mixture involved contained dibenzofuran as an impurity [102]. In 1969 an accidental discharge of 2,3,7,8-tetrachlorodibenzo *p*-dioxin (TCDD) occurred at a factory in Derbyshire, England, and resulted in 79 cases of chloracne [103].

The compound TCDD is formed as a by-product in the manufacture of trichlorophenol which itself is an intermediate in the manufacture of hexachlorophene and 2,4,5-trichlorphenoxyacetic acid (2,4,5-T), a herbicide [104]. TCDD is a potent teratogen (an agent which causes birth defects). It has an LD_{50} (the dose that kills 50% of a population within 14 days) when fed to male guinea-pigs of 0·5–1·0 $\mu g\ kg^{-1}$ of body weight; the LD_{50} is 22·3 $\mu g\ kg^{-1}$ for male rats and 45 $\mu g\ kg^{-1}$ for female rats. By comparison strychnine is two orders of magnitude less toxic. As little as 20 μg applied to human skin is sufficient to cause chloracne. In the Italian Seveso disaster in the summer of 1976 TCDD was released into the atmosphere to contaminate the local population and the countryside in the vicinity of the chemical plant where the accident happened. In view of the teratogenic properties of the pollutant much anxiety has been expressed regarding the offspring of pregnant women who were exposed. Only time will tell the full consequences of the accident. If malformations do occur

TABLE 1.14 *Concentrations of chlorinated hydrocarbons present in human blood taken from residents in Sweden and Japan*

		Concentration			
Nationality	Place of residence	BHC	DDE ($ng\ g^{-1}$)	DDT	PCB
Swedish	Stockholm	0·5	6·9	2·1	10·3
Swedish	Stockholm, 1970 Tokyo, 1972	4·7	10·0	4·5	11·0
Japanese	Stockholm, 1968–70 Tokyo, 1972	7·8	7·2	3·8	4·9
Japanese	Tokyo	7·5	10·7	3·6	4·9
Japanese	Southern Japan	30·3	20·8	7·9	21·8
American	Japan, 1964–68 Sweden, 1968–72	9·4	24·0	6·0	14·0

Table adapted from Jensen [105].

positive proof that TCDD was responsible may be difficult to obtain. Inhuman as it may seem, a statistical analysis which compares births before and after the accident may be the only way of demonstrating a cause and effect relationship in such instances.

Owing to their relative stability and global distribution, chlorinated hydrocarbons are present in human blood. Jensen [105] has shown by comparing the blood concentrations of subjects resident in Sweden and Japan that once these compounds are present in the bloodstream then they exist in the blood for several years (Table 1.14). The case of the American citizen given in Table 1.14 who lived in Japan from 1964 to 1968 and then moved to Sweden from 1968 to 1972, well illustrates this point. Even after this time period his blood BHC concentration of 9·4 ng g^{-1} still remains at the higher level in keeping with Japanese populations.

For additional information and reviews on polychlorinated hydrocarbons references 106 and 107 are recommended.

1.5 Gaseous pollutants

Table 1.15 lists the TLVs of the most important gaseous pollutants together with their toxic features.

1.5.1 Carbon monoxide

Carbon monoxide is generated during the incomplete combustion of organic matter. About 80% of the carbon monoxide present in urban atmospheres is believed to originate from motor vehicle exhausts. Because the gas is colourless and odourless, with the same density as air, it is particularly insidious, causing asphyxiation by forming meta-stable compounds with haemoglobin in the blood. Carboxyhaemoglobin levels of the order of 60% COHb saturation are normally fatal. This is equivalent to an exposure of about 10^3 parts 10^{-6} for 1 h. Lower concentrations of 40% saturation can also be lethal for very young or elderly people. The health effect of exposure is actually dependent upon the carbon monoxide content of the blood, the partial pressure of oxygen in the air breathed, the duration of exposure, the temperature, the working rate or the exertion of the individual, and general fitness. The equilibrium concentration of carbon monoxide reacting with haemoglobin of the blood for men at work is substantially complete in 6–8 h. Equilibrium concentrations in the blood in relation to exposure of carbon monoxide are given in Table 1.16. Carbon monoxide has an affinity for haemoglobin some 200 times that of oxygen, so that besides being stable COHb only dissociates relatively slowly. This effect restricts oxygen supply in the blood to body tissue.

TABLE 1.15 *Threshold limit values of some gaseous pollutants* [28]

Compound	TLV	Toxicity	Comments
CO	50 parts 10^{-6} 55 mg m^{-3}	Combines to form carboxyhaemoglobin with blood causing restricted availability of oxygen to cellular systems of the body.	Kerbside concentrations of 10 parts 10^{-6} exceeded for 7% and 25% of the time in summer and winter, respectively, for flow of 2000 vehicles h^{-1}. Exceptional values as high as 77 parts 10^{-6} have been reported. Background levels <1 parts 10^{-6}.
SO_2	5 parts 10^{-6} 13 mg m^{-3}	Irritation of mucous membranes, may impair ciliary movements in respiratory tract.	Values >1000 μg m^{-3} of smoke and SO_2 for >2 days indicative of disastrous pollution episodes. Winter values within conurbations range between 55–168 μg m^{-3}, corresponding smoke values fall in the range 48–139 μg m^{-3}.
NO	25 parts 10^{-6} 30 mg m^{-3}	High exposure (2500 parts 10^{-6}) in animals induces fatal paralysis and convulsions. Forms methaemoglobin with blood and thereby reduces oxygen carrying capacity.	Values around 1 part 10^{-6} in congested motor traffic. Kerbside concentrations 0·1–0·2 parts 10^{-6}. 0·02–0·04 parts 10^{-6} typical urban background.
NO_2	5 parts 10^{-6} 9 mg m^{-3}	13 parts 10^{-6} causes irritation of respiratory mucous membrane, 100–150 parts 10^{-6} after 30–60 min fatal due to oedema of the larynx.	At concentrations of NO 50 parts 10^{-6} the atmospheric oxidation to NO_2 occurs very slowly, Urban daytime NO_x consists mainly of NO.
O_3	0·1 parts 10^{-6} 0·2 mg m^{-3}	Influences lung function causing pulmonary congestion, oedema and haemorrhage, bronchitis and emphysema.	0·15–0·25 parts 10^{-6} (300–500 μg m^{-3}) hourly average in some USA cities

Urban air contains higher concentrations of carbon monoxide than any other gaseous pollutant. Peak concentrations of 120 parts 10^{-6} in close proximity to motor traffic have been detected, but 10–20 parts 10^{-6} are frequently found in city centres and less than 2 parts 10^{-6} are typical of residential areas. Elevated concentrations can occur in crowded confined places like inns and taverns [108]; the origin of this carbon monoxide is cigarette smoke. In fact, inhaled cigarette smoke contains 200–800 parts 10^{-6} of carbon monoxide, while smoke from pipes and cigars contains more. Smokers consuming around 20 cigarettes per day show elevated COHb levels of between 5 and 7% in their blood. During light activity the half-life of COHb is about 3 h but this rises to about 7 h during sleep. According to Castleden and Cole [109] this is the reason smokers awaken with high COHb levels in their blood. Inhalation of tobacco smoke is by far the biggest cause of personal pollution by carbon monoxide [110]. The resulting COHb elevation in the blood exceeds that which arises from breathing carbon monoxide in community air derived from vehicle exhaust emissions.

TABLE 1.16 *Relation between carbon monoxide concentration in air, equilibrium carboxyhaemoglobin blood content and carboxyhaemoglobin blood content with exposure time*

Concentration of CO in air (parts 10^{-6})	Equilibrium carboxyhaemoglobin blood content %COHb	COHb in blood (%) exposure time			
		30 min		60 min	
		Rest	Work	Rest	Work
30	5	0·27	1·00	0·42	2·00
50	8	0·45	1·65	1·06	3·30
125	20	1·12	4·12	2·64	8·24
250	40	2·25	8·24	4·50	16·48

Symptoms of carbon monoxide poisoning occur when COHb levels in the blood reach 10–20%. Nausea, headaches and fatigue occur at 30–40%, loss of memory and muscular control result at 50–60%, 70–89% causes death within a few hours, 80–90% within an hour and greater than 90% within minutes. In the USA, about one-half of the fatal cases of poisoning are due to carbon monoxide.

1.5.2 Sulphur dioxide

Dry air at sea level contains more carbon monoxide, hydrogen, methane and nitrogen dioxide than sulphur dioxide (Table 1.17). However, since sulphur is present in naturally occurring fossil fuels, when the fuel is burnt this sulphur is converted into sulphur dioxide. The severest air pollution episodes summarized in Table 1.1 have all been associated with elevated concentrations of sulphur dioxide and smoke particulate. The maximum concentration of sulphur dioxide of 3830 $\mu g\ m^{-3}$ (1·46 parts 10^{-6})* attained in the London air pollution disaster of 1952 is less than the 5 parts 10^{-6} TLV. It is worthwhile to point out that in extreme cases of high pollution it is possible that particulate concentrations become so high that the filters are blocked and this reduces the flow of air through the filter. In such circumstances the recorded value estimated by gas absorption (section 3.3 and Table 3.4) will be lower than the true value. Nevertheless, repeated occurrences of elevated sulphur dioxide and smoke concentrations have been blamed for increased mortality in the London incident and elsewhere (Table 1.18). Accordingly, it has been suggested that periods of high

TABLE 1.17 *Minor constituents of clean dry air*

Compound	Concentration (parts 10^{-6})
CH_4	1·5
H_2	0·5
CO	0·1
O_3	0·02
NO_2	0·001
SO_2	0·0002

TABLE 1.18 *Concentrations of smoke and sulphur dioxide in London during air pollution episodes* [111]

		Concentration in air $\mu g\ m^{-3}$			
		SO_2		Smoke	
Date	Duration	Highest day	Highest hourly	Highest day	Highest hourly
1952 Dec. 5–9	5	3830		4460	
1953 Jan. 20–21	2	1670		1480	
Mar. 3–4	2	2179		1660	
Mar. 24–25	2	1687		1370	
1956 Jan. 4–6	3	1430	—	2830	9700
1957 Dec. 3–5	3	1335	4200	2417	7200
1959 Jan. 28–30	3	1850	4570	1723	3980
Feb. 16–19	4	1584	4460	1486	2690
Nov. 12–13	2	1467	3260	1280	3230
1962 Dec. 3–7	5	3834	5650	3244	4700

At 1 atm pressure and 298 K, parts $10^{-6} = \dfrac{\mu g m^{-3} \times 0{\cdot}0244}{\text{molecular weight}}$.

pollution should be defined as periods of two or more consecutive days during which the daily mean concentrations of smoke and sulphur dioxide exceed 10^3 $\mu g\,m^{-3}$ [111]. Analysis of deaths for New York metropolitan region [112] with daily sulphur dioxide measurements have shown that pollution significantly contributed to the prediction of deaths. By using regression analysis it was established that mortality was 1·5% less than expected on 232 days with sulphur dioxide levels below 30 $\mu g\,m^{-3}$ and 2% greater than expected on 260 days with sulphur dioxide levels above 500 $\mu g\,m^{-3}$. It was also shown in this study that the coefficient of haze could be used instead of sulphur dioxide as a measure of atmospheric pollution.

Although epidemiological and circumstantial evidence implicates sulphur dioxide as a hazard under the circumstances outlined in the previous paragraph, definite experimental proof is more difficult to obtain. In the first place it is known that the solubility of sulphur dioxide ensures that lung penetration is restricted to the upper respiratory tract. The physiological response of the lungs to sulphur dioxide as reflected by measurements of pulmonary flow resistance (PFR) have been reported by Frank [113]. Exposure through the mouth at the 5 parts 10^{-6} level for 10 min produces an average increase in PFR of 39%, and at the 13 parts 10^{-6} level gives a 72% increase in PFR. At the same time it was shown that the effect on PFR was dependent upon the time of exposure and whether sulphur dioxide was inhaled through the nose or mouth. The maximum increase in PFR occurs after 10–15 min exposure; after that, the response decreases. The effect of inhalation through the nose produces an increase in PFR of rather less than 20%, much less than through the mouth.

A later study by Wolff *et al.* [114] on sulphur dioxide and tracheobronchial clearance in man shows, at least for healthy adults at rest, that exposure to 5 parts 10^{-6} had little effect. These experiments suggest that the real toxicological danger from sulphur dioxide experienced in urban atmospheres may arise not from sulphur dioxide itself but through its oxidation to sulphur trioxide and conversion to sulphuric acid and sulphate aerosols [115].

1.5.3 Nitrogen oxides

The presence of oxides of nitrogen, NO_x (i.e. a mixture of NO_2 and NO), in town air is very largely derived from internal combustion engines. When released into the atmosphere nitric oxide oxidizes to nitrogen dioxide, but at concentrations of less than 50 parts 10^{-6} this oxidation takes place slowly so that NO_x present in polluted air is mainly nitric oxide. Nitrogen dioxide is not very soluble in water; during respiration the gas passes through the relatively dry trachea and bronchi before reaching the moist alveoli where it is converted into a mixture of nitrous and nitric acids. Information from epidemiological studies among infants and school children suffering from high incidence of acute bronchitis indicates that exposure to nitrogen dioxide levels of 0·062–0·109 parts 10^{-6} (117–205 $\mu g\, m^{-3}$), 24 h mean, for a 6 month period could well be responsible. As an air quality standard for health an annual arithmetic mean value of 100 $\mu g\, m^{-3}$ should not be exceeded. Nitric oxide occurs in cigarette, pipe and cigar tobacco smoke where concentrations of 300, 950 and 1200 parts 10^{-6}, respectively, have been reported from these sources.

Photolytic reactions between the oxides of nitrogen and hydrocarbons in polluted air form the basis of photochemical "smog" reactions. These are discussed in some detail in Chapter 5. From the occupational hazard point of view, oxides of nitrogen always arise whenever strong nitric acid contacts organic matter and many metals. Perhaps less obviously, oxides of nitrogen are produced during welding from oxy-gas burners and from processes which involve the "shrinking on" of one metal component onto another, especially when such operations are carried out in a confined or poorly ventilated space [116].

1.5.4 Ozone

In experimental animals 4 h exposures to 1–3 parts 10^{-6} of ozone proves fatal. At lower concentrations a whole range of ailments develop including bronchitis, broncho-pneumonia, abscesses of the lung, and lung tumours. Photochemical oxidant concentrations of 0·1 parts 10^{-6} (200 $\mu g\, m^{-3}$) produce eye irritation and hourly average values of 0·15 parts 10^{-6} (300 $\mu g\, m^{-3}$) increases asthmatic attacks in humans.

Like the oxides of nitrogen, ozone is always present in photochemical smog episodes. Ozone damages plants where, after entering the leaf through the stomata, it specifically attacks the pallisade cells and causes them to collapse. This sensitivity of plants to ozone has been used in the USA as an indicator of ozone pollution. Tobacco plants are particularly prone to damage and were used in the experiment of Menser and Heggestad [117] to show that different air pollutants, in this case sulphur dioxide and ozone, can have a synergistic influence on each other (Table 1.19).

Few experiments have been performed on humans. An exception is the work of Hazucha *et al.* [118] where normal subjects, smokers, and non-smokers were exposed to 0·37 or 0·75 parts 10^{-6} of ozone for 2 h in an environmental chamber. Total lung capacity was not significantly affected by the exposure, but the residual volume increased. This increase is closely related to the changes in the closing volume and indicates an early effect in the small airways. It was concluded that a concentration of 0·37 parts 10^{-6} of ozone for a period of 2 h is unacceptably high if impairment of pulmonary function is to be avoided in a normal active population. In line with the conclusions summarized in Table 1.19 that ozone and sulphur dioxide together are more harmful than either gas alone, Hazucha and Bates [119] have recently demonstrated similar trends in human pulmonary function after exposure to these gases.

TABLE 1.19 *Ozone and sulphur dioxide synergism in damage to tobacco plants after 2 h of exposure*

Toxicant parts $\times 10^{-8}$		Percentage number of leaves damaged	
		Tobacco variety	
O_3	SO_2	Bel-W3	Bel-B
3·0	—	0	0
—	24·0	0	0
2·7	24·0	38	25

1.6 Radioactivity

Man has always been exposed to natural background radioactivity. This comes from:

(1) isotopes of uranium and thorium which are distributed in trace amounts in rocks and soils;

(2) cosmic radiation;

(3) the radionuclides, tritium, beryllium-7, carbon-14, sodium-22, silicon-32, phosphorus-32, phosphorus-33, phosphorus-35, sulphur-35, chlorine-36, sulphur-38, chlorine-38 and chlorine-39 produced in the earth's atmosphere by cosmic ray bombardment [120] and from potassium-40 which is of terrestrial origin. Additional exposure at the present time comes from man-made radiations associated with:

(1) the production, the processing of nuclear fuels and their use for power generation;

(2) the use of radioactive isotopes in industry, medical and scientific research;

(3) nuclear weapon testing;

(4) medical X-rays for diagnostic and treatment purposes; and
(5) high-altitude air travel.

Before considering the relative importance of these radiations from the health hazard aspect, it is pertinent to summarize the units currently used to measure radiation.

1.6.1 Units

The dose of radiation received by tissue depends on the quantity of energy absorbed per unit mass of tissue. This quantity of energy will depend on the number of atoms that have disintegrated and the energy released by each disintegration within the tissue. The fundamental unit of radioactivity, the curie, is defined as the amount of an element that produces $3{\cdot}7\times10^{10}$ disintegrations per second (dps). The amount of radiation absorbed by tissue is measured in rad, where 1 rad corresponds to 100 erg of energy per gram of tissue. This quantity is approximately equal to the energy absorbed when 1 roentgen (R), the X-ray dose unit, is received by one gram of soft tissue. The roentgen is conventionally defined as that amount of X-radiation which produces 1 esu of ions in 1 cm^3 of air at s.t.p., where 1 esu = $3{\cdot}3\times10^{-10}$ coulomb.

The extent of damage induced by radiation in biological systems will depend upon the type of radiation as well as the tissue irradiated. To take these factors into account the roentgen-equivalent-man (rem) has been introduced. This specifies that the ionization must be equivalent to X-rays of 250 keV energy and that the radiation must be measured by absorption and ionization within the tissue concerned. Under certain circumstances and for certain biological effects 1 rad of α-radiation may be about 10 times as damaging as 1 rad of X-radiation. In such cases the α-radiation is said to have a relative biological efficiency of 10 and 1 rad of α-radiation will be equivalent to 10 rem of radiation exposure. For β-, γ-, and X-radiation the relative biological efficiency is normally unity and the dose in rem is equal to that in rad.

1.6.2 Effects of radiation

Table 1.20 shows the effects of acute ionizing radiation over the whole body. A 50% chance of fatality occurs for a dose of 400 rem. Obviously, this is an extreme situation and represents an upper undesirable level to which few would be exposed except in the case of accident or nuclear war. Of more relevance to the population at large is information on the extent of exposure from unavoidable natural sources already mentioned, and any additional contributions likely to be encountered.

1.6.3 Internal radiation

Carbon-14 and potassium-40 are two unavoidable sources of radioactivity which find their way internally into the body. Carbon-14 is formed by nitrogen-14 capture of neutrons produced in the upper atmosphere by cosmic ray bombardment from galactic sources. The incident cosmic ray neutron flux is about 1 neutron $cm^{-2} s^{-1}$ over the earth's surface. This will lead to a natural production of 1·6 carbon-14 atoms $cm^{-2} s^{-1}$, a value which is believed to have remained unchanged for at least 15 000 years, until in 1954 the nuclear weapon test programmes interfered with this natural process. Carbon-14 eventually diffuses from the stratosphere into the troposphere from where it becomes taken up by plants and living organisms. These carbon-14 exposures account for an internal dose to human tissue of about 1 mrem $year^{-1}$. Cosmic radiation itself will contribute about 24 mrem $year^{-1}$ at sea level. The other naturally occurring radioactive isotope potassium-40 is a residual species from primordial nuclear synthesis. Since potassium is an essential element, potassium-40 is found in all living tissue and is one of the main contributors to the radiation

TABLE 1.20 *Effects of acute ionizing radiation over the whole body. (Adapted from USAEC report "The effects of nuclear weapons", 1962)*

Dose (rem)	Effect	Symptoms	Delay time and critical time after	Organs affected	Death
0–25					
25–50	no obvious injury	none	none	blood cell changes	none
50–100					
100–200	recovery within several weeks	moderate leukopaenia	3h	haematopoietic tissue	none
200–400	certain disability possibly death	severe leukopaenia, purpura, haemorrhage, infection	2h 4–6 weeks	haematopoietic tissue	often within 2 months
400–600	fatal to 50%			haematopoietic tissue	
600–1000	fatal		1 h 4–6 weeks		
1000–1500	fatal	diarrhoea, fever	30 m 5–14 days	gastrointestinal tract	90–100% within 2 weeks
>5000	fatal	convulsions, tremor, ataxia, lethargy	30 min 1–48 h	central nervous system	90–100% within 2 days

dose received by humans. This is clearly seen in the background spectrum shown in Fig. 1.3. It accounts for an estimated dosage of 21 mrem $year^{-1}$ to the gonads and 15 mrem $year^{-1}$ to bone [121].

1.6.4 External radiation

Populations are exposed to external radiation originating from rocks and soils which in ambient outside atmospheres ranges from 50 mrem $year^{-1}$ over sedimentary rocks to about 200 mrem $year^{-1}$ in granite districts. Inside houses and buildings the dose rate will depend upon the materials of construction. This is illustrated by Table 1.21, which shows the mean dose rate inside and outside houses in Scotland, where the influence of external radiation of 104 mrad $year^{-1}$ in Aberdeen is more than twice the mean dose of 48·5 mrad $year^{-1}$ in Edinburgh. Also, dose rate inside houses constructed of granite blocks in Aberdeen of 85·3 mrad $year^{-1}$ is higher than the 60 mrad $year^{-1}$ of homes built of sandstone in Edinburgh [121].

As mentioned at the beginning of this section, medical X-rays which amount to 20–50 mrem $year^{-1}$ must be included so that a final dose of around 200 mrem $year^{-1}$ is inescapable. This value is slightly in excess of the maximum permissible exposure set by the International Commission on Radiation Protection for the population at large, but is less than the 500 mrem $year^{-1}$ set for a single person. For occupationally exposed individuals the maximum upper limit is 5 rem $year^{-1}$, but these special cases require continuous monitoring. It has been pointed out by Schaefer [122] that crews of supersonic aircraft who complete 480 h $year^{-1}$ at an altitude with an average radiation level of 1 mrem $year^{-1}$ could fall into the latter category.

TABLE 1.21 *Radiation from natural sources in Scotland* [121]

Local γ-radiation measured in air	Mean dose rates (mrad $year^{-1}$)			
	Edinburgh	Dundee	Aberdeenshire	Aberdeen
Out-of-doors	48·5	63·0	69·5	104·0
In houses	60·0	67·2	81·5	85·3
24 h average	57·1	66·3	78·5	90·0

Fortunately, significant amounts of the three principal radionuclides strontium-90, iodine-131 and caesium-137 encountered as fission products from nuclear reactions (Table 1.22) are not likely to enter the body by inhalation. Most of the debris from atmospheric nuclear explosions returns to the earth by way of rainfall and consequently these elements enter the body through the food chain. In contrast to strontium-90 and iodine-131 which become fixed in bone and the thyroid gland, respectively, caesium-137 is very soluble and is eliminated from the body with a biological half-life of 100 days. Wrenn *et al.* [123] have demonstrated that the main dose from nuclear fall-out that enters the lungs is by inhalation of zirconium-95 and cerium-144. The amount, however, is small—something like 1% of the natural radiation background. A potentially more serious problem, perhaps for the future, would be the failure to contain plutonium isotopes from nuclear power operations (see section 4.5.2).

TABLE 1.22 *Maximum permissible concentrations of radionuclides encountered in air and produced by nuclear reactions*

Isotope	Radiation emitted	Half-life: Radioactive	Half-life: Biological	Non-occupational exposure	Occupational exposure
			Days	μCi ml^{-1}	μCi ml^{-1}
			Target organ		
^{3}H	β	12·26 years	12 tb[a]	2×10^{-7}	5×10^{-6}
^{14}C	β	5730 years	12 fat	1×10^{-7}	4×10^{-6}
^{85}Kr	$\beta\gamma$	10·76 years	—	3×10^{-7}	1×10^{-5}
^{89}Sr	β	52·7 days	18 000 bone	3×10^{-10}	3×10^{-8}
^{90}Sr	β	27·7 years	18 000 bone	3×10^{-11}	1×10^{-9}
^{95}Zr	$\beta\gamma$	65·5 days	450 tb		3×10^{-8}
^{131}I	$\beta\gamma$	8·05 days	138 thyroid	1×10^{-10}	9×10^{-9}
^{137}Cs	$\beta\gamma$	30 years	70 tb	2×10^{-9}	1×10^{-8}
^{140}Ba	$\beta\gamma$	13 days	65 bone		4×10^{-8}
^{144}Ce	$\beta\gamma$	284 days	1500 bone		6×10^{-9}
^{239}Pu	$\alpha\gamma$	24 390 years	73 000 bone		2×10^{-12}
^{240}Pu	$\alpha\gamma$	6580 years	73 000 bone		2×10^{-12}
^{242}Pu	αx	$3{\cdot}79\times10^{5}$ years	73 000 bone		2×10^{-12}

[a] tb = total body.
Source: US Atomic Energy Commission (10 C FR 20) 1969.

These remarks concerning radioactive atmospheric pollutants refer to community air. There are occasions, of course, in which inhaled radioactivity presents a serious problem. These are generally of an occupational nature, normally associated with accidents involving nuclear reactors or nuclear fuel processing plants. An exception concerns the recognition that miners in uranium mines [124] and in iron ore mines [125, 126] can contract lung cancer through inhalation of radon-222 which is naturally present in these underground atmospheres.

References

1. Heimann, H. (1961). Effects of air pollution on human health. *In* WHO Air pollution monograph. Ser. No. 46, pp. 159–220.
2. Schrenk, H. H., Heimann, H., Clayton, G. D., Gafefer, W. M. and Wexler, H. (1949). U.S. Public Health Service Bulletin 306.
3. Martin, A. E. (1964). Mortality and morbidity statistics and air pollution. *Proc. Roy. Soc. Med.* **57**, 969–975.
4. Greenburg, L., Field, F., Erhardt, C. L., Glasser, M. and Reed, J. I. (1967). Air pollution, influenza and mortality in New York City. *Arch. Environ. Health* **15**, 430–438.
5. Greenburg, L., Glasser, M. and Field, F. (1967). Mortality and morbidity during periods of high levels of air pollution. *Arch. Environ. Health* **15**, 684–694.
6. Hecker, L. H., Allen, H. E., Dinman, B. D. and Neel, J. V. (1974). High metal levels in acculturated and unacculturated populations. *Arch. Environ. Health* **29**, 181–185.
7. Butt, E. M., Nusbaum, R. E., Gilmour, T. C., Didio, S. L. and Sister Mariano (1964). Trace metal levels in human serum and blood. *Arch. Environ. Health* **8**, 52–57.
8. Kubota, J., Lazar, V. A. and Losee, F. (1968). Copper, zinc, cadmium and lead in human blood from 19 locations in US. *Arch. Environ. Health* **16**, 788–793.
9. Imbus, H. R., Cholak, J., Miller, L. H. and Sterling, T. D. (1963). Boron, cadmium, chromium and nickel in blood and urine. *Arch. Environ. Health* **6**, 286–295.
10. Sumino, K., Hayakawa, K., Shibata, T. and Kitamura, S. (1975). Heavy metals in normal Japanese tissues. *Arch. Environ. Health* **30**, 487–494.
11. Winell, M. (1975). An international comparison of hygiene standards for chemicals in the work environment. *Ambio* **4**, 34–36.
12. Roschin, A. V. and Timofeevskaya, L. A. (1975). Chemical substances in the work environment: Some important aspects of USSR and US hygienic standards. *Ambio* **4**, 30–33.
13. Zenz, C. and Berg, B. (1967). Human responses to controlled vanadium pentoxide exposure. *Arch. Environ. Health* **14**, 709–712.
14. Tanaka, S., and Lieben, J. (1969). Manganese poisoning and exposure in Pennsylvania. *Arch. Environ. Health* **19**, 674–684.
15. Davies, T. A. L. (1946). Manganese pneumonitis. *Brit. J. Indust. Med.* **3**, 111–135.
16. Oyanguren, H. and Pérez, E. (1966). Poisoning of industrial origin in a community. *Arch. Environ. Health* **13**, 185–189.
17. Holmberg, R. E. and Ferm, V. H. (1969). Inter-relationships of selenium, cadmium and arsenic in mammalian teratogenesis. *Arch. Environ. Health* **18**, 873–877.
18. Hammond, A. L. (1971). Mercury in the environment: natural and human factors. *Science* **171**, 788–789.
19. Jalili, M. A. and Abbasi, A. H. (1961). Poisoning by ethyl mercury sulphonanilide. *Brit. J. Indust. Med.* **18**, 303–308.
20. Berlin, M., Nordberg, G. F. and Serenius, F. (1969). On the site and mechanism of mercury vapour resorption in the lung. *Arch. Environ. Health* **18**, 42–50.

21. Kleinfeld, M., Messite, J., Shapiro, J., Kooyman, O. and Levin, E. (1968). A clinical, roentgenological, and physiological study of magnetite workers. *Arch. Environ. Health* **16**, 392–397.
22. Brief, R. S., Blanchard, J. W., Scala, R. A. and Blacker, J. H. (1971). Metal carbonyls. *Arch. Environ. Health* **23**, 373–384.
23. Lancet **2**, 268–270 (1968). Editorial comment.
24. Sullivan, R. J. (1969). Air pollution aspects of nickel (PB188 070) and chromium (PB188 075) and their compounds. Bethesda, Maryland, Sept. 1969. Litton Systems Inc.
25. Sunderman, F. W. and Donnelly, A. J. (1965). Studies of nickel carcinogenesis: metastizing pulmonary tumors in rats induced by the inhalation of nickel carbonyl. *Am. J. Path.* **6**, 1027–1041.
26. Baumslag, N. and Keen, P. (1972) Trace elements in soil and plants and antral cancer. *Arch. Environ. Health* **25**, 23–25.
27. Kazantis, G. (1972) Chromium and nickel carcinogenicity. *Ann. Occup. Hyg.* **15**, 25–29.
28. "Documentation of the Threshold Limit Values for Substances in Workroom Air." Amer. Conf. of Governmental Industrial Hygienists, 3rd Ed. 1971. Cincinnati, Ohio 45201.
29. Patterson, C. C. (1965). Contaminated and natural lead environments of man. *Arch. Environ. Health* **11**, 344–360.
30. Bryce-Smith, D. (1971). Lead pollution, a growing hazard to public health. *Chem. in Brit.* **7**, 54–56 and Lead pollution from petrol, ibid. **7**, 284–286.
31. Bryce-Smith, D. and Waldron, H. A. (1974). Lead, behaviour and criminality. *The Ecologist* **4**, 367–377.
32. Lob, M. and Desbaume, P. (1971). Étude de la plombémie et de la plomburie chez deux groupes de détenus, les uns intonés à la campagne, les autres à proximité immédiate d'une autoroute. *Schweiz. Med. Wschr.* **101**. 357–361.
33. Lob, M. and Desbaume, P. (1976). Lead and criminality. *Brit. J. Indust. Med.* **33**, 125–127.
34. Byers, R. K. and Lord, E. E. (1943). Late effects of lead poisoning in mental development. *Amer. J. Dis. Child.* **66**, 471–494.
35. Perlstein, M. A. and Attala, R. (1966). Neurologic sequelae of plumbism in children. *Clin. Pediat.* **5**, 292–298.
36. Moncrieff, A. A., Koumides, O. P., Clayton, B. E., Patrick, A. D., Renwick, A. G. C. and Roberts, G. E. (1964). Lead poisoning in children. *Arch. Dis. Child.* **39**, 1–13.
37. Pueschel, S. M., Kopito, L. and Schwachman, H. (1972). Children with an increased lead burden: A screening and follow up study. *J. Amer. Med. Assoc.* **222**, 462–466.
38. Betts, P. R., Astley, R. and Raine, D. N. (1973). Lead intoxication of children in Birmingham. *Brit. Med. J.* **1**, 402–406.
39. Hammond, P. B. (1976). Air standards for lead and other metals. *J. Occup. Med.* **18**, 351–355.
40. Barltrop, D., in Barltrop, D. and Burland, W. L., Eds. (1969). "Mineral Metabolism in Paediatrics," pp. 135–151. Blackwell Scientific Publications, Oxford.
41. Lancranjan, I., Popescu, H. I., Găvănescu, O., Klepsch, I., and Serbanescu, M. (1975). Reproductive ability of workmen occupationally exposed to lead. *Arch. Environ. Health* **30**, 396–401.

42. Grundy, R. D. (1969). Toxicologic and epidemiologic basis for air quality criteria: Carbon monoxide and lead. *J. Air Pollut. Contr. Assoc.* **19**, 729–732.
43. Rabinowitz, M. B., Wetherill, G. W. and Kopple, J. D. (1973). Lead metabolism in the normal human: Stable isotope studies. *Science* **182**, 506–508.
44. McLaughlin, M., Linch, A. L. and Snees, R. D. (1973). Longitudinal studies of lead levels in a US population. *Arch. Environ. Health* **27**, 305–311.
45. Goldsmith, J. R. and Hexter, A. C. (1967). Respiratory exposure to lead. Epidemiological and experimental dose-response relationships. *Science* **158**, 132–134.
46. Kehoe, R. A. (1966). Criteria for human safety from the contamination of the ambient atmosphere with lead. Proc. XV Congr. Internat. Med., Travail, Vienna, Vol. III, p. 83.
47. Chamberlain, A. C., Clough, W. S., Heard, M. S., Newton, D., Scott, A. N. B. and Wells, A. C. (1975). Uptake of lead by inhalation of motor exhaust. *Proc. Roy. Soc.* B **192**, 77–110.
48. Knelson, J. H., Johnson, R. J., Coulston, F., Golberg, L. and Griffen, T. (1972). Kinetics of respiratory lead uptake in humans. International Symposium on Environmental Health Aspects of Lead, pp. 391–401, Amsterdam.
49. Beasley, W. H., Jones, D. D., Megit, A. and Lutkins, S. G. (1973). Blood lead levels in a Welsh rural community. *Brit. Med. J.* **4**, 267–270.
50. Butler, J. D. and MacMurdo, S. D. (1974). Blood levels in Aberystwyth. *Brit. Med. J.* **2**, 502.
51. Kehoe, R. A., Cholak, J., Spence, J. A. and Hancock, W. (1963). Potential hazard of exposure to lead. I. Handling and use of gasoline containing tetramethyl lead. *Arch. Environ. Health* **6**, 239–254.
52. Kehoe, R. A., Cholak, J., McIlhinney, J. G., Lofquist, G. A. and Sterling, T. D. (1963). Further investigation in the preparation, handling and use of gasoline containing tetramethyl lead. *Arch. Environ. Health* **6**, 255–272.
53. Symposium on lead, reported in *Arch. Environ. Health* **8**, 202–354 (1964).
54. Cotzias, G. C., Borg, D. C. and Selleck, B. (1961). Virtual absence of turnover in cadmium metabolism: ^{109}Cd studies in the mouse. *Amer. J. Physiol.* **201**, 927–930.
55. Schroeder, H. A. (1965). Cadmium as a function in hypertension. *J. Chron. Dis.* **18**, 647–656.
56. Thind, G. S. (1972). Role of cadmium in human and experimental hypertension. *J. Air Pollut. Contr. Assoc.* **22**, 267–270.
57. Macaull, J. (1971). Building a shorter life. *Environment* **13**, 2–15.
58. Nandi, M., Gick, H., Sloane, D., Jick, H., Shapiro, S. and Lewis, G. P. (1969). Cadmium content of cigarettes. *Lancet* **2**, 1329–1330.
59. Lewis, G. P., Jusko, W. J., Coughlin, L. L. and Hartz, S. (1972). Contribution of cigarette smoking to cadmium accumulation in man. *Lancet* **1**, 291–292.
60. Carroll, R. E. (1966). The relationship of cadmium in the air: Cardiovascular disease death rate. *J. Amer. Med. Assoc.* **198**, 267–269.
61. Parizek, J. and Zahor, A. (1956). Effect of cadmium salts on testicular tissue. *Nature* **177**, 1036.
62. Hickey, R. J., Schoff, E. P. and Clelland, R. C. (1967). Relationship between air pollution and certain chronic disease death rates. *Arch. Environ. Health* **15**, 728–738.

63. Whittaker, E. J. W. (1953). The structure of chrysotile. *Acta Cryst.* **6**, 747–748.
64. Whittaker, E. J. W. (1955). A classification of cylindrical lattices. *Acta Cryst.* **8**, 571–574.
65. Whittaker, E. J. W. (1956). The structure of chrysotile II, clin-chrysotile. *Acta Cryst.* **9**, 855–862.
66. Whittaker, E. J. W. (1956). The structure of chrysotile III, ortho-chrysotile. *Acta Cryst.* **9**, 862–864.
67. Whittaker, E. J. W. (1956). The structure of chrysotile IV, para-chrysotile. *Acta Cryst.* **9**, 865–867.
68. Whittaker, E. J. W. (1957). The structure of chrysotile V. Diffuse reflection and fibre texture. *Acta Cryst* **10**, 149–156.
69. Speil, S. and Leineweber, J. P. (1969). Asbestos minerals in modern technology. *Environ. Res.* **2**, 166–208.
70. Gloyne, S. R. (1935). Two cases of squamous carcinoma of the lung occurring in asbestosis. *Tubercle* **17**, 5–10.
71. Lynch, K. M. and Smith, W. A. (1935). Pulmonary asbestosis: Carcinoma of the lung in asbestos-silicosis. *Amer. J. Cancer* **24**, 56–64.
72. Doll, R. (1955). Mortality from lung cancer in asbestos workers. *Brit. J. Indust. Med.* **12**, 81–86.
73. Knox, J. F., Holmes, S., Doll, R. and Hill, I. D. (1968). Mortality from lung cancer and other causes among workers in an asbestos textile factory. *Brit. J. Indust. Med.* **25**, 293–303.
74. Pooley, F. D. (1973). Asbestos fibre in the lung and mesothelioma: A re-examination of the Malmo material. *Acta Pathol. Microbiol. Scand. Sect. A. Pathol.* **81**, 390–400.
75. Holt, P. F. and Young, D. K. (1973). Asbestos fibres in the air of towns. *Atmos. Environ.* **7**, 481–483.
76. Gross, P. (1974). Is short-fibred asbestos dust a biological hazard? *Arch. Environ. Health* **29**, 115–117.
77. Timbrell, V., Griffiths, D. M. and Pooley, F. D. (1971). Possible biological importance of fibre diameters of South African amphiboles. *Nature* **232**, 55–56.
78. Newhouse, M. L., Berry, G., Wagner, J. C. and Turok, M. E. (1972). A study of mortality of female asbestos workers. *Brit. J. Indust. Med.* **29**, 134–141.
79. Mancuso, T. F. and El-Attar, A. A. (1967). Mortality pattern in a cohort of asbestos workers. A study based on employment experience. *J. Occup. Med.* **9**, 147–162.
80. Elmes, P. C. and Simpson, M. J. C. (1971). Insulation workers in Belfast, 3. Mortality 1940–1966. *Brit. J. Indust. Med.* **28**, 226–236.
81. Stocks, P. (1936). Distribution in England and Wales of cancer of various organs. *Ann. Rep. Brit. Emp. Cancer Campgn.* **13**, 240–274.
82 Curwen, M. P., Kennaway, E. L. and Kennaway, N. M. (1954). The incidence of cancer of the lung and larynx in urban and rural districts. *Brit. J. Cancer* **8**, 181–198.
83. Dean, G. (1966). Lung cancer and bronchitis in Northern Ireland. *Brit. Med. J.* **1**, 1506–1514.
84. Lambert, P. M. and Reid, D. D. (1970). Smoking, air pollution and bronchitis in Britain. *Lancet*, **1**, 853–857.

85. Henry, S. A. (1946). "Cancer of the Scrotum in Relation to Occupation", p. 112. Oxford University Press, London.
86. Hendricks, N. V., Berry, C. M., Lione, J. G. and Thorpe, J. J. (1959). Cancer of the scrotum in wax pressmen. I. Epidemiology. *Arch. Indust. Health* **19**, 524–529.
87. Henry, S. A. (1947). Occupational cutaneous cancer attributable to certain chemicals in industry. *Brit. Med. Bull.* **4**, 389–401.
88. Southam, A. H. and Wilson, S. R. (1922). Cancer of the scrotum. The etiology, chemical features and treatment of the disease. *Brit. Med. J.* **2**, 971–973.
89. Leitch, A. (1924). Mule-spinners cancer and mineral oils. *Brit. Med. J.* **2**, 941–943.
90. Cruickshank, C. N. D. and Gourevitch, A. (1952). Skin cancer of the hand and forearm. *Brit. J. Indust. Med.* **9**, 74–79.
91. Mastromatteo, E. (1955). Cutting oils and squamous cell carcinoma. I. Incidence of a plant report of six cases. *Brit. J. Indust. Med.* **12**, 240–243.
92. Eastcott, D. F. (1956). The epidemiology of lung cancer in New Zealand. *Lancet* **1**, 37–39.
93. Dean, G. (1961). Lung cancer among white South Africans. *Brit. Med. J.* **2**, 1599–1605.
94. Dean, G. (1962). Lung cancer in Australia. *Med. J., Aust.* **49**, 1003–1006.
95. Haenszel, W. (1961). Cancer mortality among the foreigh born in the United States. *J. Nat. Cancer Institut.* **26**, 37–132.
96. Butler, J. D. (1976). Air pollution, smoking and lung cancer. *Chem. in Brit.* **11**, 358–363.
97. Carnow, B. W. and Meier, P. (1973). Air pollution and pulmonary cancer. *Arch. Environ. Health* **27**, 207–218.
98. Harris, C. C., Sporn, M. P., Kaufman, G., Smith, J. M., Baker, M. J. and Saffioti, U. (1971). Acute ultrastructural effects of benzo(a)pyrene and ferric oxide on the hamster tracheobronchial epithelium. *Cancer Res.* **31**, 1977–1989.
99. Laskin, S., Kuschner, M. and Drew, R. T. (1970). US Atomic Energy Commission Symposium "Inhalation Carcinogenesis", p. 321.
100. Stanley, C. W., Barney, J. E., Helton, M. R. and Yobs, A. R. (1971). Measurement of atmospheric levels of pesticides. *Environ. Sci. and Technol.* **5**, 430–435.
101. Wade, N. (1971). Hexachlorophene: F.D.A. temporized on brain-damaging chemical. *Science* **174**, 805–807.
102. Hammond, A. L. (1972). Chemical pollution: polychlorinated biphenyls. *Science* **175**, 155–156.
103. May, G. (1973). Chloracne from the accidental production of tetrachlorodibenzodioxin. *Brit. J. Indust. Med.* **30**, 276–283.
104. Milnes, M. H. (1971). Formation of 2,3,7,8-tetrachlorodibenzodioxin by thermal decomposition of sodium 2,4,5-trichlorophenate. *Nature* **232**, 395–396.
105. Jensen, S. (1972). The PCB story. *Ambio* **1**, 123–134.
106. Woodwell, G. W., Craig, P. P. and Johnson, H. A. (1971). DDT in the biosphere: Where does it go? *Science* **174**, 1101–1107.
107. Edwards, R. (1971). The polychlorobiphenyls: Their occurrence and significance, a review. *Chem. Indust.* **47**, 1340–1348.

108. Cuddeback, J. E., Donovan, J. R. and Burg, W. R. (1976). Monitoring of the atmosphere in crowded taverns at peak hours. *Amer. Indust. Hyg. Assoc. J.* **37**, 263–267.
109. Castleden, C. M. and Cole, P. V. (1974). Variations in carboxyhaemoglobin levels in smokers. *Brit. Med. J.* **4**, 736–738.
110. Cole, P. V. (1975). Carboxyhaemoglobin levels in man: Comparative effects of atmospheric pollution and cigarette smoking. *Nature* **255**, 699–701.
111. Waller, R. E. and Commins, B. T. (1966). Episodes of high pollution in London, 1952–66. *In* Internat. Clean Air Congress Proc. **1**, 228–231, London.
112. Buechley, R. W., Riggan, W. B., Hasselblad, V. and Van Bruggen, J. B. (1973). Sulfur dioxide levels and perturbations in mortality. A study in New York, New Jersey metropolis. *Arch. Environ. Health* **27**, 134–137.
113. Frank, N. R. (1964). Studies on the effect of acute exposure to sulphur dioxide in human subjects. *Proc. Roy. Soc. Med.* **57**, 1029–1033.
114. Wolff, R. K., Dolovich, M., Rossman, C. M. and Newhouse, M. T. (1975). Sulfur dioxide and tracheobronchial clearance in man. *Arch. Environ. Health* **30**, 521–527.
115. Amdur, M. O. (1971). Aerosols formed by oxidation of sulfur dioxide. *Arch. Environ. Health* **23**, 459–468.
116. Buchanan, W. D. (1974). Industrial hazards. *In* "Some gaseous Pollutants in the Environment". Report of seminar of Inter-Research Council Committee on Pollution Research, p. 23. NERC, London; May.
117. Menser, H. A. and Heggestad, H. E. (1966). Ozone and sulfur dioxide synergism. Injury to plants. *Science* **153**, 424–425.
118. Hazucha, M., Silverman, F., Parent, C., Field, S. and Bates, D. V. (1973). Pulmonary function in man after short exposure to ozone. *Arch. Environ. Health* **27**, 183–188.
119. Hazucha, M. and Bates, D. V. (1975). Combined effect of ozone and sulphur dioxide on human pulmonary function. *Nature* **257**, 50–51.
120. Perkins, R. W. and Nielson, J. M. (1965). Cosmic ray produced radionuclides in the environment. *Health Phys.* **11**, 1297–1304.
121. MRC (1960). "The Hazards to Man of Nuclear and Allied Radiations." 2nd Report of the Medical Research Council. Her Majesty's Stationery Office, London.
122. Schaefer, H. J. (1971). Radiation exposure in air travel. *Science* **173**, 780–783.
123. Wrenn, M. E., Monafy, R. and Laurer, G. R. (1964). ^{95}Zr–^{95}Nb in human lungs from fall-out. *Health Phys.* **10** 1051–1058.
124. Holaday, D. A. (1969). History of the exposure of miners to radon. *Health Phys.* **16**, 547–552.
125. Boyd, J. T., Doll, R., Faulds, J. S. and Leiper, J. (1970). Radon in a haematite mine in England causes lung cancer. *Brit. J. Indust. Med.* **27**, 97–105.
126. Renard, K. G. St. C. (1974). Respiratory cancer mortality in an iron ore mine in northern Sweden. *Ambio* **3**, 67–69.

2

Sources, Sinks and Removal Mechanisms of Emissions to the Atmosphere

2.1 Introduction

In the last chapter where the health aspects of air pollution were discussed we were concerned primarily with the interaction of pollutants that cause problems to life on earth. This treatment tends to restrict our attention to a very limited layer of air: a layer lying within a few metres of the surface and important because most populations spend most of their time within its confines. In practice, this concept is rather artificial, since this layer of gas does not exist in isolation. There is in reality exchange of air between adjacent layers so that mixing and diffusion ultimately extends the system beyond the troposphere into the stratosphere. This means that in order to build up a total picture of atmospheric contamination by chemicals a mass balance on a global scale must be considered. The mass balance must include an appreciation of the atmospheric input concentrations from anthropogenic and natural sources, coupled with knowledge of the manner in which they are eventually removed from the atmosphere.

2.2 Global aspects of pollution—emissions to atmosphere

As with galactic radiation and natural radioactivity, there are a number of natural processes which inject chemicals into the atmosphere. These are of geological and biological origin.

Volcanic eruptions are responsible for emissions of carbon dioxide, sulphur dioxide and particulates. The sulphate aerosol concentration in the lower stratosphere increases typically to around 10 $\mu g\ m^{-3}$ within 1 year of a major volcanic disturbance. Estimates of global volcanic emissions of sulphur dioxide and halogen hydracids entering the troposphere and stratosphere annually according to Cadle [1] are given in Table 2.1. Other pollutants, previously unsuspected, such as mercury have been detected in the vicinity of volcanic regions. Eshleman *et al.* [2] reported values from 0·04 to 23·3 $\mu g\ m^{-3}$ for atmospheric Hg in various sites in Hawaii. Emissions from volcanoes, however, although large (see Tables 2.1 and 2.2), are small compared with those of a less spectacular nature, the much more subtle biological processes.

TABLE 2.1 *Estimates of volcanic global emission of SO_2, HCl and HF in kg year^{-1}* [1]

	SO_2	HCl	HF
Troposphere	$7{\cdot}5\times10^{9}$	$7{\cdot}5\times10^{8}$	$3{\cdot}8\times10^{7}$
Stratosphere	$2{\cdot}8\times10^{8}$	$2{\cdot}8\times10^{7}$	$1{\cdot}4\times10^{6}$

TABLE 2.2 *Summary of global particulate emission from natural sources* [3]

Source	Percent of total	Production (kg $\times 10^9$ year^{-1})
Sea salt spray	44	908
Nitrate from NO and NO_2	18	390
Ammonium	11	245
Sulphate from H_2S	9	182
Terpenes	9	182
Soil dust	9	182
Volcanoes		4
Forest fires		3
Total	100	2096

The massive emissions of particulates summarized in Table 2.2 indicate that aerosols formed from nitrates, sulphates, ammonium compounds and terpenes account for about 47% of the natural emission to the atmosphere. Sea spray aerosol is the largest single contributor, amounting to 44%, with soil dust making up most of the remaining 9% of the total 2096×10^{9} kg year^{-1}. In contrast, global particulate emissions of anthropogenic origin given in Table 2.3 come to around 269×10^9 kg year^{-1}. The total particulate emission rate, therefore, comes to $(2096+269) \times 10^9 = 2365 \times 10^9$ kg year^{-1}, of which $269 \times 10^{11}/2365 \times 10^9 = 11\%$ comes from anthropogenic sources. As Table 2.3 shows, these arise from the direct emission of pollutant particulates amounting to 84×10^9 kg year^{-1} but more than twice as much as this, 185×10^9 kg year^{-1} is produced in the atmosphere by chemical reaction between gaseous pollutants.

TABLE 2.3 *Summary of global particulate emission from anthropogenic sources* [3]

Source	Production ($kg \times 10^9$ year^{-1})
Pollutant particulate materials	84
Particles formed from gaseous pollutants	
(i) Sulphate from SO_2	133
(ii) Nitrate from NO_x	27
(iii) Photochemical from hydrocarbons	25
Total	269

2.2.1 The carbon cycle

Biological processes produce the Earth's oxygen atmosphere through the fundamental photosynthesis reaction. The reaction occurs in green plants in the presence of sunlight and chlorophyll (a magnesium porphyrin compound), according to the reaction:

$$H_2O + CO_2 + h\nu \rightarrow CH_2O + O_2 \qquad (2.1)$$

The CH_2O units build up into carbohydrates like cellulose and starch which, given sufficient time, become converted into fossil fuels. In the absence of light the process is reversed somewhat, so that plants take in oxygen: plant respiration. Carbon dioxide also enters the atmosphere from biological decay, from terrestrial and marine sources. Subsequent movement of air masses over land and sea result in an exchange of carbon dioxide between the atmosphere and the oceans of the world. Altogether, a vast system exists in which interchange and distribution of carbon dioxide takes place between land, sea and air (Fig. 2.1).

Figure 2.1 shows that there are two separate carbon cycles, one terrestrial and the other marine. These are dynamically coupled at their interface with the atmosphere. The carbon cycle in the sea concerns the assimilation of carbon dioxide dissolved in sea water by phytoplankton with release of oxygen. Zooplankton and fish use this dissolved oxygen for respiration and, thus, indirectly consume the carbon dioxide fixed by the phytoplankton. Eventually, the decomposition of marine organic matter replaces the carbon dioxide taken up by the phytoplankton. On land, as already mentioned, carbon dioxide exchanges with the atmosphere through respiration of plants and soil, photosynthesis and the combustion of fossil fuels.

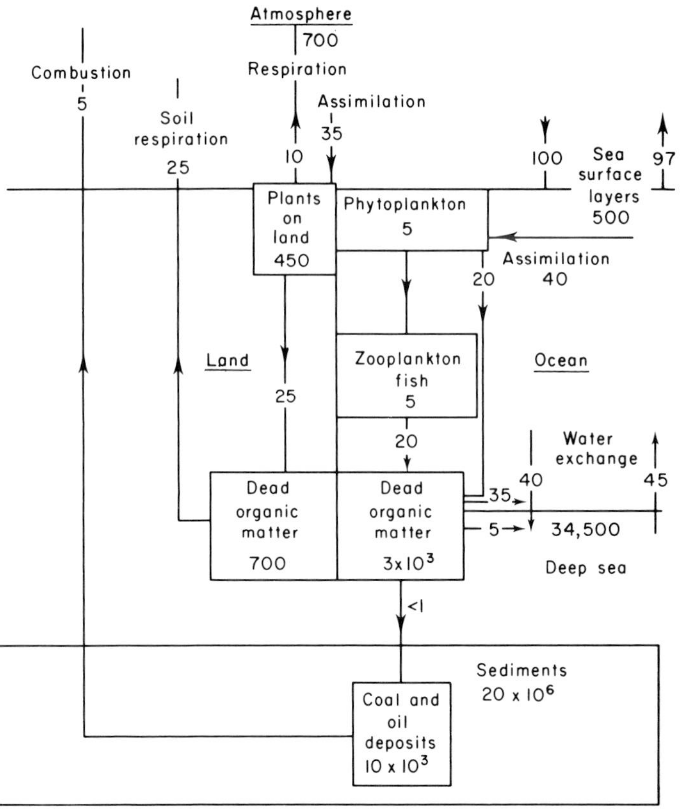

FIG. 2.1 Carbon circulation in the biosphere ($kg \times 10^{12}$ $year^{-1}$) [4]

Carbon dioxide is the fourth major constituent of the earth's atmosphere (Table 2.4). The amount in air varies slightly according to the amount of vegetation, the time of day, and the proximity of urbanization. In open countryside the lowest concentrations tend to occur in the afternoon when photosynthesis is at a maximum, and the highest values occur at night, when photosynthesis is minimal. Under these conditions carbon dioxide concentrations near the ground may approach 350 parts 10^{-6} (see Fig. 2.2). This high value reflects partly the release of carbon dioxide from the decomposition of organic matter in the soil and partly the tendency of air

TABLE 2.4 *Principal components of normal dry air*

Compound	Concentration (parts 10^{-6})
N_2	780 900
O_2	209 5000
Ar	9 300
CO_2	320 ± 10

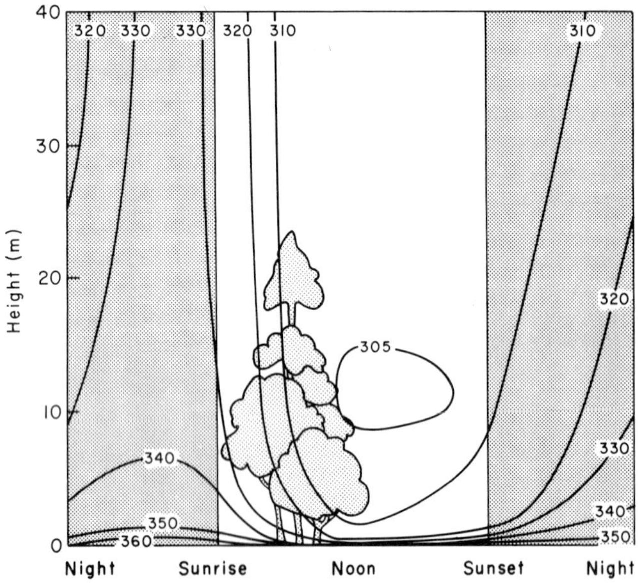

FIG. 2.2 Vertical distribution of carbon dioxide in the air around a forest varies with time of day. At night, when photosynthesis is shut off, respiration from the soil can raise the carbon dioxide at ground level to as much as 400 parts 10^{-6}. By noon, owing to photosynthetic uptake, the concentration at treetop level can drop to 305 parts 10^{-6} [4]

to remain stationary near the ground at night when there is no solar radiation to produce convection currents. Seasonal variations also exist because of the proliferation or lack of vegetation (Fig. 2.3). Between April and September the atmosphere in the northern hemisphere, north of 30° N latitude, loses nearly 3% of its carbon dioxide content through uptake by forests and tundra.

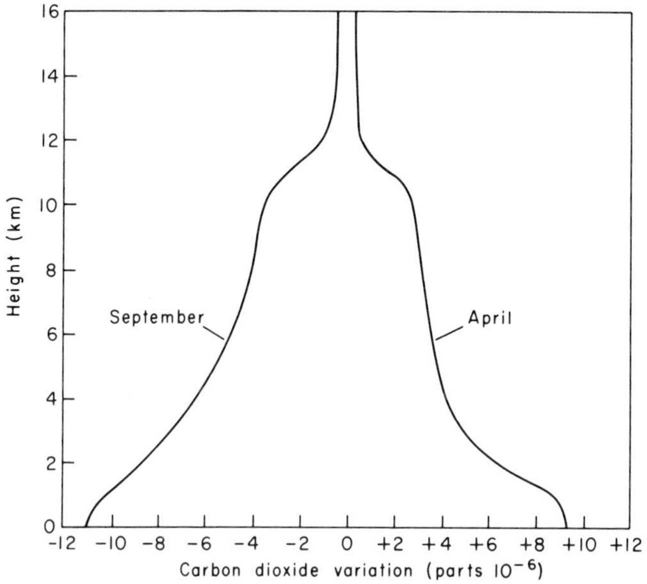

FIG. 2.3 Seasonal variations in the carbon dioxide content of the atmosphere reach a maximum in September and April for the region north of 30 degrees north latitude. The departure from a mean value of about 320 parts 10^{-6} varies with altitude as shown by these two curves [4]

From a global point of view, there is evidence to show that the concentration of atmospheric carbon dioxide is increasing by about 1 part 10^{-6} $year^{-1}$. It is possible that this could cause an increase in the temperature of the earth. Carbon dioxide absorbs some of the out-going infrared radiation from earth that normally would escape into space. This has been called the "greenhouse effect". On the other hand, particulate matter from combustion sources, if diffused into the stratosphere in sufficient quantity, causes atmospheric turbidity. Atmospheric turbidity influences the extent of scattering and reflection of in-coming solar radiation, so that this may lead to a decrease in atmospheric temperature. These effects are currently in equilibrium, but if upset, by either of the mechanisms outlined, there could be disastrous climatic and meteorological consequences.

Any appreciation of the global atmospheric carbon oxide balance, CO as well as CO_2, must include all the processes outlined in Fig. 2.1 and should also account for all the production and removal routes, both natural and anthropogenic. An important contribution to the understanding of this problem has been made by Weinstock and Niki [5], who recognized that tropospheric carbon monoxide was produced at a rate of about 5×10^{12} kg year^{-1}. This value is over 20 times greater than the rate of carbon monoxide production from combustion, e.g. the value of $7{\cdot}45 \times 10^{10}$ kg year^{-1} given in Table 2.10 is roughly one-third the world total so that 5×10^{12} kg/$7{\cdot}45 \times 3 \times 10^{10}$ kg ~ 22. The value of 5×10^{12} kg of CO produced annually was derived by considering a steady-state model for tropospheric CO.

Let P_N be the unknown natural emission of CO to the atmosphere and P_F the rate of production of CO from fossil fuel sources. If this CO is removed from the atmosphere at a rate of $k[\mathrm{CO}]$, where k is the first-order removal rate constant and [CO] is the atmospheric concentration of CO, the equation

$$\frac{\mathrm{d}}{\mathrm{dt}}[\mathrm{CO}] = P_N + P_F - k[\mathrm{CO}] = 0 \tag{2.2}$$

can be written for the steady state. Since interaction of cosmic radiation with nitrogen-14 naturally produces carbon-14 in the stratosphere and if this rate of production is given by P_R, another equation can be written to take this into account

$$\frac{\mathrm{d}}{\mathrm{d}t}[^{14}\mathrm{CO}] = nP_N + P_R - k[^{14}\mathrm{CO}] = 0 \tag{2.3}$$

where n is the known mole fraction of ^{14}CO, viz. $1{\cdot}17 \times 10^{-12}$. Setting P_F, the rate of production of CO from "dead" carbon as in fossil fuels equal to 7×10^{12} mol year^{-1}, P_R equal to 290 mol year^{-1}, the steady state concentration of [CO] and [^{14}CO] equal to $1{\cdot}71 \times 10^{13}$ mol and 45 mol, respectively, equations (2.2) and (2.3) can be solved simultaneously for P_N and k. When this is done,

$$P_N + 7 \times 10^{12} = 1{\cdot}71 \times 10^{13} k \qquad \text{(from 2.2)}$$

$$1{\cdot}17 \times 10^{-12} P_N + 290 = 45k \qquad \text{(from 2.3)}$$

P_N comes to $1{\cdot}86 \times 10^{14}$ mol year^{-1} or $5{\cdot}2 \times 10^{12}$ kg year^{-1}, and $k = 11{\cdot}3$ year^{-1}. Hence, from this analysis P_N/P_F, the ratio of CO produced naturally to that of anthropogenic origin, is $1{\cdot}86 \times 10^{14}/7 \times 10^{12} = 26{\cdot}6$.

To summarize, the calculation shows that very much more carbon monoxide enters the troposphere from natural sources than it does from combustion sources. Yet despite this, the concentration in the troposphere remains sensibly constant. A removal mechanism must be operating to account for this anomaly. There are several possibilities:

(1) CO may be transferred into the stratosphere where it rapidly oxidizes to CO_2;

(2) CO may be consumed by living organisms;
(3) CO may be oxidized by some suitable atmospheric species, as for example OH radicals.

The first possibility is untenable on the basis of the work by Junge *et al.* [6], who showed that this mechanism alone would give a residence time of 9 years and a depletion rate for CO of only 11% per year. Some estimates of the removal of CO by acidic surface soils by Inman *et al.* [7] suggest that this sink could account for the removal of about 6·5 times the amount of CO produced in the USA by combustion. Although helpful, this removal rate is inadequate to satisfy the overall requirement, so that the atmospheric route involving the reaction

$$CO + OH \xrightarrow{k_{(2\cdot4)}} CO_2 + H \tag{2.4}$$

was investigated, where $k_{(2.4)} = 2.1 \times 10^{-13} \exp -115/T$ cm^3 molecule s^{-1}, or for $T = 300K$, $k_{(2.4)} = 1.4 \times 10^{-13}$ cm^3 molecules s^{-1} [8]. For this reaction to be feasible sufficient OH radicals must be present. This can be estimated, since the required OH concentration will be given by the ratio of the first-order rate constant k derived from equation (2.2) and (2.3) for CO removal divided by $k_{(2.4)}$ the bimolecular rate constant for reaction (2.4). Taking $k_{(2.4)} = 1{\cdot}4 \times 10^{-13}$ cm^3 molecule^{-1} s^{-1} gives an OH concentration necessary to maintain CO in the steady state of $2{\cdot}5 \times 10^6$ molecules cm^{-3}. This value is in broad agreement with that derived by McConnell *et al.* [9] for the tropospheric OH concentration of between 3×10^6 molecules cm^{-3} up to altitudes of about 6 km, reducing to 1×10^6 molecules cm^{-3} at the tropopause (10 km).

OH radical concentrations of the order of 10^6 molecules cm^{-3} can also account for the natural production rate of CO represented by P_N in equation (2.2), through their interaction with atmospheric methane,

$$CH_4 + OH \xrightarrow[\text{slow}]{k_{(2.5)}} CH_3 + H_2O \qquad \Delta H^0 = -63 \text{ kJ} \tag{2.5}$$

followed by further rapid reactions that convert organic free radical species to CO, e.g.

$$CH_2O + OH \rightarrow CHO + H_2O \tag{2.6}$$

$$CHO + OH \rightarrow CO + H_2O \tag{2.7}$$

The rate constant $k_{(2.5)}$ of equation (2.5) is $2.95 \times 10^{-12} \exp(-1770/T)$ cm^3 molecules^{-1} s^{-1} or 8×10^{-15} cm^3 molecules^{-1} s^{-1} at $T = 300$ K [10]. For a global methane concentration of 1·5 parts 10^{-6} and an OH concentration in the range 1–3×10^6 molecules cm^{-3}, or twice this estimate in daylight, the CO produced by methane oxidation will be of the order $4–5\times10^{12}$ kg year^{-1}. This agrees with the value P_N and led Weinstock and Niki [5] to conclude that OH radicals play a major role in both the formation and removal of tropospheric CO.

One of the consequences of this model that should be noted, is that since methane is being removed from the atmosphere at a rate of about $2{\cdot}3\times10^{12}$ kg year^{-1}, then a source of the same order of magnitude must be identified to sustain the steady-state equilibrium concentration. By adding together estimated emissions of methane from rice paddy fields, swamp lands, and humid tropical areas Robinson and Robbins [11] estimated natural methane production to be about $1{\cdot}45\times10^{12}$ kg year^{-1}. This is the correct order of magnitude demanded by the model and gives a life-time of about 1·5 years for methane in the atmosphere.

Some further refinements in this model have been made by considering other organic molecules such as ethene that can also participate in the scheme [12] and that the oceans are saturated with CO which can exchange with atmosphere. These processes, however, do not alter the fundamental concept that global CO is primarily dependent upon the existence of methane and hydroxyl radicals in the troposphere.

Besides carbon dioxide, natural biological degradation will produce oxides of nitrogen and sulphur, ammonia, hydrogen sulphide and methane, each of which distribute themselves throughout the biosphere with different atmospheric residence times.

Natural global emissions of nitrogen compounds from biological action are given in Table 2.5, and those for sulphur compounds produced by biological action or sea spray are shown in Table 2.6. Global emission estimates from combustion and industrial operations for nitrogen and sulphur compounds are contained in Tables 2.7 and 2.8, respectively.

2.2.2 The nitrogen cycle

In a similar manner to that outlined for the carbon cycle shown in Fig. 2.1, cycles indicating the distribution and exchange of N-compounds

TABLE 2.5 *Natural global emissions of nitrogen compounds produced by biological action* [13]

Compound	Emission ($kg\times10^9$ year^{-1})	Emission expressed as N ($kg\times10^9$ year^{-1})
NO	455	212
N_2O	537	342
NH_3	1053	867

TABLE 2.6 *Natural global emissions of sulphur compounds produced by biological action and sea spray expressed as S in kg × 10^9 year* [14]

Compound	Emission expressed as S ($kg \times 10^9$ $year^{-1}$
H_2S	90
Sea spray	40

TABLE 2.7 *Global emissions of nitrogen compounds produced from combustion, expressed as NO_2 in kg × 10^9 $year^{-1}$* [13]

Compound	Source	Emission ($kg \times 10^9$ $year^{-1}$)	Emission expressed as N
NO_2	Coal combustion	24·4	7·4
	Petroleum refining	0·6	0·2
	Gasoline combustion	6·8	2·1
	Other oil combustion	12·8	3·9
	Natural gas combustion	1·9	0·6
	Other combustion	1·4	0·4
NH_3	Combustion	3·8	3·1

TABLE 2.8 *Global emission of sulphur dioxide from pollutant emissions in kg × 10^9 $year^{-1}$* [14]

Source	Emission ($kg \times 10^9$ $year^{-1}$)	Emissions expressed as S
Coal	92	46
Petroleum:		
Combustion and refining	26	13
Smelting:		
Cu	12	6
Pb	1	0·5
Zn	1	0·5

and S-compounds can also be written. The environmental N-cycle as illustrated in Fig. 2.4 is in reality three separate cycles, one involving N_2O, one NH_3 and the other NO_2.

The atmospheric nitrous oxide background concentration on a global scale is 0·25 parts 10^{-6} or 1362×10^9 kg expressed as N. This represents 97% of the total nitrogen compounds present in the atmosphere calculated on the basis of their N-content. The N_2O cycle shown in Fig. 2.4 is essentially independent of the rest of the system, except that it draws on soil nitrogen for formation through the agency of bacteria. About 6% of this tropospheric N_2O diffuses into the stratosphere where it undergoes

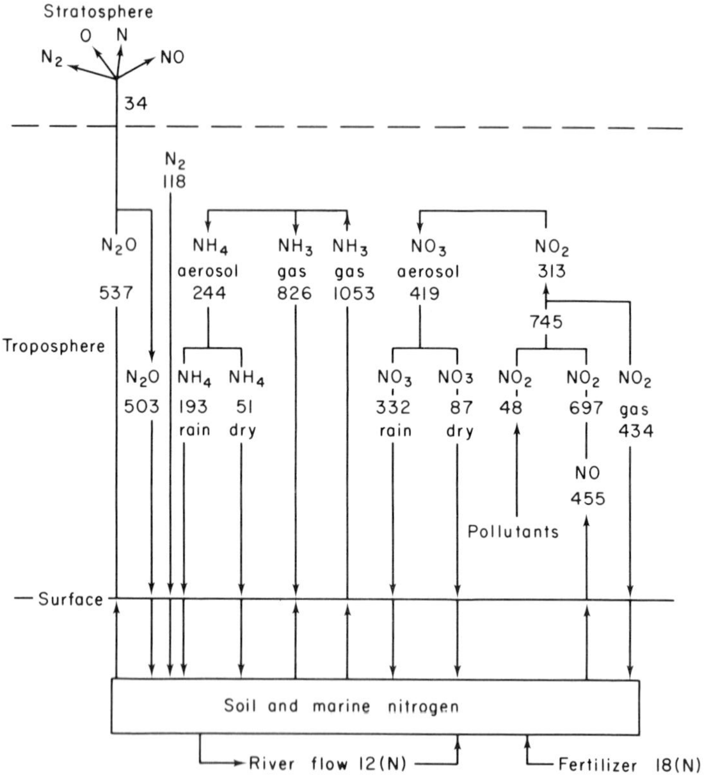

FIG. 2.4 The nitrogen cycle in the biosphere in kg $\times 10^9$ year^{-1} [13]

conversion to nitric oxide, nitrogen and oxygen. The remainder returns to the earth's surface. Nitrous oxide diffusion into the stratosphere mainly accounts for the presence of nitric oxide at these high altitudes through the reaction

$$N_2O + O(^1D) \rightarrow NO + NO \tag{2.8}$$

This aspect of the global nitrous oxide budget and its influence on stratospheric ozone is discussed in Chapter 5. Bacterial fixation of nitrogen by soil has been estimated at 118×10^9 kg year^{-1} and again this portion of the cycle does not relate to the remainder, except in the sense that nitrogen in the soil acts as a source of NH_3, NO, NO_2 and N_2O.

According to Robinson and Robbins [13], the amount of NH_3 in the atmosphere is about 33×10^9 kg. Since NH_3 is very soluble and originates near the surface, the atmospheric life-time is probably rather short, of the order of a week. If this is so, then an annual production of about 50 times the atmospheric mass or 1650×10^9 kg is indicated. In Fig. 2.4, the annual production rate of 1053×10^9 kg has been used in order to compensate the calculated scavenging mechanisms responsible for returning the ammonium aerosol back to earth, either as ammonia or as ammonium salts. Evidence exists that precipitation both of NH_4 and NO_3 does not occur uniformly, but that the majority falls out in the northern hemisphere in the region between 30° and 55° latitude.

The third cycle involving the oxides of nitrogen NO_x as NO and NO_2 stem mainly from combustion. Robinson and Robbins [13] estimate that NO_x emissions come to around 48×10^9 kg year^{-1}, i.e. the sum total of the emission expressed in terms of NO_2 given in Table 2.7. As shown in Fig. 2.4, the main natural NO_x emission is estimated to be NO at a rate of 455×10^3 kg year^{-1}. This is equivalent to an atmospheric NO residence time of 4 days. Gaseous deposition of NO is considered to be unimportant; most of the atmospheric NO becomes oxidized to NO_2 and eventually to nitrate. The nitrate aerosol is returned to earth by both rain precipitation and dry deposition. The dry deposition rate of 87×10^9 kg year^{-1} is believed to be about one-quarter of that returned by rainfall.

The cycle shown in Fig. 2.4 is completed by including an estimate of 18×10^9 kg year^{-1} for fertilizer addition to soil expressed as N, and 12×10^9 kg year^{-1} again expressed as N for the transfer of nitrogen, mainly as nitrates from land to sea by rivers. Finally, it will be noted that atmospheric fixation of nitrogen during electrical storms, although important in evolutionary times, now plays little part in the nitrogen cycle.

2.2.3 The sulphur cycle

The S-cycle shown in Fig. 2.5 describes the source concentrations and the circulation of SO_2, H_2S and sulphates in the biosphere. Sulphur dioxide emissions come from anthropogenic pollutant sources. Hydrogen sulphide is mainly derived from the decomposition of organic matter in swamps, bogs and tidal flats, but some also comes from industrial activity. The sulphate aerosol consists principally of either ammonium sulphate or sulphuric acid.

The total global emission of SO_2, about 132×10^9 kg year^{-1}, can be estimated by addition of the itemized source concentrations given in Table

2.8. Most of this SO_2, some 93%, is emitted in the northern hemisphere. The largest single contribution to the total, 71%, is accounted for by coal combustion. Robinson and Robbins [14] maintain that SO_2 emissions have

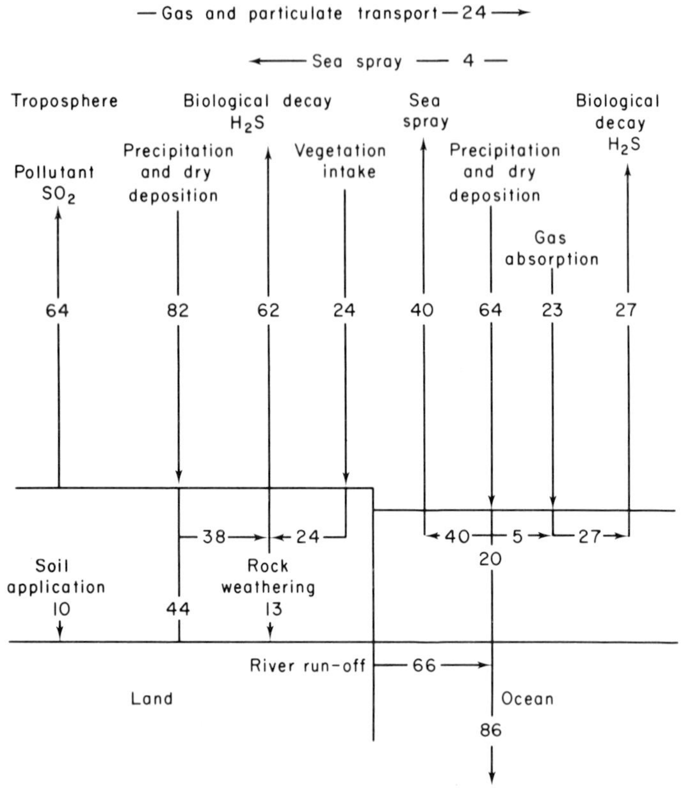

FIG. 2.5 The sulphur cycle in the biosphere ($kg \times 10^9\ year^{-1}$) [14]

approximately doubled in the period between 1940 and 1965. The average tropospheric concentration of SO_2 is about 0·2 parts 10^{-9} or if expressed as sulphur 0·25 $\mu g\ m^{-3}$. Considerably more than half of this background SO_2 will be derived from H_2S which has undergone oxidation to SO_2 after being released into the atmosphere. This reaction occurs in the troposphere by ozone oxidation in a heterogeneous surface reaction. Aerosol particulates provide the necessary surface. The residence time of H_2S varies from about 2 h in urban locations to 2 days in rural unpolluted areas. The average trophospheric H_2S concentration is 0·2 parts 10^{-9} or 0·14 $\mu g\ m^{-3}$ when expressed as S. Ultimately, SO_2 becomes oxidized and finishes up in an aerosol sulphate form, usually ammonium sulphate or sulphuric acid mist. This oxidation stage is enhanced by moist foggy conditions and the presence of ammonia [15]. Under these conditions the residence time of SO_2 may be as short as 1 h. The other principal oxidation route for SO_2 is photochemical, especially when oxides of nitrogen and hydrocarbons are also present. These processes are discussed in some detail in Chapter 5. The average tropospheric sulphate aerosol concentration is 2 $\mu g\ m^{-3}$ or 0·7 $\mu g\ m^{-3}$ expressed as S.

Several processes exist for scavenging sulphur compounds from the atmosphere. Among them are uptake of SO_2 by vegetation [16] and, of course, rainfall and clouds provide an effective removal mechanism, as does dry deposition (see section 2.4.1). All these processes are summarized in Fig. 2.5. In order to maintain atmospheric S-content constant at 0·25, 0·14 and 0·7 $\mu g\ m^{-3}$ expressed as S for SO_2, H_2S and SO_4, the scheme requires a sink of 86×10^9 kg year^{-1} into the oceans of the world.

2.3 Details of anthropogenic emissions

In the last 200 years or so the growth in world population and the industrial revolution have resulted in an increased demand for energy. Up till now these energy requirements have been supplied largely by the combustion of fossil fuels. The planet's resources of convenient carbonaceous fuels, coal and oil, have been used for heating purposes, to power industry, for transport and synthesis of chemicals. The by-products of these operations—particulates, the oxides of carbon, nitrogen and sulphur—have been committed to the atmosphere in enormous quantities. Some idea of the type and scale of emission from these operations is listed in Table 2.9 under the headings: (A) industrial manufacturing, (B) food and agriculture, (C) fossil fuel combustion in furnaces, (D) fossil fuel combustion in engines, (E) evaporative losses, and (F) solid waste disposal. [17].

The US emissions given in Table 2.10 represent the main single producer and contribute about $\frac{1}{3}$ of the carbon monoxide, $\frac{3}{8}$ of the sulphur oxides, $\frac{1}{4}$ of the hydrocarbons, $\frac{1}{6}$ of the oxides of nitrogen and $\frac{1}{2}$ of the particulates to the northern hemisphere.

2.3.1 *Primary fossil fuels*

2.3.1.1 Coal composition and combustion

As an alternative to burning coal directly as an energy source it can be processed into gas or liquid fuels or kept in the solid form but solvent refined. In these modified forms it is generally cleaner and easier to handle,

TABLE 2.9 *General emission sources*[a]

Operation[b]	Emission factor [kg $(10^3\ kg)^{-1}$ unless stated otherwise] Particulates	CO	NO_x	SO_x	HC
A HEAVY INDUSTRY AND MANUFACTURING					
1. *Chemical processing*					
(i) C—black	5–1150	2250–16 750			200–5750
(ii) Wood charcoal	160				50
(iii) Explosives—TNT			80		
(iv) HNO_3			0·1–27·5		
(v) S from $2H_2S+O_2 \rightarrow 2S+H_2O$				62–84	
(vi) Synthetic rubber, butudiene + acrylonitrile or styrene					up to 41·5
(vii) Terephthalic acid			6·5		
(viii) Nylon					11
(ix) Spray drying detergents	1·5–45				
(x) Printing inks					up to 295
(xi) Plastics—PVC	24				0·35–8·5
(xii) Phthalic anhydride					up to 21·5
(xiii) Paint and varnish	1				15–285
(xiv) Phosphoric acid	8·95				
2. *Petroleum refining*		kg $(10^3 l)^{-1}$ of fresh feed			
(i) Fluidized bed catalytic cracking	0·695	39·2	0·20	1·41	0·68
(ii) Moving bed cracking units	0·049	10·8	0·01	0·17	0·28
3. *Metallurgical processes*					
(i) Cu—smelter	65·5			625	
(ii) Pb—smelter	6·0–37·5			330	
(iii) Zn—smelter	4·0–60·0			550	
(iv) Fe—blast furnace	65–100			700–1050	
(v) $Al(OH)_3$ calcining	100				
4. *Mineral products*					
(i) Asphaltic concrete plants	up to 22·5				
(ii) Bricks (coal fired)	0·5 of ash in coal	0·95	0·45	3·6 × %S in fuel	0·3
(iii) Portland cement (excluding heat source)	130–170		1·3	5·1	
B FOOD AND AGRICULTURE					
1. *Coffee roasting*	2·1–3·8		0·05		0·55
2. *Fermentation*					
(i) Grain handling and drying	4·0				

Operation[b]	Emission factor kg $(10^3\ kg)^{-1}$ unless stated otherwise				
	Particulates	CO	NO_x	SO_x	HC
3. *Grain handling*					
(i) Drying	3·0–3·5				
(ii) Screening and cleaning	2·5–4·0				
(iii) Transporting	1·0–1·5				
(iv) Processing					
corn meal	2.5				
soya bean	3·5				
barley milling	1·5				
4. *Stubble burning*	8.5	50	1		10
C FOSSIL FUEL COMBUSTION IN FURNACES FOR POWER OR HEAT GENERATION					
1. *Bituminous coal*					
Boiler size					
>100 BTU h^{-1}	(1–8) × % ash in coal	0·5	9–27·5	19 × % S in coal	0·15
10–100 BTU h^{-1}	6 × % ash in coal	1	7·5	19 × % S in coal	0·5
<10 BTU h^{-1}	1 × % ash in coal	5	3	19 × % S in coal	1·5
2. *Anthracite coal*	(1–8·5) × % ash in coal	0·5–45	1·5–9	19 × % S in coal	0.1–1·25
3. *Wood*	12·5–15	1	5	up to 1·5	1
			kg $(10^3\ l)^{-1}$		
4. *Fuel oil*	1–2·75	0·4–0·6	1·9–12·6	(17–19) × % S in fuel	0·5
5. *L.P.G.*	0·22	0·19–0·23	0·8–1·45	0·09 × % S in fuel	0·04–0·1
			kg $(10^6\ m^3)^{-1}$		
6. *Natural gas*	240–302	270–320	1920–9600	9·6	16–128
			g $(km)^{-1}$		
D FOSSIL FUEL CONSUMPTION IN INTERNAL COMBUSTION ENGINES					
1. *Motor vehicles (1977)*					
Average all vehicles	0·36	22	2.6 as NO_2	0·12	2·82
2. *Aircraft* (during a typical landing-take-off cycle including idle, climbout and approach)					
			kg $(engine)^{-1}$		
(i) Jumbo jet, 4 engines	0·59	21·2	14·2	0·83	5·5
(ii) Long-range jet, 3 and 4 engines	0·55	21.5	3·6	0·71	18·7

TABLE 2.9 *General emission sources*[a]—*cont.*

Operation[b]	Emission factor [kg $(10^3\ kg)^{-1}$ unless stated otherwise] Particulates	CO	NO_x	SO_x	HC
(iii) Medium-range jet, 2 or 3 engines	0·19	7.7	4·6	0·46	2·2
(iv) Piston transport, 4 engines	0·25	138	0·18	0·13	18·5
(v) Light piston transport, 1 engine	0·01	5·5	0·02	0·006	0·18
3. *Locomotives*					
Average all vehicles	3	16	44	6·8	12·5
4. *Transport marine or inland water*					
			kg $(km)^{-1}$		
(i) Steam ships (fuel oil)					
(a) cruise	0·098	5×10^{-4}	1·13	1·73	0·06
			kg day^{-1}		
(b) in dock	6·8	0·036	90·7	138·0	5·0
			kg $(km)^{-1}$		
(ii) Motor ships (diesel)					
(a) cruise	0·49		0·34	0·37	0·24
			kg day^{-1}		
(b) in dock	7·5		22·7	19·5	16·1
			kg $(10^3\ l)^{-1}$		
4. *Stationary sources*					
(i) Gas turbine (fuel oil)	1·0		14	17	
			kg $(10^6\ m^3)^{-1}$		
(ii) Gas turbine (natural gas)			9600	9·6	
E EVAPORATIVE LOSS					
1. *Dry cleaning*					17·5–152·5
2. *Painting*					420–770
			kg $(10^3\ l)^{-1}$		
3. *Petroleum storage/marketing/filling tanks*					44·1
F SOLID WASTE DISPOSAL					
1. *Refuse incineration*					
(iii) Municipal	7–15	17·5	1·5	1·25	0·75
(ii) Industrial/commercial	17·5–18·5	5	1·5	1·25	1·5
(iii) Domestic	3·5–17·5	150	0·5	0·25	50
2. *Open burning*					
(i) Municipal refuse	8	42·5	3	0·5	15
(ii) Wood refuse	8·5	25	1		2
(iii) Automobile components	50	62·5	2		15

[a] The emission factor used in this table is defined as the amount of pollutant material produced as a consequence of the operation named being performed, e.g. in the production of wood charcoal, every 10^3 kg of wood charcoal manufactured, 160 kg of particulate and 50 kg of hydrocarbon on average are relased to the atmosphere.

[b] Omitted from the table: Crop dressing, cotton ginning and airborne allergens.

Source: Table adapted from reference 17.

TABLE 2.10 *Nationwide emission USA 1970* ($kg\ year^{-1}$)

Pollutant	Stationary combustion	Solid waste disposal	Mobile combustion	Industrial processes	Miscellaneous	Total
Particulates	$3{\cdot}4\times10^9$	$7{\cdot}0\times10^8$	$4{\cdot}0\times10^8$	$6{\cdot}6\times10^9$	$2{\cdot}0\times10^9$	$13{\cdot}1\times10^9$
CO	$4{\cdot}0\times10^8$	$3{\cdot}6\times10^9$	$55{\cdot}5\times10^9$	$5{\cdot}7\times10^9$	$9{\cdot}2\times10^9$	$74{\cdot}5\times10^9$
SO_x	$13{\cdot}2\times10^9$	$5{\cdot}0\times10^7$	$5{\cdot}0\times10^8$	$3{\cdot}2\times10^9$	$1{\cdot}0\times10^8$	$17{\cdot}1\times10^9$
NO_x	$5{\cdot}0\times10^9$	$2{\cdot}0\times10^8$	$5{\cdot}8\times10^9$	$1{\cdot}0\times10^8$	$3{\cdot}0\times10^8$	$11{\cdot}4\times10^9$
Hydrocarbons	$3{\cdot}0\times10^8$	$1{\cdot}0\times10^9$	$9{\cdot}8\times10^9$	$2{\cdot}8\times10^9$	$3{\cdot}7\times10^9$	$17{\cdot}5\times10^9$
As	$3{\cdot}81\times10^5$	$1{\cdot}32\times10^5$	—	$3{\cdot}164\times10^6$	$1{\cdot}602\times10^6$	$5{\cdot}278\times10^6$
Asbestos	—	—	—	$3{\cdot}34\times10^6$	$1{\cdot}4\times10^3$	$3{\cdot}354\times10^6$
Be	81×10^3	—	—	$5{\cdot}2\times10^3$	—	$86{\cdot}2\times10^3$
Cd	—	49×10^3	$0{\cdot}5\times10^3$	$1{\cdot}03\times10^6$	3×10^3	$1{\cdot}082\times10^6$
Fluorides	$5{\cdot}25\times10^6$	—	—	$7{\cdot}77\times10^7$	—	$8{\cdot}295\times10^7$
Pb	$1{\cdot}11\times10^6$	$1{\cdot}36\times10^6$	$1{\cdot}07\times10^8$	$5{\cdot}35\times10^6$	$2{\cdot}1\times10^4$	$1{\cdot}1485\times10^8$
Mn	$1{\cdot}015\times10^6$	88×10^3	—	$7{\cdot}85\times10^6$	—	$8{\cdot}95\times10^6$
Hg	$1{\cdot}34\times10^5$	90×10^3	—	74×10^3	$1{\cdot}3\times10^5$	$4{\cdot}28\times10^5$
Ni	$3{\cdot}136\times10^6$	—	—	$5{\cdot}18\times10^5$	—	$3{\cdot}654\times10^6$
V	$1{\cdot}0055\times10^7$	—	—	$1{\cdot}16\times10^5$	—	$1{\cdot}017\times10^7$

Source: Compilation of Air Pollutant Emission Factors 2nd Ed. US Environmental Protection Agency (1973).

especially as gas or liquid since it can then be piped. The organic component of coal is mainly organized into polyaromatic structures in association with oxygen, sulphur, nitrogen and a very wide range of other metallic and non-metallic elements. Some mean analytical values for chemical elements present in coal are given in Table 2.11 [18]. The value

TABLE 2.11 *Elements commonly present in coal* [18]

Element	Mean (parts 10^{-6})	Standard deviation
Al	12 900	4500
As	14·02	17·7
B	102·2	54·65
Be	1·61	0·82
Br	15·42	5·92
Ca	7700	5500
Cd	2·52	7·60
Co	9·57	7·26
Cl	1400	1400
Cr	13·75	7·26
Cu	15·16	8·12
F	60·94	21·00
Fe	19 200	7900
Ga	3·12	1·06
Ge	6·59	6·71
Hg	0·20	0·20
K	1600	600
Mg	500	400
Mn	49·4	40·15
Mo	7·54	5·96
Na	500	400
Ni	21·07	12·35
P	71·10	72·81
Pb	34·78	43·69
Sb	1·26	1·32
Se	2·08	1·10
Si	24 900	8000
Sn	4·79	6·15
Ti	700	200
V	32·7	12·03
Zn	272·29	694·23
Zr	72·46	57·78

of coal as a fuel is governed mainly by the carbon content. A classification on this basis showing a range of quality, or rank as it is called, from lignites to anthracites is given in Table 2.12. Coals with high carbon content are low in oxygen. The presence of oxygen affects the calorific value of the fuel as this oxygen is combined with carbon and hydrogen of the coal. Hence, the heat of combustion has already been utilized and will not be available during combustion. The amount of nitrogen in coal varies between about 1

TABLE 2.12 *Classification of coal*

	Composition				
Coal	C	H	O, N and S content	% Moisture	% Volatiles
Lignites	60–70	5	20–35	20	45
Bituminous	75–80	5	15–20	10–20	40–45
	80–85	5·5	10–15	5–10	30–40
	85–90	5·5	5–10	5	25–30
	88–90	~5	~6	5	18–25
Semi-bituminous	90–92	~4	~4	~3	10–20
Semi-anthracites	92–94	~3	3·5	~2	10–15
Anthracites	92–94	~3	~3·5		~8

and 2·25% and the sulphur content between 0·5 and 2·5%. The sulphur in coal is in three forms: as iron pyrites FeS_2, as gypsum $CaSO_4$, and bound organically with carbon compounds. During carbonization to coke, about 20% of this sulphur is released as hydrogen sulphide. Most of the remainder is retained in the coke produced, although a little can be found in the tar and aqueous liquor. Most of the nitrogen released at the same time comes off as ammonium compounds, cyanides and various pyridine bases.

After combustion, the non-volatile inorganic residues, mainly oxides, are left as an ash. The principal oxides are silica, alumina and iron(III) oxide, they range between about 37–58%, 26–38% and 6–25%, respectively. Smaller quantities of alkaline earth oxides and alkali oxides, 0·4–6%, are also produced. It has been found [19] that the elements present in fly-ash fall into three categories dependent upon their mass-size distributions. The first group of elements all belong to the rock fraction of coal. They include Mg, Al, Ca, Sc, Ti, Cr, Fe, Co, Rb, Ba, Hf, the rare earths and Th, and are associated with larger particles, greater than 1 μm in size. A second group of elements—Cl, Cu, Zn, As, Se, Br, In, Sb, I and Hg—which are less than 1 μm in size, belong to the organic fraction of coal. For some of the elements there appears no clear preference for either small or large particles. These elements include V, Ni, Mo, Au, Ag, Cd, W, Na, K, Cs, Mn and Ga and comprise the third group. In comparison to the abundances of elements in the original coal, the first group are depleted in the fly-ash and in the second group they are enriched.

2.3.1.2 Oil composition, distillation and metal content

Oil differs from coal in that it contains much less inorganic matter and hence produces less ash on combustion. The hydrocarbon content of crude oils is present in a range of molecular weights. The lower molecular weight material constitutes the dissolved gases, and the higher molecular weight the solid waxes. Four basic types can be identified:

(1) paraffins;
(2) olefins;
(3) aromatics;
(4) cyclic compounds.

TABLE 2.13 *Distillation and product usage of petroleum oil*

Fraction	Boiling range (K)	Composition	Usage
Natural gas		Methane	Fuel gas and town gas
Liquefied petroleum gas (LPG)		Propane and butane	Domestic and industrial fuel, synthetic chemicals
Gasoline	300–450	Complex mixture of hydrocarbons (no sulphur)	Motor fuel
Kerosene (paraffin)	410–575	Complex mixture of paraffinic hydrocarbons with aromatics (low sulphur content)	Aviation gas turbines, tractor fuel, lighting and heating
Gas oil (diesel fuel)	450–650	Paraffinic hydrocarbons (high sulphur content)	Diesel fuel, heating and furnace fuel
Fuel oil	500–700	Paraffinic hydrocarbons (high sulphur content)	Large-scale industrial heating
Lubricating oil	Reduced pressure distillation	Wide range can be mainly aromatic, mainly paraffinic or mixed	Lubrication from greases to light machine oil
Wax	Reduced pressure distillation	High molecular weight paraffins	Candles, cosmetic preparation, petroleum
Bitumen	Residue from distillation	Complex	Road surfacing, roof waterproofing

The paraffins form the main group. The lower members of the series give liquefied petroleum gas and primary flash distillate; they consist of propane and butane, i.e. the C_3 and C_4 members. Upon distillation (Table 2.13), the various fractions named gasoline, kerosene, gas oil, diesel oil, fuel oil, lubricating oil, waxes and bitumen are obtained. Paraffins above $C_{16}H_{34}$ are present as solids and are isolated as waxes. Aromatic and cyclic components such as benzene, toluene, cyclopentane, tetrahydronaphthalene are present in these fractions in varying amounts dependent upon the source of the original crude oil.

Hydrocarbons in combination with oxygen, sulphur and nitrogen also exist in crude oil. Russian crudes generally contain significant amounts of naphthenic acids. These are carboxylic acids attached to a naphthene ring, e.g.

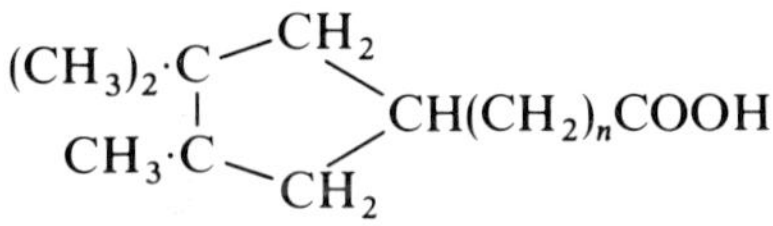

Sulphur and nitrogen compounds are often a feature of some oils. Sulphur compounds like mercaptans, thioethers and thiophenes are associated with Middle East crudes. Oils from California contain nitrogen compounds belonging to pyridine and quinoline base derivatives.

Naturally occurring petroleum hydrocarbons contain measurable quantities of many metals (Table 2.14). Ni and V are frequently the most abundant with Fe, Zn, Cr, Cu, Mn, Co, As, Sb and Hg present in amounts ranging from 1 $\mu g\ g^{-1}$ to more than 100 $\mu g\ g^{-1}$. Rather more is known

TABLE 2.14 *Metal content of crude oils* [21, 22]

Element	Concentration range (parts 10^{-6})	Mean concentration (parts 10^{-6})
As	0·05–1·11	0·26
Co	0·03–12·75	1·71
Cu	0·13–6·33	1·32
Cr	0·002–0·02	0·008
Fe	3·36–120·8	40·67
Hg	0·023–30	3·24
Mn	0·63–2·54	1·17
Mo	0·008–0·053	0·031
Ni	49·1–344·5	165·8
Pb [23]	0·17–0·31	0·24
Se	0·026–1·39	0·34
Sb	0·03–0·11	0·05
V	4·0–1100	88·5
Zn	3·57–85·8	29·8

about how these metals are incorporated into oil compared with metals in coal. The metal components of petroleum can be classified into four categories:

(1) metalloporphyrin chelates (Fig. 2.6);
(2) transition metal complexes of tetradentate ligands;
(3) organo-metallic compounds (Hg, Sb and As);
(4) carboxylic acid salts of the polar functional groups of resins.

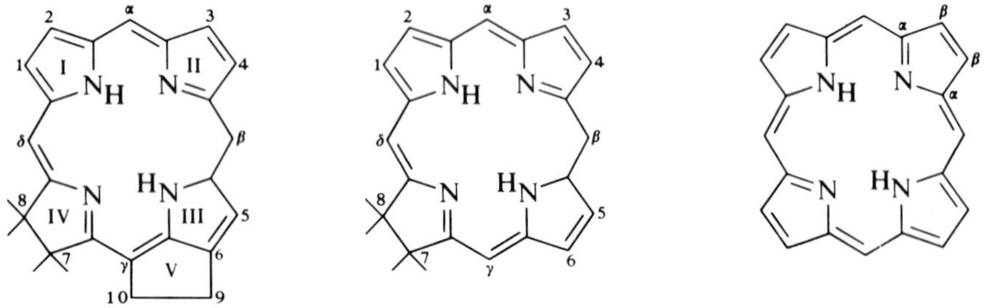

Phorbin Chlorin Porphin

FIG. 2.6 Fundamental structures. Chlorin is the basic skeleton of natural chlorophylls. The 7,8- positions carry dihydro-groups and some derivatives have a penta cyclic ring V as in phorbin [25]

The metalloporphyrins which are extractable from oil are identified as a series of alkyl homologues of cycloalkanoporphyrin. These are the etio-type porphyrins and deoxophylloerythroetioporphyrins (DPEP) [20].

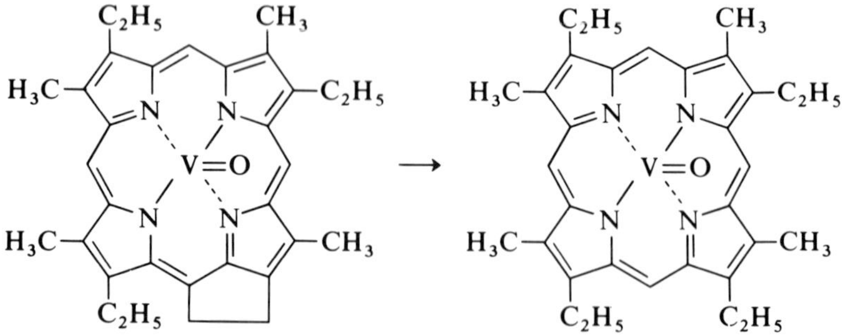

DPEP Etioporphyrin III

The combinations and ratios of their structures are related to geological factors. As the age of the petroleum increases, for example, the petroporphyrin changes from DPEP to etio.

Besides alkyl porphyrins, monobenzo and cycloalkanomonobenzoporphyrins are known. The benzo group may be fused to ring I, II, III or IV, e.g. [24]

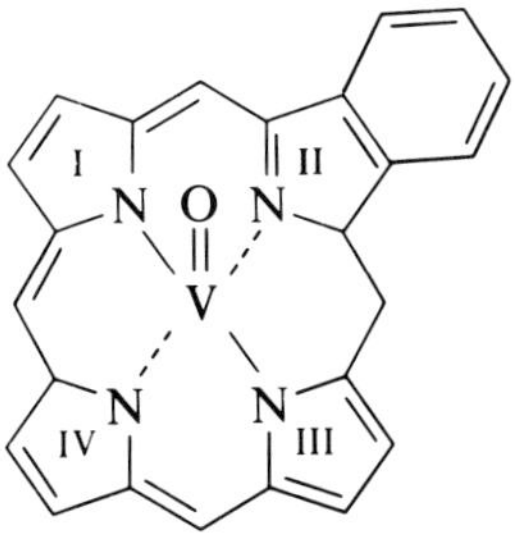

Transitional metal complexes of tetradentate ligands include porphyrin degraded compounds, arylporphyrins and hydroporphyrins. In many cases these derivatives lose their porphyrin identity. It has been suggested by Yen [25] that these ligands can be any combination of four atoms from, nitrogen, sulphur and oxygen (Fig. 2.7). The petroporphyrins are of biogenic origin and are derived from chlorophylls by replacement of the central magnesium atom with other metals [25]. If MeP represents a

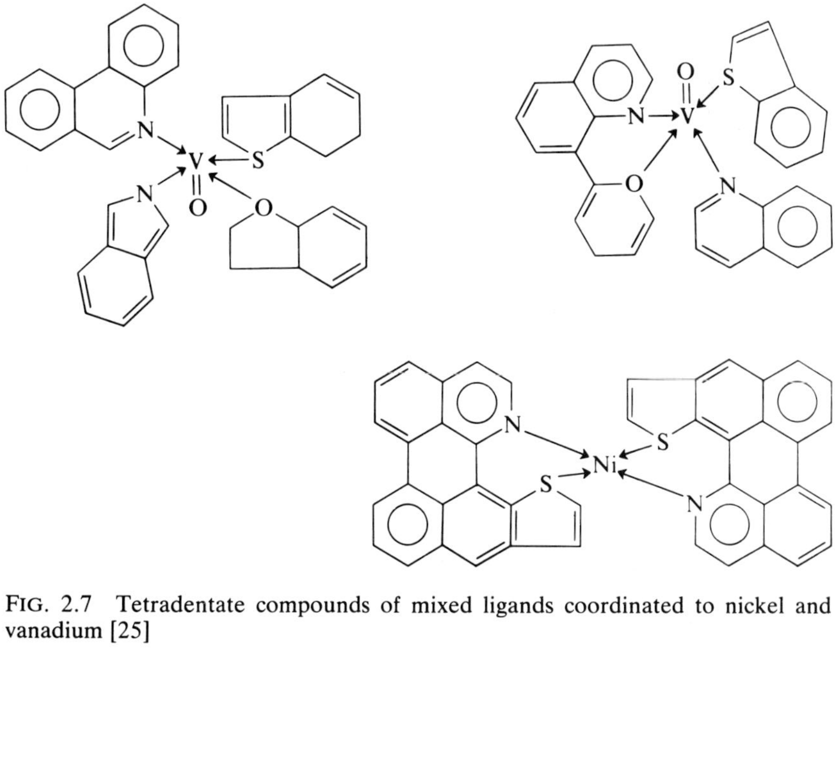

FIG. 2.7 Tetradentate compounds of mixed ligands coordinated to nickel and vanadium [25]

metalloporphyrin and PH_2 is the porphyrin, then the exchange may be indicated by the equation

$$MeP + 2H^+ \rightleftharpoons PH_2 + Me^{2+} \qquad (2.9)$$

The more stable the metalloporphyrin, the more the equilibrium will lie to the left-hand side of the equation.

Ag(I), K, Na, Pb(II) and Mg(II) can be replaced by water. Zn(II), Cd(II), Hg(II), Fe(II) require dilute hydrochloric acid. Cu(II), Mn(II), Co(II), Ni(II), Fe(III) need concentrated sulphuric acid, whereas VO(IV) requires trichloroacetic acid. Hence, once the vanadyl complex is formed, considerable stability results.

2.3.2 Mechanism of formation of pollutants during combustion—the internal combustion engine

Basically, the internal combustion engine draws a mixture of fuel and air into a combustion chamber or chambers which is subsequently ignited. The hot combustion product gases are then expelled before a new charge is introduced into the chamber, in preparation for repetition of the cycle. In the spark ignition engine, the ignition of compressed gas mixture is initiated by a sparking plug and in the diesel engine the mixture ignites spontaneously on compression. Besides the complete combustion products, carbon dioxide and water vapour, both types of engine emit:

(1) carbon monoxide
(2) unburned and partially oxidized hydrocarbons as well as polynuclear hydrocarbons (PNHs) which can either be present in the original fuel or are synthesized in the combustion chamber;
(3) oxides of nitrogen, NO_x—mainly NO;
(4) oxides of sulphur, SO_2 and SO_3;
(5) soot and solid particulate matter.

In addition, petrol-fuelled engines emit compounds of lead. The extent of emissions varies in a complex manner determined by the fuel-to-air ratio, quality of fuel as for example the lead content, and the load put on the engine. Characteristic emission patterns as a function of air-to-fuel ratio are given in Fig. 2.8. It will be noted from the diagram that when NO_x emissions are at a maximum, hydrocarbon and carbon monoxide emissions are minimal. The engine load is reflected in the case of engines used in transport systems by the driving mode of the vehicle. Through the years a great deal of work has been done to discover precisely what goes on during combustion. Although some questions still remain unresolved the following sub-sections summarize current knowledge.

2.3.2.1 Carbon monoxide

Carbon monoxide is not a stable end-product of balanced combustion, but it represents the penultimate stage through which all the carbon species must pass on their route to the formation of carbon dioxide. The gas

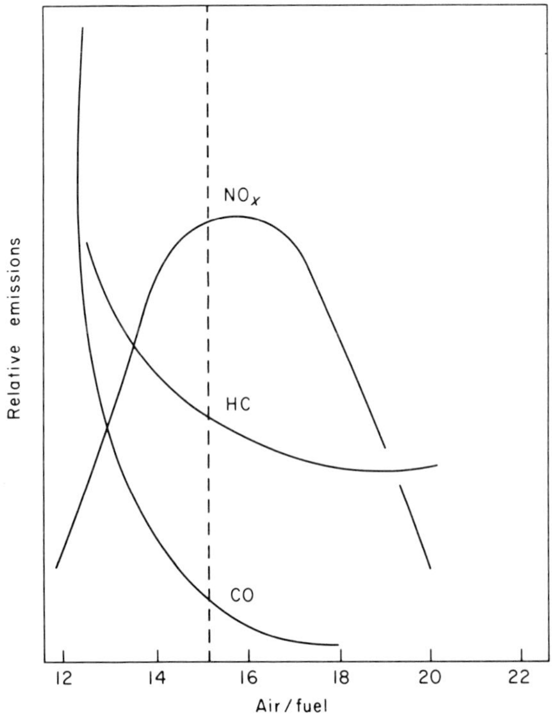

FIG. 2.8 Effect of air/fuel ratio on emissions from internal combustion engines

reaches a high temperature equilibrium concentration inside the combustion chamber. A perplexing aspect which is not entirely understood, is why does this not fall off to an equilibrium level in the exhaust lower than that actually observed. The principal removal reaction of CO during the exhaust cycle is

$$CO + OH = CO_2 + H \tag{2.4}$$

for which the equilibrium should move to the right as the temperature falls. One explanation that has been offered [26] suggests that the Arrhenius activation energy of this reaction is negligible up to 10^3 K but rapidly increases at higher temperatures. The possibility that the supply of OH radicals limits the extent of reaction (2.4) seems unlikely. These OH radicals are formed by the following chain reactions:

$$M+H+O_2 \rightarrow OH+O+M \tag{2.10}$$

$$O+H_2 \rightarrow OH+H \tag{2.11}$$

$$OH+H_2 \rightarrow H_2O+H \tag{2.12}$$

The largest activation energies in the cycle $E_{(2.11)}$ 37 kJ and $E_{(2.12)}$ 21·5 kJ appear insufficient to give rise to energy barriers that would restrict the supply of OH radicals. Actually, in an expanding system as would obtain in the exhaust stage, there will be a further enhancement of OH radicals because the 3-body recombination reaction

$$H+OH+M \rightarrow H_2O+M \tag{2.13}$$

will "freeze" during expansion.

2.3.2.2 Oxides of nitrogen

Nitric oxide is an endothermic molecule and consequently enjoys a greater stability at high rather than low temperatures. The compound is formed in the wake of the flame front as it spreads through the combustion zone (Fig. 2.9). At the temperatures existing in this zone of around 1900–2500 K, atomic oxygen is available for reaction either by dissociation of molecular oxygen:

$$O_2 \rightarrow 2O \tag{2.14}$$

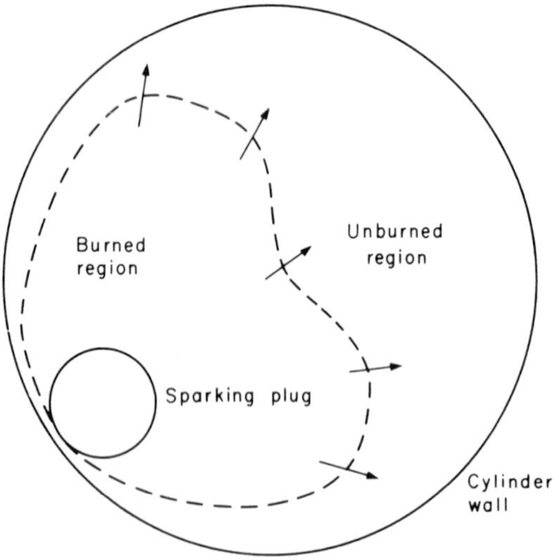

FIG. 2.9 Propagation of the flame front within the combustion chamber

or by the interaction of carbon monoxide with OH radicals:

$$CO + OH \rightarrow CO_2 + H \tag{2.4}$$

followed by

$$H + O_2 \rightarrow OH + O \tag{2.15}$$

Once atomic oxygen has been generated in the system, this initiates the Zeldovich chain reaction responsible for formation of nitric oxide:

$$O + N_2 \rightarrow NO + N \tag{2.16}$$

$$N + O_2 \rightarrow NO + O \tag{2.17}$$

These reactions are also accompanied by:

$$N + NO \rightarrow N_2 + O \tag{2.18}$$

and

$$O + NO \rightarrow N + O_2 \tag{2.19}$$

As the temperature begins to fall after combustion and during the exhaust cycle, nitric oxide formation ceases abruptly because of the high activation energy of reaction (2.16), $E_{(2.6)}$ 316 kJ [10]. At the same time processes which would destroy the nitric oxide already formed, for example by reaction (2.19), similarly have a high activation energy $E_{(2.19)}$ 162 *kJ* and are ineffective. The low atomic nitrogen concentration throughout the system prevents reactions (2.17) and (2.18) from taking place, so the overall effect is to preserve the nitric oxide concentration that existed at the high temperature stage of the process.

Nitric oxide emissions are always accompanied by nitrogen dioxide. These combined emissions of oxides of nitrogen define the term NO_x:

$$NO_x = NO + NO_2 \tag{2.20}$$

The nitrogen dioxide content of NO_x is generally not more than about 10%. It is produced during combustion by a fast reaction between hydroperoxyl radicals and nitric oxide

$$HO_2 + NO \rightarrow HO + NO_2 \tag{2.21}$$

Other things being equal, the production of NO_x depends on combustion chamber temperature. The harder the engine works, by increasing the load or the number of revolutions per unit time, the higher the engine operating temperature becomes and the greater the amount of nitric oxide formed. Any device which reduces temperature within the combustion chamber will reduce nitric oxide formation. Several procedures have been used to achieve this objective:

(1) exhaust gas recycle;
(2) injection of moisture to the combustion mixture;
(3) engine operation biased towards lean fuel-to-air ratios;
(4) engine combustion chamber design.

Haynes and Weaving [27] report a reduction in NO emissions of up to 60% achieved by returning about 13% of the exhaust gas to the intake region of the carburettor. One disadvantage of the device is that, at least under some operating conditions, there is a falling off in the engine performance. Alternatively, Moore [28] has shown that about 1% by weight of water incorporated into the charge to the combustion chamber lowers the flame temperature by some 20 K and this is sufficient to lower nitric oxide formation by about 25%.

Figure 2.8 shows that air/fuel ratios greater than stoichiometric, in the range of 18–20 (i.e. lean mixtures), result in low emissions of NO_x and hydrocarbons. Normally, engines will not run reliably under these conditions but developments in engine design have demonstrated the feasibility of this solution to the problem. One concept which has been tried with some success, is to vaporize the fuel in the induction manifold to ensure a homogeneous air/fuel charge distribution to the cylinders of the engine. This can be done by using hot exhaust gases to heat the inlet manifold system. The other method of ensuring lean mixture operation makes use of the "stratified charge" principle. In this case, two air/fuel feeds are supplied to each cylinder. A rich mixture is introduced into the combustion chamber near the sparking plug and an over-lean mixture fills the remainder of the chamber. The burning zone initiated in the rich region propagates through this lean zone, burning the hydrocarbons efficiently and at a lower temperature, so that nitric oxide production is minimized.

An extension of this technique studied by Newhall and El-Messiri [29] makes use of the pre-combustion chamber concept of engine design. In this system, outlined in Fig. 2.10, combustion is initiated in a fuel-rich environment of a small pre-combustion chamber connected to the main chamber. As burning commences gases expand into the main chamber. Under these conditions of chamber configuration, combustion proceeds in the main chamber at a lower temperature. This lower temperature will restrict the amount of nitric oxide produced and under favourable conditions can be only 5–10% of that from a conventional engine.

2.3.2.3 Oxides of sulphur

Combustion emissions of sulphur dioxide arise from the existence of sulphur compounds naturally present in fossil fuels. The nature of some of these compounds found in petroleum has already been mentioned in Fig. 2.7, section 2.3.1.2. Typically sulphur is present in the ligands that are attached to vanadium and nickel atoms. When emitted in exhaust gas this sulphur is converted to SO_2 in concentrations of about 70 parts 10^{-6} accompanied by smaller amounts (2–3 parts 10^{-6}) SO_3.

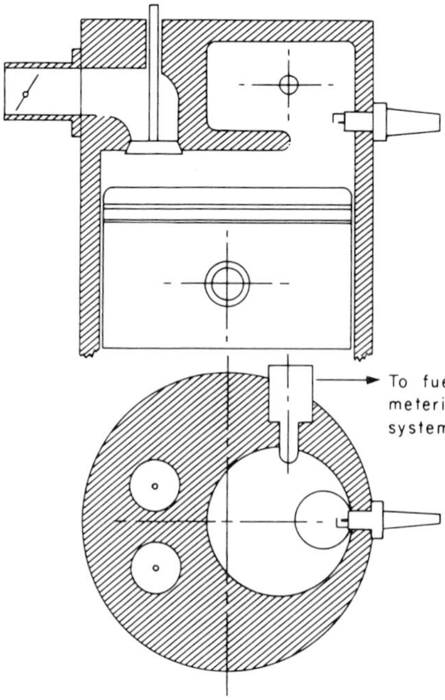

FIG. 2.10 Pre-combustion chamber design for reducing NO_x emissions [29]

2.3.2.4 Hydrocarbons and soot

Organic products from combustion can be divided into three separate categories, those that are present in the fuel that pass through the engine without reaction, belong to one. Partially oxidized hydrocarbons such as aldehydes and carboxylic acids belong to another. These partially oxidized compounds are the cause of odour and the unsaturated compounds can account for photochemical smog. A typical engine exhaust gas comprises 0·2% hydrocarbons, 0·004% aldehydes and a particulate emission of 4000 $\mu g\ m^{-3}$. The hydrocarbon abundances shown in Table 2.15 indicate that they are mainly paraffinic in character.

The other category describes those compounds which are believed to be formed by pyrosynthesis routes from the initial cracking of higher molecular weight material into radicals. These radicals are very reactive and recombine to form polynuclear hydrocarbons. Badger *et al.* [30, 31] have proposed a mechanistic scheme for the pyrosynthesis of aromatics and polynuclear hydrocarbons during combustion. This is outlined in Fig. 2.11,

TABLE 2.15 *Approximate composition of hydrocarbon exhaust emissions from an internal combustion engine*

Compound	Approximate abundance (parts 10^{-6})
CH_4	200
C_2H_4	250
C_2H_2	100
Propene + isobutane	100
Isobutene + butene + butadiene	50
Benzene	25
Dimethylpentane + toluene	50
2,3,4-Trimethylpentene + *m*- and *p*-xylene	25

Source: "Combustion Generated Pollution", Science Research Council 1976.

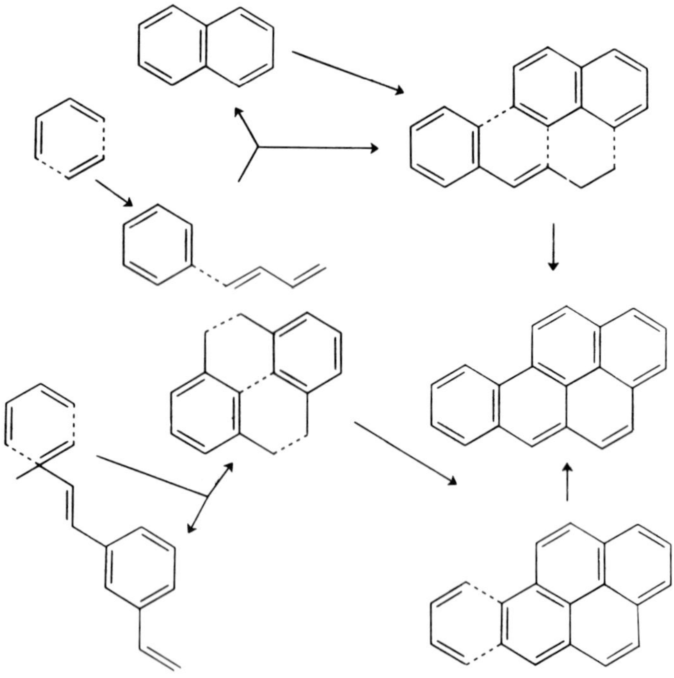

FIG. 2.11 Formation of benzo(a)pyrene by pyrosynthesis according to Badger *et al.* [30, 31]

where the formation of benzo(a)pyrene is illustrated. Pyrolysis of organic matter ruptures C—C and C—H bonds to give lower molecular weight precursor free radicals. These radicals orientate themselves and recombine in the style shown. Acetylene plays a fundamental role and since acetylene is also associated with soot production, the two processes are closely related mechanistically.

As a rough guide, the lower the engine efficiency or the lower the compression ratio, either by design or through engine wear, the more pronounced hydrocarbon emissions become. The reasons for this are associated with wall effects in the combustion chamber. It is a well known feature of combustion that a large surface-to-volume ratio enhances hydrocarbons in the exhaust. Although inefficient mixing of fuel and air may be an important additional factor, the steep temperature gradient near the walls of the chamber are held primarily responsible. Gas temperatures in this region will be cooler than in the remainder of the chamber so that burning will not proceed efficiently. Furthermore, films of higher molecular weight lubricating oils will be present on the cylinder walls. The older and the more worn the engine the greater this contribution will be.

Begeman and Colucci [32] have shown that some of the benzo(a)pyrene in fuel itself can survive the combustion process. This was established by adding 1.1 parts 10^{-6} of labelled benzo(a)pyrene to fuel and recovering 0·1–0·2% of the compound uncharged in the exhaust gas after combustion. Begeman and Colucci also investigated polynuclear hydrocarbon emissions from commercial petrol in comparison to indolene petrols and found that about five times as much benzo(a)pyrene was present in the exhaust from the indolene petrols as in that from ordinary commercial petrol. Undoubtedly, on this evidence polynuclear hydrocarbons are formed during combustion and the fuel aromatic content has some bearing on how much is produced.

Candeli *et al.* [33] have extended these observations by comparing the polynuclear hydrocarbon emission from four different unleaded fuels labelled A, B, C and D, containing up to 46% aromatic compounds. Table 2.16 shows the percentage aromatic hydrocarbon content of the fuels in the C_9–C_{10} and C_{10+} ranges. Their results expressed in terms of the amount of polynuclear hydrocarbons analysed in the fuel to that found in the exhaust gas are shown in Table 2.17. It can be seen from the table that the total emissions of polynuclear hydrocarbons, viz. $D > C > B > A$ are in the same order as the amounts of aromatics in the fuels used. This order is the same, in fact, as that which shows an increase in the amount of benzo(a)pyrene present in the exhaust of $30\ \mu g\ l^{-1}$, $14\ \mu g\ l^{-1}$, $3\ \mu g\ l^{-1}$ and $2\ \mu g\ l^{-1}$. Although this relationship between the aromatic content of fuel and the emission of polynuclear hydrocarbons has been demonstrated, the precise mechanism of formation is not known for certain. It is usually assumed that the pyrosynthesis route suggested in Fig. 2.11 occurs. It seems reasonable that partially consumed aromatic or polyaromatic molecules at least have

TABLE 2.16 *Fuel compositions used by Candeli* et al. [33]

Fuel	Composition	(C_9+C_{10+}) Hydrocarbons (%)
A	Fuel blend characteristic of catalytic reforming.	0
B	Main aromatic components are benzene and toluene.	1·2
C		15·6
D	Main components are alkyl aromatics in the C_9, C_{10} and C_{10+} region.	46·3

the correct orientation for synthesizing similar skeletal structures. It is interesting to note in this connection that the formation of benzo(a)pyrene appears to be favoured. This can be seen by comparing the ratios of polynuclear hydrocarbons in the fuel to polynuclear hydrocarbons in exhaust gas. The values for benzo(a)pyrene for the fuels A, B, C and D are 4/2 = 2; 26/3 = 8·66; 273/14 = 19·5 and 3587/30 = 119·6, respectively. These are the highest ratios for any of the compounds given in Table 2.17.

Polynuclear hydrocarbon emissions from diesel engines are associated with smoke formation, which with this type of engine is critically dependent upon engine maintenance and correct tuning of the fuel injection

TABLE 2.17 *Concentration of polynuclear hydrocarbons in fuels* ($\mu g\ l^{-1}$) *and in exhaust gas* ($\mu g\ l^{-1}$) *of fuel burnt* (33)

	A		B		C		D	
Compounds	Fu	Ex	Fu	Ex	Fu	Ex	Fu	Ex
Phenanthrene + anthracene	370	384	3669	1561	95 144	4704	370 690	11 710
Fluoranthene	44	174	334	454	4781	578	3350	2348
Pyrene	91	143	1341	333	25 596	1224	65 116	2387
Benzofluoranthenes	10	38	76	70	1289	170	10 063	658
Benz(a)anthracene	9	17	40	57	487	72	12 228	642
Chrysene	12	35	75	109	926	136	10 543	767
Benzo(e)pyrene	8	6	73	8	945	49	6815	225
Benzo(a)pyrene	4	2	26	3	273	14	3587	30
Indeno(1,2,3-cd)pyrene	2	12	1	10	61	27	1756	37
Benzo(ghi)perylene	11	20	74	35	1476	126	5815	395
TOTAL	561	831	5709	2640	130 978	7 100	489963	19 199

equipment. An engine employing liquefied petroleum gas is cleaner from this point of view than a conventionally powered engine. Liquefied natural gas may be advantageous in certain industrial operations, for example where engines in fork-lift trucks are used in confined or poorly ventilated areas.

The mechanism of soot and luminous carbon formation in flames involves reactions of the type:

$$C_2 + C_2H_2 \rightarrow C_4H + H \qquad (2.22)$$

$$C_4H + C_2H_2 \rightarrow C_6H, \quad \text{etc.} \qquad (2.23)$$

The existence of C_4H_2 and higher polyacetylenes in the region prior to soot formation in pre-mixed hydrocarbon flames, has been demonstrated by mass spectrometry [34] and by visible and u.v. absorption spectral evidence [35]. Little attention has been taken in combustion studies in flames of the type or the amount of polynuclear hydrocarbon formed with soot. An exception to this is the work of Chakraborty and Long [36]. They showed that polynuclear hydrocarbons produced in soots from ethylene and ethane diffusion flames under similar conditions generate about 2·6 times as much polynuclear hydrocarbon with ethane as with ethylene, although 3·5 times as much soot was formed from the ethylene-fuelled flame.

The reaction attributed to the removal or suppression of soot formation involves hydroxyl radicals.

$$C_n + OH = CO + C_{n-1} + H \qquad (2.24)$$

This is a very fast reaction which can consume OH radicals and deplete their concentration to below the equilibrium value. Their replacement through the reaction

$$H_2O + M \rightarrow H + OH + M \qquad (2.25)$$

is limited by the large activation energy of reaction (2.25). The reaction is relevant for an understanding of the role of alkaline earth metal additives in the suppression of soot from diesel engines. Cotton *et al.* [37] consider that metal additives, especially barium, assist in maintaining an adequate supply of hydroxyl radicals through the following chain reaction:

$$BaOH + H_2O = Ba(OH)_2 + H \qquad (2.26)$$

equilibrium of this stage lies mainly on the right)

$$Ba(OH)_2 + M \rightarrow BaO + H_2O + M \qquad (2.27)$$

$$BaO + H_2 \rightarrow BaOH + H \qquad (2.28)$$

$$H + H_2O \rightarrow OH + H_2 \qquad (2.29)$$

2.3.2.5 Lead

The story of lead additives in petrol goes back to 1919 when Charles Kettering, who was director of research at General Motors at the time, appointed a young engineering graduate named Thomas Midgley to look into the problem of engine "knock". Previously, Sir Henry Ricardo in England had established that engine "knock" was intimately associated with the chemical structure of hydrocarbon molecules of the fuel. Although the octane scale was not invented until 1926–27 by Dr Graham Edgar, Ricardo appreciated that the best paraffinic aviation fuel, around 55–56 octane number, as used by the Allies in 1917, was inferior to the dirty benzene petrol (90–100 octane number) burnt off as waste in the Borneo oil fields. On 9 December 1921, Midgley reported to Kettering that the problem was solved: addition of 1 part in 2000 of a compound called tetraethyl lead was sufficient to prevent "knocking" in internal combustion engines. Subsequently, lead alkyls were introduced in 1923 in the USA and in 1926 in the UK and have been used ever since. Internal combustion engines and the fuels they use today have been designed to be compatible with lead additives in the fuel. Besides controlling the fuel octane number, lead additives act as lubricants for valve guides and seats. Most modern engines will quickly seize up if run on lead-free petrol, unless of course special provision has been made to allow for this factor.

Once the combustion stage of the cycle is complete, the lead residues need to be exhausted as efficiently as possible. This can be accomplished by the addition of ethylene dichloride and ethylene dibromide to the fuel to act as lead scavenging agents. As a consequence of their use, exhausted lead compounds are principally lead bromochlorides; PbBrCl, lead bromochloride is associated with ammonium chloride, for example as $NH_4Cl{\cdot}2PbBrCl$ or 2NH4Cl·PbBrCl and an oxide double salt 2PbO·PbBrCl [38]. Of the total particulate matter exhausted by petrol propelled vehicles between 40 and 60% contains lead, 30–35% iron(III) oxide and the remainder soot. According to Lee *et al.* [39] some 95% of the lead emitted has a particle size diameter of less than about 5 μm. The lead size distribution curve shows a maximum at around 0·25 μm and has 25-percentile points at 0·15 μm and 0·40 μm.

The utilization of lead in the UK and the USA shown in Table 2.18 indicates that the production of lead alkyls accounts for about 16% of the total lead usage in these countries. On a global scale lead emissions to the atmosphere amount to $5{\cdot}45 \times 10^8$ kg year^{-1} of which some $3{\cdot}36 \times 10^8$ kg or 62% is derived from automative sources and the remainder, $2{\cdot}09 \times 10^8$ kg comes from lead smelting [3]. Patterson [41] has calculated a pre-industrial value of lead in air of 5×10^{-14} μg m^{-3} as an indication of the background level derived from natural processes. This may be compared with airborne lead concentrations in cities of between 1 and 4 $\mu g\,m^{-3}$, in rural areas something like 0·1 $\mu g\,m^{-3}$, and in remote areas around 10^{-3} $\mu g\,m^{-3}$ as found in Novaya Zemlaya, USSR [42] and in air above the mid-Pacific ocean [43].

TABLE 2.18 *Utilization of lead in the UK and USA in 1974* [40]

Use	UK Amount $kg \times 10^6$	UK % Fraction of total	USA Amount $kg \times 10^6$	USA % Fraction of total
Batteries	80·1	24·6	772·7	53·3
Lead alkyls	56·1	17·3	227·3	15·7
Ammunition	7·3	2·3	79·0	5·4
Cable covering	44·4	13·7	39·4	2·7
Sheet, foil and pipe	47·6	14·6	38·2	2·6
Collapsible tubes	1·4	0·4	2·3	0·2
Pigments	36·7	11·3	105·4	7·3
Alloy	32·3	9·9	121·1	8·3
Miscellaneous	19·3	5·9	65·5	4·5
TOTAL	325·2	100·0	1451·0	100·0

The atmospheric lifetime of lead aerosols is dependent upon a number of factors which include particle size, meteorological conditions and the concentration of total particulate matter (see also section 2.4.1). The larger coarse particles 0·3–3 mm in diameter will be deposited by gravitational settling. Lead emissions from motor vehicles which fall into the 5–50 μm diameter range will be precipitated by turbulent deposition near the kerb-side [44]. In both of these instances the atmospheric lifetime of lead particles will be short—of the order of minutes at most. Coagulation represents the main mechanism responsible for the removal of the smallest particles, by adding them to the range of particles from 0·1 to 1·0 μm in size. The rate of growth will depend on the total aerosol concentration, so parodoxically the removal rate will increase at higher pollution concentrations [45]. Estimates of the residence time of airborne particulate lead vary from several weeks [46] to 9·6 ± 20% days [47].

It seems on present evidence that the residence time of atmospheric lead which is composed of lead particles less than about 1 μm in size, cannot be measured precisely because too many extraneous influences can affect the determination. If more exact times are required it may well be necessary to specify the conditions so that urban and rural locations are considered separately and measurements of windspeed, temperature and relative humidity are taken into account.

2.3.3 Gas turbine engines

Figure 2.12 shows the conventional layout of a turbine combustor having design requirements which satisfy the following criteria: high combustion efficiency, good combustion stability, low pressure loss, easy ignition, and

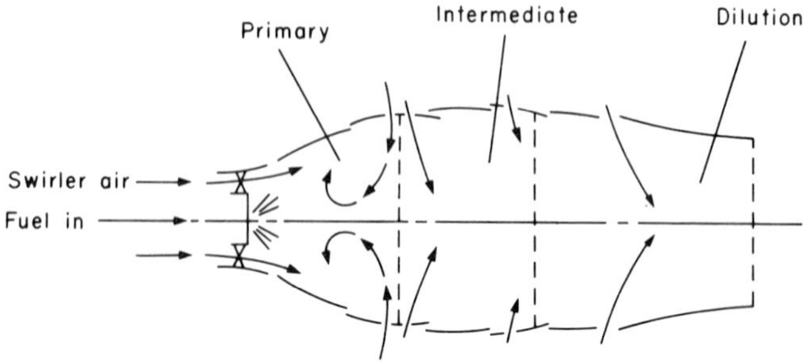

FIG. 2.12 Cross-section of a gas turbine combustor showing air flow pattern (Source: "Combustion Generated Pollution". Science Research Council 1976)

low emisson of pollution. Pollution characteristics shown in Fig. 2.13 follow the same pattern as internal combustion engines, in that maximum NO_x is associated with minimum carbon monoxide and hydrocarbon emissions.

With this type of engine carbon monoxide is formed at the flame front in concentrations that greatly exceed equilibrium values. It is converted to carbon dioxide only if sufficient time and temperature are available in the

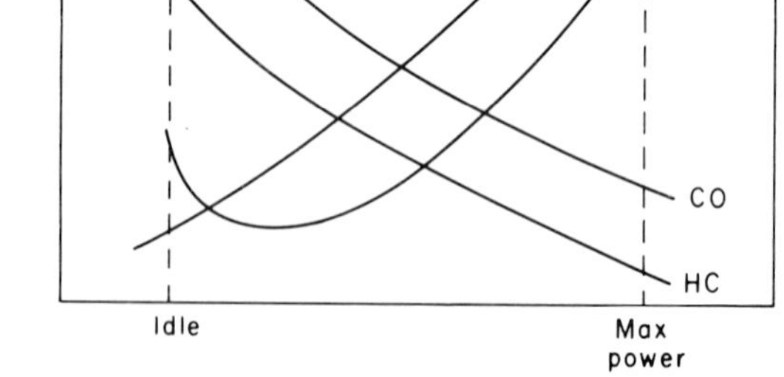

FIG. 2.13 Pollution emission characteristics of a gas turbine engine (Source: a Fig. 2.12)

gas zone. At full load operation the combustion chamber walls and the adjacent gases are at higher temperatures than during idle, fulfilling these conditions. Thus, carbon monoxide and hydrocarbon emissions are at a minimum at full power. Poor fuel atomization and excessive quantities of air in the primary and intermediate zones all serve to increase CO levels. Annular chambers, because of their lower surface/volume ratio, tend to give lower CO emissions than tubular systems.

Unburned hydrocarbons and partially burned hydrocarbons are normally a consequence of poor atomization of fuel, inadequate burning rates and too low temperatures during combustion. Higher inlet air pressures will enhance chemical reaction rates in the primary combustion zone and lower emissions of CO and hydrocarbons.

Measured concentrations of nitric oxide in combustor and engine exhausts are appreciably lower than calculated values based on thermal equilibrium temperatures in the primary zone. This indicates that the production of nitric oxide is limited by the kinetics of formation and that for any given combustor, the exhaust concentration will depend on the pressure, temperature and composition of gas in the combustion zone coupled with residence time in this zone.

Smoke emissions are caused by finely divided soot particles generated in the fuel-rich region of the flame. This is often the result of poor mixing of fuel and air. With pressure atomizers the main soot-forming region lies inside the fuel spray at the centre of the combustor. In this region the recirculating burned products move upstream towards the fuel spray. Local pockets of fuel vapour are enveloped in oxygen-deficient gases at high temperature. Most of the soot produced in the primary zone is consumed in the high temperature region downstream. The soot concentration actually observed in exhaust gases reflects the dominance of one or other of these zones within the combustor.

Soot concentrations are critically dependent on pressure. Higher pressures tend to delay evaporation of fuel droplets and this provides more time for soot formation in the liquid phase. Increase in pressure accelerates chemical reaction rates so that combustion is initiated earlier and a greater proportion of this fuel is burned in the fuel rich region adjacent to the atomizer spray.

2.3.4 Stationary sources: power stations and industrial plants

Increasing demand for electrical power, roughly doubling in every decade, has had two principal effects. These are:

(1) the capacity of individual boilers within a power station have increased to around 600 to 1300 MW;

(2) thermal efficiencies of boilers must be high to offset the increasing cost of fuel.

As with other combustions involving fossil fuels the main pollutants emitted from large power stations are carbon monoxide, oxides of nitrogen and sulphur, partially combusted organic matter, and fly-ash particulates. In the case of pulverized coal, emissions are influenced by the degree of fineness of the coal, the amount of excess air, the aerodynamics of the combustion chamber and the rate of heat absorption by the boiler. In general these factors apply to oil-fired burners as well, except that these units operate with low (i.e. less than 2%) excess air. One of the consequences of operation with so little excess air means that the emission of solids such as soot, very strongly depends upon the quality of the atomizer. To limit the formaton of smoke the mean drop size of the fuel from the atomizer should be less than 200 μm. Good aerodynamic design of the burners is essential for oil-fired units, if solid emissions are to be minimized. Elimination of solid particulates is improved by importing "swirl" to the air fed into the chamber. This affects the turbulence and recirculation patterns of gases in the chamber. Under conditions of low swirl, large quantities of carbon in the range 10–40 μm diameter escape from the combustion chamber. Increasing the swirl to optimum values reduces the mass of solid carbon production by a factor of 4 [48]. This is achieved by reduction in the number of particles in the 10 μm to 40 μm size range. Increasing swirl to beyond this optimum setting generates more soot particles in the sub-micron (<1 μm) range. It is believed that these sub-micron particles arise from vapour phase cracking of hydrocarbons, and the latter originate in their inability to follow the changes in air direction adequately.

As a general rule, the following guidelines are helpful [49].

(1) Emission of solids is mainly determined by boiler design, combustion chamber and atomization characteristics. When properly maintained, and correct operating procedures employed, solid emissions should remain constant during the life of the plant.

(2) For industrial boilers conversion of SO_2 to SO_3 is usually in the range 0·2–2·5%. For gas turbines the range is 1·5–3·5% and up to 30% may be obtained from brick kiln furnaces.

(3) The production of NO_x depends mainly on the scale of fuel consumption, ranging from 100 parts 10^{-6} for small boilers to 500 parts 10^{-6} for large power-station boilers. Significant reductions are possible when the equipment is operated on part load.

Industrial boilers of this type have reasonably low emissions of carbon monoxide, less than 0·01%. When emissions are not of this order, combustion conditions are not efficient and must be adjusted accordingly.

Frequently, optimization of conditions can be controlled automatically by monitoring of carbon monoxide in the flue gases using an infrared analyser (see Chapter 4). This instrument can be coupled to the boiler control system so that as far as possible ideal conditions are continuously maintained by the unit.

2.3.4.1 Emissions of polynuclear hydrocarbons from coal, oil, gas burning units, incinerators and open burning

Emissions of polynuclear hydrocarbons from selected sources have been specifically studied. These include:

(1) industrial and domestic boilers consuming solid fuel, oil and gas;
(2) incineration of municipal and commercial waste;
(3) open burning of waste;
(4) asphalt air blowing and asphalt hot-road-mix manufacture [50].

Emissions from heat generation sources are summarized in Table 2.19 under three categories dependent on the fuel used—coal, oil or gas.

Coal-burning units. The table shows that polynuclear hydrocarbon emissions from coal-burning units varies considerably depending upon the efficiency of combustion. The small domestic furnaces employed for home heating have the highest emissions and the large fully controlled pulverized coal power plants show the lowest. Among the polynuclear hydrocarbons studied by Hangebrauck *et al.* [51] pyrene and benzo(a)pyrene were the only compounds detected in all the coal-fired units. In all cases the ratio of pyrene to benzo(a)pyrene was greater than unity, ranging from 1·5 to 23. Ratios of benzo(a)pyrene to benzo(ghi)perylene were found to vary between 1·3 and 6·5 and ratios of benzo(a)pyrene to coronene ranged from 1·0 to 30. Commins [52] has commented on ratios of polynuclear hydrocarbons from coal combustion as distinct from an internal combustion engine. The equation $1{\cdot}5x+0{\cdot}25(1-x)=y$ was devised where y is the ratio of the concentration of coronene to benzo(e)pyrene and x is the fraction of the total benzo(e)pyrene present that is contributed by motor vehicles. Thus, by substituting the monitored values for y at an urban site, the amount of benzo(e)pyrene that is derived from traffic can be assessed. Because the ratio of benzo(e)pyrene to benzo(a)pyrene is approximately the same from coal and traffic sources, this value of x will also be the amount of benzo(a)pyrene that has been contributed by traffic.

Oil burning units. These units emit less polynuclear hydrocarbons than coal-fired boilers. Again, pyrene was found from all units tested. In cases where benzo(a)pyrene, benzo(ghi)perylene and coronene were also present the ratios were pyrene to benzo(a)pyrene about 6·6, benzo(a)pyrene to benzo(ghi)perylene between 2 and 3, and for benzo(a)pyrene to coronene

TABLE 2.19 *Summary of polynuclear hydrocarbon emissions from heat generation sources* [51]

Fuel used	Firing method	Benzo(a)pyrene		Pyrene	Benzo(e)-pyrene	Benzo(a)-pyrene	Benzo(ghi)-perylene	Perylene	Benz(a)-anthracene	Anthan-threne	Fluor-anthene	Coronene
		ng m^{-3}	$\mu g\ (kg)^{-1}$				$\mu g(10^6\ BTU)^{-1}$					
Coal	Pulverized	58·5	0·71	195	92	25·5					365	
	Chain grate stoker	71	0·97	390	130	37					680	
	Spreader stoker	49	0·77	590	350	26					360	26
	Underfeed stoker	5650	211	11 850	6650	6900	4500	1600	3900	290	38 000	330
	Hand-stoker	340 000	13 216	600 000	100 000	138 000	300 000	60 000		90 000	10^6	30 000
Oil	Steam atomized	39	1·31	174					27		270	
	Low pressure air atomized	1900	40	6100			300				1900	2100
	Vaporized	34	4	1200							15 000	
Gas	Pre-mix burners	108	3·3	4610	254						860	

from 0·4 to greater than 5. An air-atomized burner gave the highest benzo(a)pyrene emission rate.

Gas-burning units. Polynuclear hydrocarbon emissions from gas-burning units indicated in Table 2.19 suggest orders of magnitude similar to oil-burning ones, certainly less than solid fuel boilers. Ratios of benzo(a)pyrene to benzo(ghi)perylene of 0·1 and of benzo(a)pyrene to coronene of 0·41 are lower than those associated with coal-fired sources and are similar to the lower ratios typically found in motor vehicle exhausts.

Incinerators and open burning. Table 2.20 shows that large municipal incinerators emit less polynuclear hydrocarbons than do smaller commercial units used for waste disposal purposes. The inclusion of a scrubber primarily for removel of fly-ash also reduced benzo(a)pyrene emission by 98%. Some details of the relative amounts of polynuclear hydrocarbons discharged from incinerators through (1) the stack gas, (2) the fly-ash residues in the electrostatic precipitator, and (3) in water containing suspended matter leaving the scrubbing tower, have been provided by Davies *et al.* [53]. The unit investigated was rated at 9×10^3 kg h^{-1} and discharged about 0·06 g, 1·9 g and 20·5 g of 4-, 5- and 6-membered ring polynuclear hydrocarbons per day in the scrubbing water, in the stack gas and in the fly-ash residues, respectively. This shows that the greatest daily emission of polynuclear hydrocarbons is the solid residue. Care should be taken, therefore, to ensure that fly-ash residues are not dumped or carelessly disposed of in such a manner that weathering can leach these compounds to cause contamination of ground and surface waters.

By the very nature of the operation measurements of emissions from open burning cannot be made with as much precision as with enclosed boilers or burners. Consequently, the best that can be done in these situations is to collect particulate material by filtration of air in the vicinity of the source and then analyse the deposition on the filter chemically. Inevitably, the results reported in Table 2.21 must be interpreted much more qualitatively. They are consistent, however, regarding trend in the sense that the burning of rubber tyres and motor vehicle bodies, as might be expected, generated more polynuclear hydrocarbons than the other sources designated in the table.

2.3.5 Organic halogen compounds

2.3.5.1 Methyl chloride

Open burning and incineration are also sources of tropospheric chlorine compounds [54]. Methyl chloride, for example, is formed during wood

TABLE 2.20 *Emission of polynuclear hydrocarbons through incineration* [51]

Type of unit		Benzo(a)-pyrene ($ng\ m^{-3}$)	$\mu g\ kg^{-1}$ of refuse charged								
			Benzo(a)-pyrene	Pyrene	Benzo(e)-pyrene	Perylene	Benzo(ghi)-perylene	anthracene	Anthan-threne	Fluor-anthene	Coronene
A. Municipal											
(i) 227×10^3 kg day^{-1} multiple chamber	Breeching (before settling chamber)	19	0·16	18	0·75			0·8		21·6	0·53
(ii) 45×10^3 kg day^{-1}	Breeching (before scrubber)	2700	13·4	114	26·4		5			10	33
	Stack (after scrubber)	17	0·2	4·6	1·3		1·4	0·3		7·3	1·4
B. Commercial											
(i) $4{\cdot}8 \times 10^3$ kg day^{-1}	Stack	11 000	117	705	99	6·8	198	10	14	484	46
(ii) $2{\cdot}7 \times 10^3$ kg day^{-1}	Stack	52 000	573	9250	573	132	1916	639	174	8590	462

TABLE 2.21 *Polynuclear hydrocarbon emissions through open burning of waste* [51]

Emissions	$\mu g\, g^{-1}$ of particulate								
	Pyrene	Benzo(a)-pyrene	Benzo(e)-pyrene	Perylene	Benzo(ghi)-perylene	Benz(a)-anthracene	Anthan-threne	Fluor-anthene	Coronene
Municipal refuse	29	11	4·5					13	4·7
Motor vehicle tyres	1300	1100	450	72	660	560	53	470	110
Motor vehicle bodies	670	270	120	33	150	40	12	450	220
Grass, leaves, tree branches	120	35	21		5·4	25		110	4·7

burning and in the burning of polyvinyl chloride (PVC). About 1% of the combustion products evolved from wood (i.e. cellulose) is methane, and since chlorine is also present in wood, it has been estimated that 2.2 mg of methyl chloride is formed from every gram of cellulose burnt. It has been calculated from available USA data for the period 1972–74, that agricultural field burning, forest fires and various category wild fires contribute 62% to the total USA emission of around $1{\cdot}905\times10^{8}$ kg year^{-1}. The remaining 38% can be attributed to anthropogenic emission derived from PVC burnt in waste and from fires in buildings (Fig. 2.14). These sources arise from the widespread use of PVC as a packaging commodity and in the building construction industry.

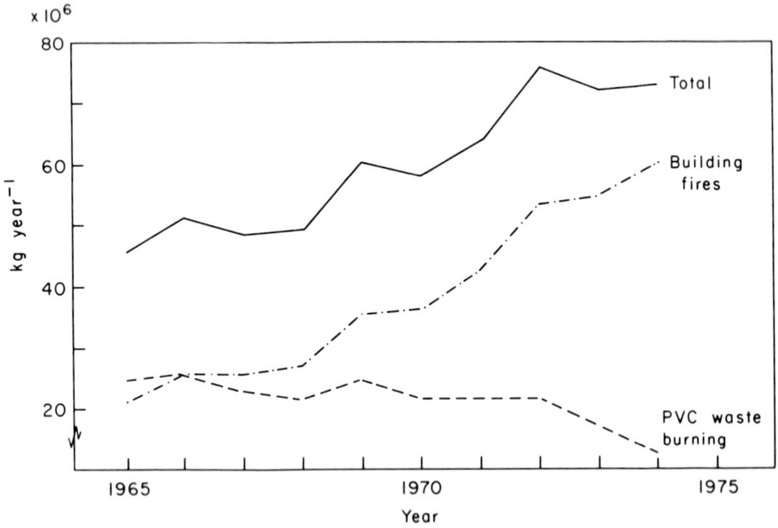

FIG. 2.14 Estimated emissions of methyl chloride from PVC burning and building fires in the USA [54]

2.3.5.2 Carbon tetrachloride

Carbon tetrachloride is used as a cleaning solvent, fumigant, fire extinguisher, and metal degreasing solvent. Estimates by Galbally [55] of emissions between 1934 and 1973 are shown in Fig. 2.15. Future emissions will probably tend to diminish because since the 1950s carbon tetrachloride has been increasingly replaced by other hydrocarbon compounds.

In order to understand the carbon tetrachloride content of the atmosphere, Galbally [55] has used a mass balance equation of the form

$$\frac{d}{dt}[CCl_4]_{atm} = A_{(t)} - R_i \tag{2.30}$$

where $[CCl_4]_{atm}$ is the mass of CCl_4 in the atmosphere, $A_{(t)}$ are anthropogenic emissions as a function of time and R_i is the rate of loss of CCl_4 from the atmosphere; the different loss mechanisms are indicated by the suffix i. These loss mechanisms fall into four categories:

(1) atmospheric reaction with active species (free radicals);
(2) hydrolysis by water;
(3) photolysis by u.v. radiation;
(4) absorption and subsequent decomposition by biological systems.

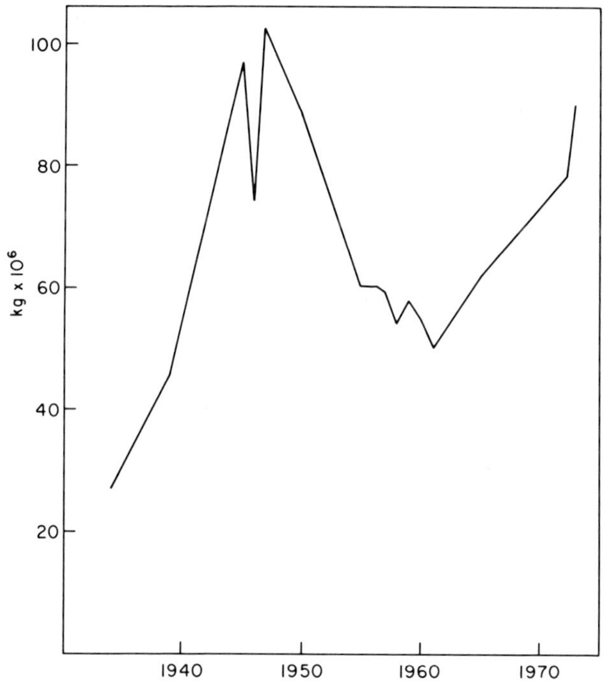

FIG. 2.15 Anthropogenic emissions of carbon tetrachloride [55]

In principle the first of these atmospheric removal routes could involve radical species, $O(^1D)$, $O(^3P)$, H, CH_3, and OH. In practice, only the $O(^1D)$ reaction is significant [see section 5.1 for definition of $O(^1D)$ and $O(^3P)$]. It gives a pseudo-first-order rate constant $k_1 \sim 10^{-3}$ year^{-1}, which is valid up to altitudes of 18 km. At higher altitudes the $O(^1D)$ reaction is less than 1% of the photolysis rate (route 3) and hence can be neglected. Consideration of the second route, hydrolysis by water, reveals that the amount of carbon tetrachloride stored in oceans is less than 1% of that present in air. The large contact between ocean and atmosphere will ensure a relatively large turnover between these phases giving an ocean lifetime for carbon tetrachloride of about 6 months. The corresponding atmospheric loss rate constant k_2 comes to 7×10^{-3} year^{-1}.

Stratospheric carbon tetrachloride absorbs u.v. radiation in the 195–225 nm region and dissociates:

$$CCl_4 + h\nu \rightarrow CCl_3 + Cl \quad (2.31)$$

The model of Molina and Rowland [56] which takes into account varying vertical eddy diffusivity, various photodissociation rates with altitude, latitudinal, and diurnal averaged global loss rates, gives a value due to photodissociation of about 10^6 molecules cm^{-3} s^{-1} at a tropospheric carbon tetrachloride concentration of 73 parts 10^{-12}. This model gives a value for k_3 of $1{\cdot}7 \pm 0{\cdot}8 \times 10^{-2}$ year^{-1}.

Finally, land biota must be considered as a sink for carbon tetrachloride. The carbon tetrachloride content of land biota comes to about 2×10^6 kg which is a negligible amount compared with the atmospheric content of about 10^9 kg of CCl_4. Evidence suggests that the half-life of carbon tetrachloride concentrations found in processed foodstuffs are in a similar range to that present in plants and plant products. If a 4-month half-life is assumed for carbon tetrachloride in land biota, then the loss rate in land biota is 4×10^6 kg of CCl_4 year^{-1}. Hence the atmospheric loss rate $k_4 = 2 \times 10^{-3}$ year^{-1}.

On the basis of these findings equation (2.30) may be solved. Galbally [55] concludes from the result that the majority of the carbon tetrachloride found in the atmosphere is man-made. Values of 75 parts 10^{-12} for carbon tetrachloride in both the northern and southern hemispheres are consistent with an equilibrium concentration due to anthropogenic emissions of about 50×10^6 kg year^{-1}. This is the same order of magnitude as that given in Fig. 2.15 during the last decade.

2.3.5.3 Chloroform

According to Yung *et al.* [57], estimates of the global production of chloroform vary between $2{\cdot}3 \times 10^8$ kg and $1{\cdot}0 \times 10^9$ kg year^{-1}. The smaller value assumes that $CHCl_3$ is of anthropogenic origin and that the measured mole fraction (v/v) of between 19×10^{-12} and 27×10^{-12} is uniformly dispersed.

Sources of chloroform reviewed by Morris [58] refer to the haloform reaction. In this reaction chloroform is obtained as a by-product during the chlorination of municipal water supplies. The following sequence of reactions has been suggested, based on the initial production of hypochlorous acid:

$$Cl_2 + HCO_3^- \rightarrow HOCl + Cl^- + CO_2 \quad (2.32)$$

$$HOCl \rightleftharpoons H^+OCl^- \quad (2.33)$$

Hypochlorous acid then reacts with organic residues present in the water, e.g.

$$CH_3COR + 3HOCl \rightarrow CCl_3COR + 3H_2O \quad (2.34)$$

Eventually hydrolysis of these chlorinated compounds liberates chloroform:

$$CCl_3COR + H_2O \rightarrow CHCl_3 + R{\cdot}COOH \quad (2.35)$$

The yield of chloroform from this chlorination reaction is of the order 1–3%, the actual value being governed by the availability of dissolved organic matter in the water. Similar reactions have been proposed for the production of chloroform from the bleaching of pulp paper by chlorine [57]. A conversion efficiency as low as 6% in this bleaching process would account for a global emission of chloroform of about 3×10^8 kg year^{-1}. As far as current knowledge is concerned, there are no known natural sources of chloroform. Once the compound has been produced and distributed in the atmosphere, the principal removal reaction is thought to be through reaction with OH radicals in an analogous reaction to that given by equation (2.36).

2.3.5.4 Bromine and organic bromine derivatives

The lower atmosphere contains bromine in various forms. This is derived (1) from ocean sources, (2) in association with lead in the exhausts of petrol-propelled vehicles, and (3) from agricultural dressing.

The oceans contain 2×10^{-2}, 7×10^{-5} and 3 to 7×10^{-8} parts by weight of chlorine, bromine and iodine, respectively [57]. Natural emissions of halogenated materials come primarily from this source. Anthropogenic contributions are derived mainly from the use of ethylene dibromide and ethylene dichloride mixtures as scavenging agents with leaded petrols and the employment of methyl bromide as an agricultural fumigant.

An estimate of the source strength for total gaseous bromine by Wofsy *et al.* [59] comes to 10^9 kg year^{-1}. Some 80% of this is derived from marine aerosols and about 10% each from methyl bromide and lead bromochloride. Methyl bromide is removed from the atmosphere by reaction with OH radicals:

$$CH_3Br + OH \rightarrow CH_2Br + H_2O \quad (2.36)$$

The bromine atom is eventually released and inorganic bromide comprising (HBr+Br+BrO) is carried out of the troposphere and returned to earth by rain. The fate of the lead bromochloride entering the atmosphere from motor vehicles is uncertain, but as discussed in Chapter 5, photodissociation has been suggested as one possibility.

2.3.5.5 The freons; fluorotrichloromethane ($CFCl_3$) and difluorodichloromethane (CF_2Cl_2)

These compounds are sometimes referred to as fluorocarbon-11 and fluorocarbon-12, respectively. The 1972 world production figures for $CFCl_3$ and CF_2Cl_2 are about 3×10^8 kg year^{-1}, respectively [60]. These rates are increasing by about 9% year^{-1}. As far as is known there are no biological processes which are capable of removing freons from the environment. Furthermore, the low water solubility of these compounds prevents their removal by rain-out in the troposphere. This feature also precludes the oceans as an acceptable sink. Photodissociation in the stratosphere, as discussed in Chapter 5, appears to be the only significant route for their eventual disappearance. Estimates of their atmospheric lifetimes due to this mechanism vary from a minimum of 30–40 years to a maximum of several centuries.

2.3.5.6 DDT in the biosphere

Peak production of DDT in the USA occurred in 1963 when approximately 8×10^7 kg were manufactured; by 1969 and 1970 the amount had fallen to 6×10^7 kg and 3×10^7 kg, respectively. Woodwell *et al.* [61] estimate DDT residues in soils to be about $0{\cdot}17$ g m^{-2} in agricultural areas and $4{\cdot}5\times10^{-4}$ g m^{-2} in non-agricultural regions. By far the highest values ($8{\cdot}18$–$14{\cdot}6$ g m^{-2}) are associated with orchards. The rate of loss of DDT residues from soils in the USA was estimated by these authors in the following manner. From a total land area of $7{\cdot}7\times10^{12}$ m^2 some 11% or $8{\cdot}5\times10^{11}$ m^2 is kept in crops on which DDT preparations are used. This agricultural land retains about $0{\cdot}17\times8{\cdot}5\times10^{11}=1{\cdot}44\times10^8$ kg. The rate of use of DDT in the USA during this period of the early 1960s was $2{\cdot}7\times10^7$ kg year^{-1} so that the mean lifetime of DDT in soils must be around $1{\cdot}44\times10^8/2{\cdot}7\times10^7=5{\cdot}3$ years. Four processes are attributed to losses of DDT from soils:

(1) volatilization,
(2) removal during harvest of organic matter,
(3) water drainage,
(4) chemical or biological degradation.

The occurrence of DDT residues in rain water suggests that transport into the atmosphere takes place either as vapour or adsorbed onto particulate matter. The vapour pressure of DDT at 20°C is $1{\cdot}5\times10^{-7}$ torr. This is equivalent to about $3\ \mu g\ m^{-3}$ or 2 parts 10^{-9} by weight. Assuming that DDT remains in the atmosphere as vapour the saturation capacity of the troposphere comes to about 10^9 kg of DDT or approximately as much as has been manufactured. This is likely to be an underestimate since DDT residues adsorbed on particulates have not been considered. This means that the atmosphere is potentially a large reservoir as well as a major method of transport of the material. On the basis of land to marine areas of the earth, rainfall will return atmospheric DDT mainly to the oceans. Since DDT is virtually insoluble in water but soluble in fatty tissue, a concentration process is observed in which marine DDT collects in organic matter—notably fish, mammals and plants. Eventually, as in the case of carbon dioxide, transport of residues from the upper layers of ocean waters 75–100 m deep to the much larger lower volumes in the depths will occur.

Calculated average DDT concentrations in the atmosphere and in the mixed layer of the oceans for the period 1940–2000 are shown in Fig. 2.16. The two possibilities envisaged are represented in the diagram by curves A and B, and are dependent upon whether or not manufacturing of DDT continues. Assuming world DDT production ceased in 1974, the lower atmospheric concentration in 1966 will attain the maximum value of $84\ ng\ m^{-3}$ and the mixed layer of the ocean will contain its maximum concentration of $17\ ng\ m^{-3}$ by 1971. The DDT content of these major reservoirs will decrease gradually, reaching 10% of the peak value in air by

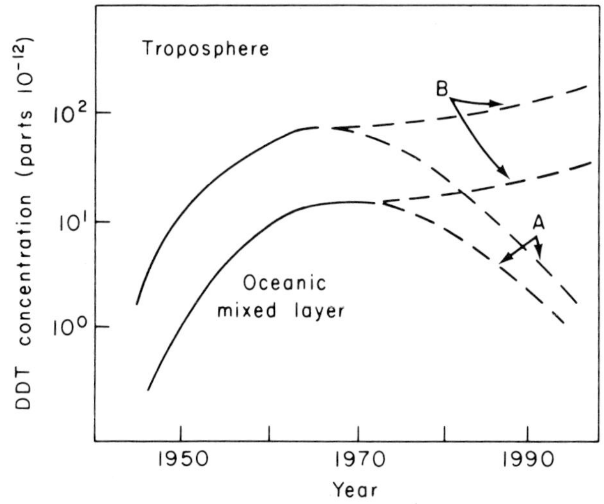

FIG. 2.16 Calculated average DDT concentrations in the atmosphere and in the mixed layer of oceans from mid-1940s to 2000 [61]

1984 and in the ocean by 1993. Curve B assumes that restrictions on the manufacture of DDT imposed in the USA will not be followed by other nations—these nations taking the view that the beneficial aspects of DDT for controlling diseases like malaria and pest control in agriculture outweigh the disadvantages. Curve B, therefore, represents an attempt to predict the consequences of such a policy.

2.3.6 Trace elements in the atmosphere

Table 2.22 gives examples of the trace element content of the atmosphere measured in non-urban regions of the northern hemisphere during 1975. Typical concentrations analysed in the UK between latitudes 51°N and 55°N and between longitude 2°E and 4°W are reported. These concentrations are compared with those measured in the marine environment of the Shetland Isles at 60°N 2°W [62].

Data of this type can be interpreted with the assistance of derived parameters. These parameters help to identify the elements into three main source categories: (1) maritime, (2) soil-derived and (3) industrial.

Two useful derived parameters defined by Chamberlain [16] are the washout factor W, where

$$W = \frac{\text{concentration in rain } (\mu\text{g kg}^{-1})}{\text{concentration in air } (\mu\text{g kg}^{-1})} \tag{2.37}$$

and the dry deposition velocity Vg, where

$$Vg \text{ (cm s}^{-1}) = \frac{\text{rate of dry deposition } (\mu\text{g cm}^{-2}\text{ s}^{-1})}{\text{concentration in air } (\mu\text{g cm}^{-3})} \tag{2.38}$$

Another parameter, the enrichment factor, takes into account the trace element content of an aerosol which has been derived from the soil by the passage of winds. For convenience all concentrations are made relative to scandium; this is because airborne scandium is generally considered to originate from soil. Hence the enrichment factor is defined by:

$$\text{Enrichment factor} = \frac{\left(\dfrac{\text{air concentration of element}}{\text{air concentration of Sc}}\right)}{\left(\dfrac{\text{average soil concentration of element}}{\text{average soil concentration of Sc}}\right)} \tag{2.39}$$

Anthropogenic influences are usually the cause of large enrichment factors due to direct discharge into the atmosphere. Among the heavy metals, the highest enrichment is shown by Cd and Pb. In this regard, the high bromine

TABLE 2.22 *Elemental concentrations in air at selected non-urban sites in the UK (1975). Data abstracted from Cawse, reference 62*

Element	Non-urban concentration (ng m^{-3})	Enrichment factor	Shetland Isles concentration (ng m^{-3})	Enrichment factor
Cl	2728		4414	
Ca	1161	8·9	821	28
Na	1021		3310	
Mg	335	8·3	478	44
Fe	328	0·9	89	1·1
K	263	2·1	172	5·5
Al	212	0·4	64	0·4
Pb	126	1295	21	960
Zn	104	197	15	140
Ti	45	0·9	10	1·0
Br	38	733	15	1370
Ni	24	17	4	<39
Cu	18	108	20	440
Mn	18	2·1	3	1·4
V	11	12	3	12
Cr	6	5·6	0·7	3·4
As	5·3	78	0·6	48
Cd	5	8823	0·8	<7200
I	5	108	4	360
Rb	3·4	3·6	1	<6
Sb	2·3	31	0·4	24
Co	2·1	5·2	0·06	3·6
Se	1·7	822	0·5	1090
Mo	0·6	29	0·2	<57
W	0·6	73	0·5	<240
La	0·4	1·6	0·2	4
Ce	0·4	0·9	0·1	0·9
Ag	0·4	377	0·07	<350
Hg	0·2	857	0·004	<580
Cs	0·1	2·7	0·04	3
Th	0·07	1·4	0·02	1·8
Sc	0·07	1·0	0·01	1·0
In	0·06	208	0·02	<410
Sm	0·05	2·6	0·01	3
Eu	0·01	3·4	0·004	3·9
Au	0·004	373	0·005	1650
Total average suspended matter	29·8 μg m^{-3}		8·9 μg m^{-3}	

value is consistent with lead input to the atmosphere through the association of lead bromochloride, as already discussed in section 2.3.2.5.

Washout factors usually fall within the range 500–2000. Higher values in this range are indicative of higher concentrations of these species in the atmosphere at the rain-forming altitudes (in the region of 3 km), compared with their concentrations at ground level. It follows that high W values are commensurate with material that is dispersed at a higher altitude and therefore with a more remote source, whereas low W values will tend to reflect less dispersed material, probably of local origin.

The dry deposition velocity represents a complex function which includes particle size, surface roughness and wind speed. The soil-derived elements Al, Fe and Th have high Vg values and low W factors, features which imply that they are confined to low altitudes and consist of particles at least as large as a few micrometres in size. Some typical industrial derived elements like V, Co, Cr, As, Pb, Cu, Zn tend to have low Vg velocities as well as low W factors.

Some assistance regarding origin can be obtained by considering ratios of selected elements in the aerosol. An obvious example is the ratio of Cl/Na, which is 1·8 in sea water. Maritime derived aerosols will have Cl/Na values close to 1·8; any significant increase in this ratio is attributable to industrial emissions. The ratio can fall to a value as low as 0·2 at inland locations far removed from oceans [63].

2.4 Elimination of atmospheric pollutants

2.4.1 Natural scavenging processes

The discussion in section 2.2 dealing with the S-cycle and the N-cycle indicated that when particulate aerosols are formed these eventually return to the Earth's surface. What goes up must come down! This can happen in a number of ways:

(1) precipitation scavenging;
(2) aerosol coagulation resulting in particle growth and gravitational deposition (some mention of this process has been made in section 2.3.2.5, which dealt with the production of lead aerosols);
(3) dry deposition.

In precipitation scavenging two distinct mechanisms can be recognized:

(1) "Rainout." In this instance aerosol particles are incorporated into cloud droplets before the formation of raindrops within the cloud; and
(2) "Washout," where raindrops falling from higher altitude scrub out aerosol particulates at lower altitudes by essentially an inertial impaction mechanism.

The "rainout" mechanism depends upon capture by cloud droplets of particles smaller than 0·1 μm radius by Brownian diffusion. Particles less than 0·1 μm radius are called Aitken particles. Dufour [64] has calculated the half-life for nuclei of 0·01, 0·05 and 0·1 μm radius in the presence of 200 droplets cm^{-3} of 10 μm radius as 35 min, 10·8 h and 32 h, respectively. These estimates show that the efficiency of incorporation of particulates into cloud droplets improves with decreasing size of particulate.

Aitken-size particles are naturally associated with sea-spray evaporations, forest evaporations, atmospheric dusts of terrestrial origin, and aerosol formation in the presence of hydrocarbons, oxides of nitrogen and sulphur. According to Went [65], their concentration in the atmosphere is subject to diurnal variation. The value is lowest in the late night hours, rises by about 20% during daylight and then falls again after sunset.

Particles in the size range between 5 and 10 μm radius are removed by inertial impaction. The rain droplet scavenging efficiency, in this case, is influenced by the raindrop cross-sectional area and the intensity of the rainfall. From this analysis of precipitation scavenging, it will be apparent that a gap exists for particle sizes of radius between about 0·05 and 5 μm in which neither Brownian diffusion nor aerodynamic impaction can remove aerosols efficiently. This gap was first described by Greenfield [66] in 1957 and has been termed the "Greenfield window".

The effectiveness of rain for removing sub-micron particles has been calculated by Junge [67] in terms of a factor defined by

$$\alpha = \frac{Eh\rho}{LH} \tag{2.40}$$

where E is the rainout efficiency, h is the amount of rainfall, ρ is the density of water,

L = liquid water content in the cloud

H = depth of the atmosphere to the top of the aerosol layer

and

α = fraction of the aerosol removed

E can be estimated from the equation,

$$E = \frac{CL}{C_o\rho} \tag{2.41}$$

where C is the aerosol concentration of precipitation and C_o is the aerosol concentration in air. Values of L/ρ may be taken as 10^{-6}.

By taking some typical values of C and C_o for aerosols of sulphate nitrate, Cu and Pb, Robinson and Robbins [3] have calculated the values of E given in Table 2.23.

TABLE 2.23 *Estimated values of* E, *the rainout efficiency for specific atmospheric aerosols* [3]

Aerosol	C (μg cm^{-3})	C_o (μg m^{-3})	E
Sulphate	2	4·2	0·48
Nitrate	0·7	0·7	1·00
Cu	0·02	0·07	0·29
Pb	0·034	0·066	0·51

Assuming that aerosol concentration is uniform in the troposphere, Junge [67] shows that where washout is negligible the amount of the aerosol left in the air after t days may be expressed by:

$$1-\alpha=\exp\left(\frac{-Eh\rho t}{LH}\right) \qquad (2.42)$$

Thus, substituting some reasonable values of $h = 0{\cdot}25$ cm day^{-1} and $H =$ 12 km, it is possible to calculate the time in days necessary to remove 25%, 50%, 90% and 99% of the atmospheric aerosol. These times are shown in Table 2.24 for various values of the parameter E, the rainout efficiency [3]. Table 2.24 shows that as E decreases in value, the atmospheric residence time of the aerosol increases. Further, even where $E = 1$ (the maximum possible value), it will still take 11 days for the removal of 90% of the aerosol. For values of E less than unity, as for example approximately 0·5 for the sulphate and lead aerosols given in Table 2.23, the calculation predicts that about half the particulate material will persist for at least a week. In time periods as long as these it is reasonable to conclude that the

TABLE 2.24 *Calculated times for the removal of given percentages of atmospheric aerosol as a function of rainout efficiency* [3]

Efficiency	Time in days to remove: 25%	50%	90%	99%
1·0	1·3	3·3	11	22
0·75	1·7	4·4	15	29
0·50	2·6	6·6	22	44
0·25	5·2	13	44	88
0·10	13	33	110	220

aerosol can be transported over considerable distances, even globally dispersed. The discovery of lead deposits in Greenland and Antarctic ice serves to emphasize this point [68].

Aerosol coagulation processes are controlled by diffusion coefficients of the particles present in the system. This dependence upon diffusion will be more important for Aitken-type particles. This is illustrated in Fig. 2.17, which shows the aerosol size distribution pattern as a function of time [69]. Clearly, particles larger than 0·1 μm radius change little with time, but particles smaller than 0·1 μm radius become progressively less during the 5-day coagulation period. Indeed, particles less than 0·01 μm radius disappear altogether after 12 h. Hence, this coagulation mechanism predicts on a mass basis an increase in total mass of particulate material between 0·1 and 1·0 μm at the expense of aerosol mass in the Aitken range.

Dry deposition accounts for about 20% of the total aerosol deposition. This sedimentation process is governed by Stokes' law:

$$v = \frac{2}{9}\frac{gr^2}{\eta}(\rho - \rho_a)\left(1 + \frac{C}{rp}\right) \tag{2.43}$$

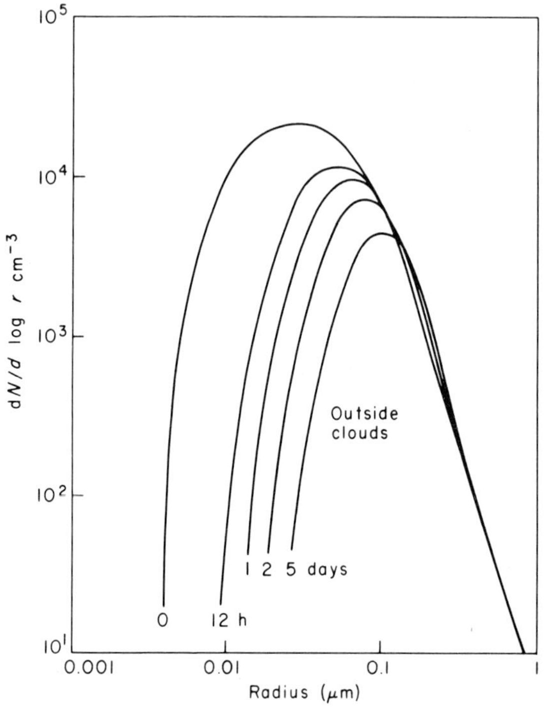

FIG. 2.17 Calculated atmospheric aerosol particle size distribution changes as a function of time without the influence of cloud droplet condensation [69]

where

v = terminal settling velocity
r = radius of particle
g = acceleration due to gravity
ρ and ρ_a = density of particle and air, respectively
η = viscosity coefficient of air
p = air pressure

and

C = constant (when p is given in millibars and r in centimetres), $C = 0{\cdot}0084$ [70]

The law predicts that the terminal settling velocity for particles in the micron and sub-micron range will be less than $0{\cdot}1 \text{ cm s}^{-1}$. These are much less than the normal eddy velocities associated with turbulence and mixing, so that on this basis, particles of this size would be expected to remain in suspension. Nevertheless, Chamberlain [71] has shown that particles of this size can be deposited on rough surfaces with greater efficiency than suspected. This is important, because it means that sub-micron particles can be removed from the atmosphere once turbulent mixing brings them close to the surface or within the laminar boundary layer of air that surrounds all obstacles. Hence, even in the absence of precipitation processes there still remains a route whereby particles can be removed from the troposphere.

2.4.2 Techniques to prevent atmospheric emission from motor vehicles

Emission inventories conducted in the USA and West Germany show that motor vehicles account for 50% by weight of all pollutants discharged to atmosphere. Transport sources, therefore, are the greatest single contributors to air pollution. If only carbon monoxide, oxides of nitrogen and hydrocarbons are considered then their contribution to the total emissions are even more significant (Table 2.25).

Legislation originally enacted in the USA called for a progressive reduction of around 95% of hydrocarbon, CO and NO_x emissions over a decade

TABLE 2.25 *Transport emissions as a percentage of total emissions* [72]

Country	CO	HC	NO_x
USA	74	53	47
Sweden	89	81	22

to 1976. This objective has proved unrealistic on technological and economic grounds so that the programme has been delayed to contain reduction to between 50 and 80%. European legislation has been less severe, demanding a 30% reduction of CO and HC and as yet no control of NO_x emissions.

When considering emissions from mobile sources it is essential to define precisely the way in which the vehicle is driven. Different driving modes encourage the formation of different pollutants. As a general rule, acceleration and heavy loading minimizes CO and HC production but favours NO_x formation. Idling, and running downhill under low-load conditions, on the other hand, give less NO_x and more HC. In order that repeatable conditions for test procedures are used, test driving cycles have been invented in which the vehicle is driven in a certain manner for a certain time. The manner and time are varied according to the requirements of the total test programme. Different countries have adopted slightly different standards so that several test cycles are currently in force (Fig. 2.18a and b). During the test the vehicle is coupled to a dynamometer through the driving wheels to simulate and measure the driving conditions required. Vehicle emissions are monitored and analysed by appropriate instruments which sample the exhaust gas (Fig. 2.19).

Some aspects of the reduction of emissions from engines have already been considered in section 2.3.2, where the mechanism of formation of noxious combustion products was discussed. In addition to these fundamental design characteristics already mentioned, two further preventive measures can be introduced without much difficulty. These are: (1) control of crank case emission; and (2) control of evaporative fuel losses. Crank-case emissions are the simplest to control and were the subject of the first stage in the original USA legislative programme. All that is required is the positive ventilation of the crank-case into the inlet manifold of the engine. This is done by a pipe connection and a one-way return valve. Evaporative losses from the fuel tank can be prevented by using an activated charcoal trap incorporated into the filler cap.

2.4.2.1 Emission control by catalytic methods

Unlike the mechanical and design concept of emission control from engines, the catalytic method recognizes that noxious gases will be produced during combustion. The method seeks to make these gases harmless by initiating further chemical reaction between these gaseous species to produce nitrogen, carbon dioxide, and water vapour. The design and development of such a catalytic system has been one of the most challenging topics presented to chemists in the last decade. Considerable effort has been expended, with, up till now, only partial success.

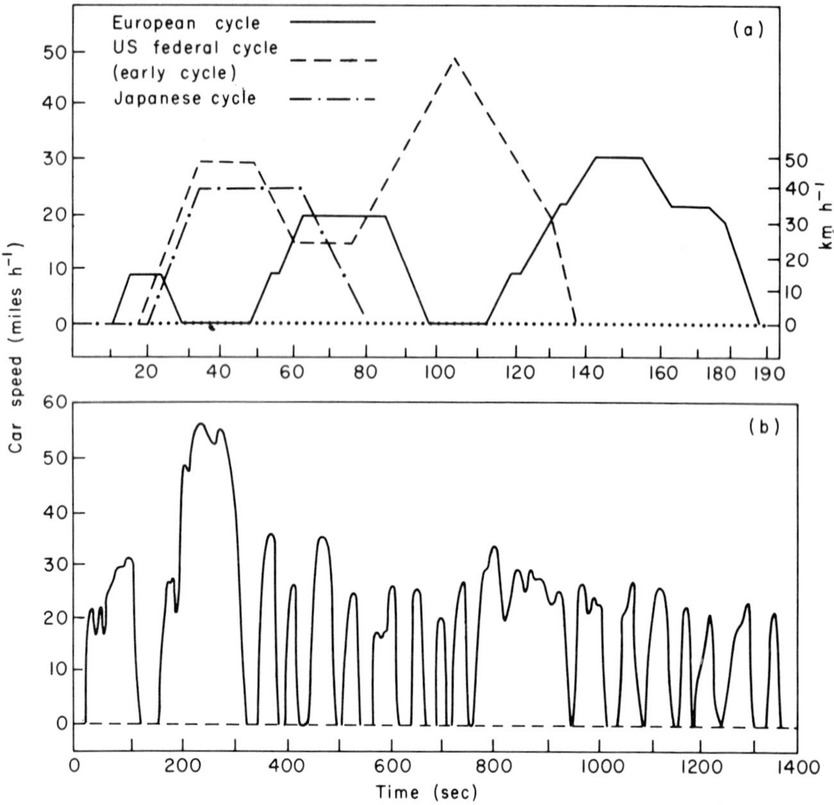

FIG. 2.18 (a) US Federal, European, and Japanese driving cycles [73]; (b) 1975 US Federal driving cycle [72]

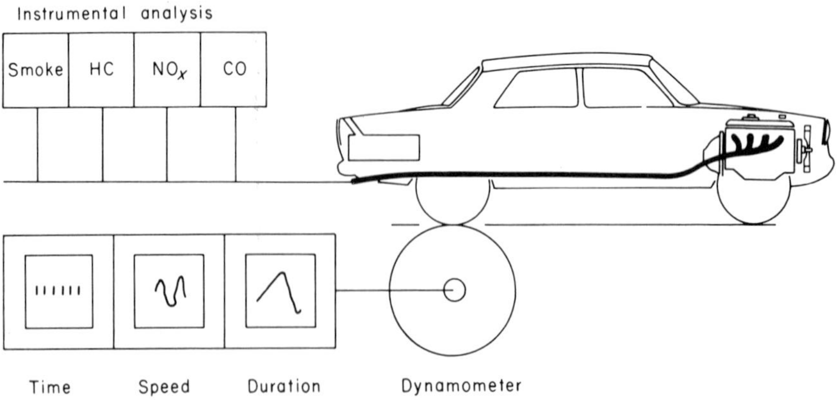

FIG. 2.19 Schematic representation of motor vehicle test procedure

The reasons for this lack of success are, primarily, the presence of lead in motor vehicle exhaust gas and the inherent complexity of exhaust gas compositions. Lead is a notorious catalyst poison so unless the catalytic reactor bed can be protected in some way, the lifetime will be limited. This does not apply to diesel systems, however, but even in this case, catalysts have not been widely used to control exhaust fumes. The difficulty lies in the incompatibility of exhaust gas compositions for the removal of hydrocarbons and carbon monoxide as well as nitric oxide. In the absence of lead, a platinum catalyst can be used to oxidize unburnt or partially burnt hydrocarbons and carbon monoxide to carbon dioxide and water vapour. To bring about this oxidation stage effectively, supplementary air is added to the exhaust gas before passage over the catalyst bed. There is no real problem in using catalysts for this limited operation.

Ideally, it should also be possible to remove the 2000–4000 parts 10^{-6} nitric oxide by reaction with the excess (1–2%) carbon monoxide that is always present in exhaust gases through the reactions:

$$2CO + 2NO = 2CO_2 + N_2 \tag{2.44}$$

or

$$CO + 2NO = CO_2 + N_2O \tag{2.45}$$

followed by

$$CO + N_2O = CO_2 + N_2 \tag{2.46}$$

Thus, it seems feasible that the two processes could be carried out by operating a dual-catalyst bed system, in which carbon monoxide is made to reduce nitric oxide to nitrogen in the first reactor and then the remaining carbon monoxide and hydrocarbons are oxidized in the second catalytic reactor. In practice, instead of carbon monoxide reducing nitric oxide, hydrogen reduces nitric oxide to ammonia:

$$2NO + 5H_2 = 2NH_3 + 2H_2O \tag{2.47}$$

in the first reactor. When this ammonia passes over the oxidizing catalyst in the second reactor it is oxidized back again to nitric oxide:

$$4NH_3 + 5O_2 = 4NO + 6H_2O \tag{2.48}$$

or

$$2NH_3 + O_2 = 2NO + 3H_2 \tag{2.49}$$

Hydrogen is present in the system at the first reactor stage because of the participation of the water gas reaction:

$$CO + H_2O \rightleftharpoons CO_2 + H_2 \tag{2.50}$$

and possibly from the steam reforming of hydrocarbons,

$$C_nH_{2n+2}+2nH_2O \rightarrow nCO_2+(3n+1)H_2 \tag{2.51}$$

Even this is not the end of the story because carbon monoxide and hydrogen under the reducing conditions that obtain in the first reactor can react by the methanation reaction, or deviants of it, to synthesize hydrocarbons, e.g.

$$CO+3H_2 \rightarrow CH_4+H_2O \tag{2.52}$$

By careful choice of catalyst in the first reactor it is possible to minimize water gas and methanation reactions, but ammonia reoxidation still remains a problem.

Base metal and noble metal catalyst preparations have been extensively studied by Shelef and Gandhi [74, 75, 76], Vartuli and Gonzalez [77], and Taylor and Klimisch [78, 79]. Most, though not all of this work has been carried out in flow systems to evaluate catalyst performance and to characterize reaction products from carbon monoxide, nitric oxide, water vapour, hydrogen and hydrocarbon mixtures. A detailed mechanistic knowledge of precisely what occurs at the catalyst surface during ammonia formation remains obscure. It is known, however, from the work of Unland [80, 81] that intermediate surface isocyanates can be formed on platinum and other noble metals during reaction between carbon monoxide and nitric oxide. The outline of a reaction mechanism in which a rate law of the form

$$\frac{dp_{CO_2}}{dt}=k\frac{p_{NO}^2}{p_{CO}} \tag{2.53}$$

was found for reaction between CO and NO over palladium which includes the formation of surface isocyanate is shown in Fig. 2.20 [82]. Briefly, the scheme outlined in Fig. 2.20 considers that the palladium surface is largely covered by CO. Occasionally, however, there will be a NO molecule adsorbed on an adjacent site of the surface to a CO molecule. If the CO molecule captures the oxygen atom belonging to the adsorbed NO, then this will give adsorbed CO_2 which will be released from the surface, and an adsorbed nitrogen atom:

$$\begin{array}{cccccc} CO & ON & & CO_2 & N \\ | & | & \rightarrow & | & | \\ S & S & & S & S \\ & & & & \\ & & & S+CO_2 & \end{array} \tag{2.54}$$

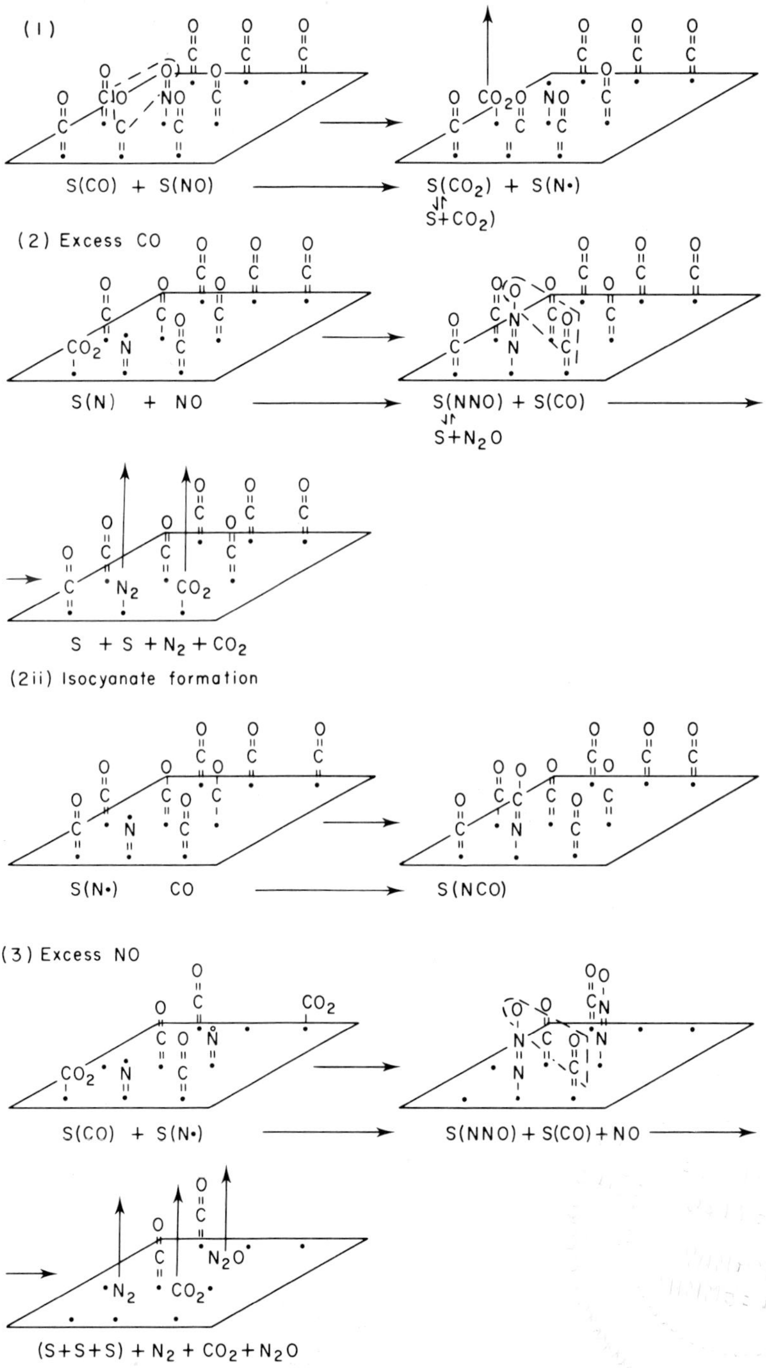

FIG. 2.20 Catalytic surface reaction scheme for the reduction of NO by CO, illustrating the formation of surface isocyanate

Once the surface nitrogen atom is generated, two possibilities are envisaged. Firstly, nitric oxide from the vapour phase reacts with adsorbed nitrogen via rapid reaction to give adsorbed nitrous oxide:

$$\begin{array}{c}N\\|\\S\end{array}\!+NO \rightarrow \begin{array}{c}N_2O\\|\\S\end{array} \tag{2.55}$$

The next stage depends upon the fraction of the surface covered by CO, θ_{CO}. If θ_{CO} is large, then an adjacent surface site is likely to be occupied by CO so that interaction between the two adsorbed species will produce N_2 and CO_2

$$\begin{array}{c}N_2O\\|\\S\end{array} \quad \begin{array}{c}OC\\|\\S\end{array} \rightarrow \begin{array}{c}N_2\\|\\S\\\rightleftharpoons\\S+N_2\end{array} \quad \begin{array}{c}CO_2\\|\\S\\\rightleftharpoons\\S+CO_2\end{array} \tag{2.56}$$

When insufficient CO is present on the surface, on the other hand, nitrous oxide will desorb into the gas phase as a reaction product:

$$\begin{array}{c}N_2O\\|\\S\end{array} \rightarrow S+N_2O \tag{2.57}$$

Secondly, instead of the surface nitrogen atom reacting with NO as indicated in equation (2.55), if reaction with gas phase CO takes place, then surface isocyanate will be formed:

$$\begin{array}{c}N\\|\\S\end{array}\!+CO \rightarrow \begin{array}{c}C=O\\||\\N\\|\\S\end{array} \tag{2.58}$$

These surface isocyanates are only obtained with noble metal catalysts when excess CO is in contact with the catalyst.

Schemes of the type represented by equations (2.54) to (2.58) are useful in the limited sense that they provide some insight into the intimate reactions that occur between the catalytic surface and the CO/NO reacting gases. In practice, all potentially interacting species in an exhaust system should be similarly treated. From the knowledge acquired of these fundamental processes it would be theoretically possible to select the most effective catalyst. In reality, because of the overall complexity of the system the empirical approach for finding the best catalyst may prove to be the most rewarding from the technological aspect.

There are three other points that warrant careful consideration regarding the widespread use of catalytic converters. Firstly, reactions shown by equation (2.45) and (2.57) imply that production of nitrous oxide from CO and NO is possible under some engine operating conditions. Additionally, nitric oxide–ammonia and nitric oxide–hydrogen interactions expressed stoichiometrically by:

$$8NO + 2NH_3 = 5N_2O + 3H_2O \tag{2.59}$$

and

$$2NO + H_2 = N_2O + H_2O \tag{2.60}$$

respectively, represent realistic processes leading to the undesirable by-product, nitrous oxide.

Secondly, it is imperative that the bed of catalytic material is effectively contained within the converter. The very nature of the operational mode of the system, which requires a free passage of exhaust gas through the catalyst before emission to the atmosphere, makes this very difficult to achieve in practice.

Finally, unless extremely stringent fail-safe provisions are made, the use of nickel catalysts must be ruled out. The excess carbon monoxide content of exhaust gas under certain operational temperature conditions could cause the formation and discharge to atmosphere of the very toxic nickel carbonyl.

2.4.2.2 Emission control legislation

Table 2.26 shows the US Federal Regulations for spark engines. Between 1973 and 1980, it will be seen that these stipulate a reduction by 1980 to about one-tenth of the 1973 value. If this is to be achieved a major effort in research and development will be needed.

TABLE 2.26 *US Federal regulations for emissions from spark ignition engines*

	CO		NO		HC	
Year	(g mile^{-1})	(g km^{-1})	(g mile^{-1})	(g km^{-1})	(g mile^{-1})	(g km^{-1})
1973	39	24·2	3·0	1·86	3·4	2·1
1975	3·4	2·1	3·0	1·86	0·41	0·25
1980	3·4	2·1	0·4	0·25	0·41	0·25

Table 2.27 shows the emission requirements for diesel engines. With these engines the main pollutants are hydrocarbons, nitrogen oxides, and smoke, and a large proportion of their hydrocarbon emissons are unburnt fuel molecules. Compared with petrol engines carbon monoxide and poly-nuclear hydrocarbons emissions are low provided the engine is well tuned

and maintained. Here again, however, if future legislation is to be met, more fundamental information on the influence of turbulence, mixing of air and fuel, and flame propagation on the production of nitric oxide, carbon monoxide and hydrocarbons will be required.

TABLE 2.27 *US Federal regulations for emissions from diesel engines*

Light duty emissions

	CO		NO_x		HC	
Year	(g mile^{-1})	(g km^{-1})	(g mile^{-1})	(g km^{-1})	(g mile^{-1})	(g km^{-1})
1975	3·4	2·1	3·1	1·9	0·41	0·25
1976	3·4	2·1	0·4	0·25	0·41	0·25

Heavy duty emissions

Year	CO (g hp^{-1} h^{-1})	$HC+NO_x$ (g hp^{-1} h^{-1})
1975	25	5

2.4.2.3 The lead trap

A method of restricting emission of lead from motor vehicles that has been tried out employs a lead trap fitted into the exhaust system. The device consists of a cylinder having a volume of about five times the engine displacement; this replaces the usual silencer in the exhaust system. The cylinder is packed with stainless steel wool of rectangular cross-section coated with alumina by steeping with sodium aluminate solution followed by calcination at 540°C. Performance characteristics of the device published by the Transport and Road Research Laboratory [83] show that on average the lead emission for a vehicle fuelled with petrol containing 0·52 g l^{-1} of lead was reduced to that which would occur for a fuel containing 0·3 g l^{-1} with a conventional exhaust system. Lead emission was found to depend on the path length of exhaust gas through the trap, engine size and engine speed. These criteria affect exhaust gas temperature and velocity.

It was established that if the trap was overheated to 850°C, a process which converts γ-alumina to α-alumina, a reduction in absorption properties resulted. This implies that the trap does not merely behave as a filter but that absorption of lead also occurs. This feature influenced the efficiency of the trap as it was found to depend on the driving mode of the vehicle. The trap was most effective in an urban environment, but lead which has accumulated in the device under these conditions can be emitted when driving conditions change radically. This purging effect continues until a new equilibrium is attained. The effect is most pronounced when a vehicle which has been operating in an urban area is subsequently driven on a motorway. The traps used in this study appeared to have a life exceeding 38 000 km (24 000 miles) on mixed routes and there was no evidence that the trap had reached the end of its useful life even after 58 000 km (36 000 miles).

Other devices that have been evaluated include cyclone traps of the type illustrated in Fig. 2.21. Some results from Fords and Chevrolets fitted with this system in comparison with the ordinary production models are given in Fig. 2.22 [84].

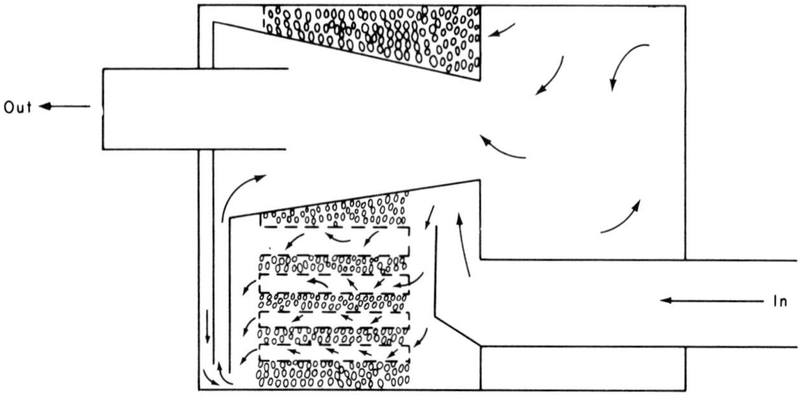

FIG. 2.21 Design of a cyclone fitted into an exhaust system to act as a lead trap

Some authorities have sought to tackle the question of leaded petrol literally at source by greatly reducing the permitted concentration of lead alkyl additive (Table 2.28). On present evidence, it is possible that such a pragmatic approach as complete removal of the lead may not be the best. Setting aside the detrimental economic and fuel utility aspect of such a policy for the moment, the known alternative of maintenance of octane number by increasing the aromatic content of the fuel should also be avoided. The proportionality between aromatic character of the fuel and the emission of carcinogenic polynuclear hydrocarbon, discussed in section 2.3.2.4, cannot be overlooked.

2.4.3 Restriction of industrial emissions—stationary sources

Industrial emissions are subject to controls to protect both the local community outside a factory and those workers involved in manufacturing processes within. This is particularly true in the case of dangerous toxic chemicals where, as we have seen in Chapter 1, TLVs or MACs are laid

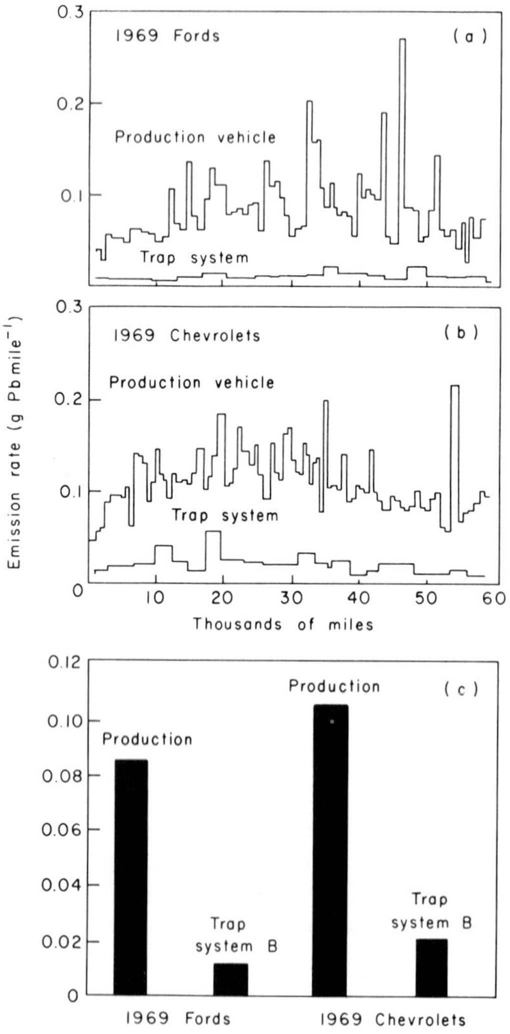

FIG. 2.22 Lead emission from Fords (a) and Chevrolets (b) fitted with cyclone trap systems compared with lead emissions from normal production models. Average lead emissions with and without trap system (c).

down by legislation to protect individuals from known health hazards. These requirements usually manifest themselves in the design of equipment to prevent or reduce the escape of pollutants during such hazardous operations.

TABLE 2.28 *Lead limitations in fuel: most severe current legislation*

Country	Concentration $g\,l^{-1}$
USA	
New York City	0·27 (current) zero (proposed)
EUROPE	
West Germany	0·15
Austria	0·4
Sweden	0·4
Norway	0·4
Italy	0·635
Portugal	0·635
France	0·64
Finland	0·7
Spain	0·74
Belgium	0·84
Greece	0·84
Netherlands	0·84
Denmark	0·84
Eire	no limit
Switzerland	0·4
UK	0·45

2.4.3.1 Particulates

Much of the equipment that has been designed and manufactured to restrict particulate emissions by filtration, electrostatic precipitation, inertial collectors, scrubbers or spray collectors, and condensers belongs to the province of engineering. The details of these methods are outside the scope of this text. Nevertheless, it is important that chemists should have at least some knowledge of the subject, and Table 2.29 has therefore been included, so that the relative merits of some of these procedures can be appreciated.

2.4.3.2 Gases

Sulphur dioxide presents the main problem regarding gaseous emissions. The "acid-rain" phenomenon falling in Scandinavia has been attributed to sulphur dioxide emissions originating in the UK and in the industrial regions of West Germany. Prevailing winds carry the acid aerosol north and east until precipitation occurs. Coal-burning or oil-burning electrical

TABLE 2.29 *Distribution of particle size and mean collection efficiencies for particle control equipment*

Type of system	Total	0–5	5–10	10–20	20–44	>44
			Particle size range Efficiency, %			
Baffled settling chamber	58·6	7·5	22	43	80	90
Multiple cyclone	93·8	63	95	98	99·5	100
Electrostatic precipitator	97·0	72	94·5	97	99·5	100
Spray tower	94·5	90	96	98	100	100
Baghouse[a]	99·7	99·5	100	100	100	100

[a] Baghouse comprises a series of filter bags fabricated from cloth, synthetic fibres or glass fibres. The weave of the cloth employed can be changed to suit specific requirements.
Source: Table adapted from "Compilation of air pollutant emission factors", 2nd Edition, US Environmental Protection Agency, 1973.

power stations and oil refineries are the largest emitters from stationary sources. An estimate of the total emissions from EEC countries is shown in Table 2.30. The table shows that the ratio of particulates to oxides of

TABLE 2.30 *Total emissions of SO_2, NO_x and particulates from stationary sources from EEC countries* [72]

Year	SO_2 (kg × 10^9)	NO_x (kg × 10^9)	Particulates (kg × 10^9)
1970	7·5	2·7	0·93
1973	10·2	3·1	0·91
1980	12·3	4·2	1·05

nitrogen to sulphur dioxide is about 1:4:12. In the USA, air quality standards for SO_2 of 0·1 parts 10^{-6} as a 24 h average, or 0·02 parts 10^{-6} as an annual ground level concentration are legally enforceable.

Preventive measures restricting sulphur dioxide discharge into the atmosphere fall into two categories:

(1) modified combustion procedures,

(2) chemical treatment and scrubbing of flue gases.

Modified combustion can be accompanied by charging the boiler unit with small coal granules and limestone. The charge is aerated from underneath to produce a heated bed of solid particles into which steam-raising coils are submerged. Sulphur oxides formed by the burning fuel react with the limestone to give calcium sulphate. The ash and calcium sulphate are withdrawn periodically for disposal or for sulphur reclamation. A somewhat similar procedure can be used for sulphur-containing fuel oil. The fuel oil is first gasified in a fluidized bed with limestone particles and with only one-fifth of the air needed for complete combustion. At a bed temperature of about 900°C sulphur is retained in the bed as calcium sulphide. The desulphurized combustible off-gas is then fed directly to a conventional boiler. The solid residue remaining in the burner is drawn off and heated to 1050°C. At this temperature calcium sulphate decomposes to sulphur dioxide and lime. The lime is recycled and the sulphur dioxide reclaimed.

Lime or limestone slurry can be used in a wet scrubber for treatment of flue gas. In either case sulphur dioxide is converted to a slurry of calcium sulphite by the reactions:

$$\text{limestone:} \quad CaCO_3 + SO_2 = CaSO_3 + CO_2 \tag{2.61}$$

$$\text{lime:} \quad CaO + SO_2 = CaSO_3 \tag{2.62}$$

Liquors from the slurry pass to a settling tank to allow fly-ash particulates and the calcium salts to precipitate.

Instead of calcium salts, sodium salts may be used as scrubbing agents, e.g.:

$$NaOH + SO_2 = NaHSO_3 \tag{2.63}$$

$$NaHSO_3 + NaOH = Na_2SO_3 + H_2O \tag{2.64}$$

$$Na_2SO_3 + SO_2 + H_2O = 2NaHSO_3 \tag{2.65}$$

The sodium sulphite can be recovered either by heating:

$$2NaHSO_3 = Na_2SO_3 + H_2O + SO_2 \tag{2.66}$$

further processing for S-reclamation, or conversion to H_2SO_4

or by addition of slaked lime:

$$2NaHSO_3 + Ca(OH)_2 = Na_2SO_3 + CaSO_3 + 2H_2O \tag{2.67}$$

Wet scrubbing of flue gases with alkaline liquors as outlined for the removal of sulphur dioxide will also assist in removing oxides of nitrogen.

As an alternative to these wet-scrubbing methods flue gases can be desulphurized by catalytic conversion to sulphuric acid or by absorption in a heated copper bed. These methods require an initial clean-up stage to remove fly-ash particulates from the gas steam. The catalytic process is merely an adoption of the contact process for the conversion of sulphur dioxide to sulphur trioxide over vanadium pentoxide:

$$2SO_2 + O_2(\text{air}) \xrightarrow{V_2O_5} 2SO_3$$

$$2SO_3 \xrightarrow[H_2O]{\text{absorption in dilute } H_2SO_4} H_2SO_4$$

The copper bed absorption method depends upon the initial production of copper(II)oxide from copper at about 400°C by air in the flue gas. Sulphur dioxide is subsequently converted to copper sulphate:

$$Cu + \tfrac{1}{2}O_2 = CuO \tag{2.69}$$

$$CuO + \tfrac{1}{2}O_2 + SO_2 = CuSO_4 \tag{2.70}$$

Copper can be regenerated by passing a stream of hydrogen over the copper sulphate:

$$CuSO_4 + 2H_2 = Cu + SO_2 + 2H_2O \tag{2.71}$$

Comparatively large supplies of hydrogen are needed for this regenerative stage, and such supplies are normally available only in oil-refinery-type operations.

References

1. Cadle, R. D. (1975). Volcanic emissions of halides and sulfur compounds to the troposphere and stratosphere. *J. Geophys. Res.* **80**, 1650–1652.
2. Eshleman, A., Siegel, S. M. and Siezel, B. Z. (1971). Is mercury from Hawaiian volcanoes a natural source of pollution? *Nature* **233**, 471–472.
3. Robinson, E. and Robbins, R. C. (1971). "Emissions, concentrations and fate of particulate atmospheric pollutants". American Petroleum Institute, March. SRI Project SCC-8507, Washington DC. Publication No. 4076.
4. Bolin, B. (1970). The carbon cycle. *Scientific American* **223** (3), 125–132.
5. Weinstock, B. and Niki, H. (1972). Carbon monoxide balance in nature. *Science* **176**, 290–292.
6. Junge, C. E., Seiler, W. and Warneck, P. (1971). The atmospheric ^{12}CO and ^{14}CO budget. *J. Geophys. Res.* **76**, 2866–2879.
7. Inman, R. E., Ingersoll, R. B. and Levy, E. A. (1971). A natural sink for carbon monoxide. *Science* **172**, 1229–1231.
8. Greiner, N. R. (1969) Hydroxyl radical kinetics by kinetic spectroscopy. V Reactions with H_2 and CO in the range 300–500 K. *J. Chem. Phys.* **51** 5049–5051.

9. McConnell, J. C., McElroy, M. B. and Wofsy, S. C. (1971). Natural sources of atmospheric CO. Nature **233**, 187–188.
10. Climatic impact assessment program of the US Department of Transportation. Chemical kinetic data survey, IV (NBSIR 73–203); VII (NBSIR 74-430). Ed. D. Garvin (1974). Appendix.
11. Robinson, E. and Robbins, R. C. (1967). "Sources, abundances, and fate of atmospheric pollutants." Stanford Research Institute Project PR-6755, prepared for American Petroleum Institute, New York.
12. Kummler, R. H. and Baurer, T. (1973). A temporal model of tropospheric carbon–hydrogen chemistry." *J. Geophys. Res.* **78**, 5306–5316.
13. Robinson, E. and Robbins, R. C. (1970). Gaseous nitrogen compound pollutants from urban and natural sources. *J. Air Pollut. Contr. Assoc.* **20**, 303–306.
14. Robinson, E. and Robbins, R. C. (1970). Gaseous sulfur pollutants from urban and natural sources. *J. Air Pollut. Contr. Assoc.* **20**, 233–235.
15. Junge, C. E. and Ryan, T. (1958). Study of SO_2 oxidation in solution and its role in atmospheric chemistry. *Quart. J. Roy. Meteorol. Soc.* **84**, 46–55.
16. Chamberlain, A. C. (1960). Aspects of the deposition of radioactive and other gases and particles. *Internat. J. Air Pollut. (London)* **3**, 63–88.
17. "Compilation of air pollutant emission factors", 2nd edition. US Environmental Protection Agency, Research Triangle Park, N. Carolina, April 1973.
18. Koppenhaal, D. W. and Manahan, S. E. (1976). Hazardous chemicals from coal conversion processes. *Environ. Sci. and Technol.* **10**, 1104–1107.
19. Block, C. and Dams, R. (1976). Study of fly-ash emission during combustion of coal. *Environ. Sci. and Technol.* **10**, 1011–1017.
20. Baker, E. W., Yen, T. F., Dickie, J. P., Rhodes, R. E. and Clark, L. F. (1967). Mass spectrometry of porphyrins II. Characterization of petroporphyrins. *J. Amer. Chem. Soc.* **89**, 3631–3639.
21. Shah, K. R., Filby, R. H. and Haller, W. A. (1970). Determination of trace elements in petroleum by neutron activation analysis I. Determination of S, Cl, K, Ca, V, Mn, Cn, Ga and Br. *J. Radioanal. Chem.* **6**, 185–192.
22. Shah, K. R., Filby, R. H. and Haller, W. A. (1970). Determination of trace elements in petroleum by neutron activation analysis II. Determination of Sc, Cr, Fe, Co, Ni, Zn, As, Se, Sb, Eu, Au, Hg, and U. *J. Radioanal. Chem.* **6**, 413–422.
23. Bratzel, M. P. and Chakrabarti, C. L. (1972). Determination of lead in petroleum and petroleum products by atomic absorption spectrometry with a carbon rod atomizer. *Anal. Chem. Acta* **61**, 25–32.
24. Yen, T. F., Boucher, L. J., Dickie, J. P., Tynan, E. C. and Vaughan, G. V. (1969). Vanadium complexes and porphyrins in asphaltenes. *J. Indust. Petrol.* **55**, 87–99.
25. Yen, T. F. (1975). "The Role of Trace Metals in Petroleum". Ann Arbor Science Publishers Inc., Michigan 48106.
26. Dryer, F., Naegeli, D. and Glassman, I. (1971). Temperature dependence of the reaction $CO + OH \rightarrow CO_2 + H$. *Combust and Flame* **17**, 270–272.
27. Haynes, C. D. and Weaving, J. W. (1971). Catalytic reduction of atmospheric pollution from the exhaust of petrol engines. *In* "Air Pollution Control in Transport Engines". Institute of Mechanical Engineers Conference, pp. 232–240.
28. Moore, J. (1971). The effects of atmospheric moisture on nitric oxide pollution. *Combust. and Flame* **17**, 265–267.

29. Newhall, H. K. and El-Messiri, I. A. (1970). A combustion chamber concept for control of engine exhaust air pollutant emissions. *Combust. and Flame* **14**, 155–158.
30. Badger, G. M., Donnelly, J. K. and Spotswood, T. M. (1965). The formation of aromatic hydrocarbons at high temperatures. XXIV The pyrolosis of some tobacco constituents. *Austral. J. Chem.* **18**, 1249–1266.
31. Badger, G. M., Donnelly, J. K. and Spotswood, T. M. (1966). The formation of aromatic hydrocarbons at high temperatures. XXVII The pyrolysis of isoprene. *Austral. J. Chem.* **19**, 1023–1043.
32. Begeman, C. R. and Colucci, J. M. (1968). Benzo(a)pyrene in gasoline partially persists in automobile exhaust. *Science* **161**, 271.
33. Candeli, A., Morozzi, G., Paolacci, A. and Zoccolillo, L., (1975) Analysis using thin layer and gas-liquid chromatography of polycyclic aromatic hydrocarbons in the exhaust products from a European car running on fuels containing a range of concentrations of these hydrocarbons. *Atmos. Environ.* **9**, 843–849.
34. Fenimore, C. P. and Jones, S. W. (1969). Coagulation of soot to smoke in hydrocarbon flames. *Combust. and Flame* **13**, 303–310.
35. Laud, B. B. and Gaydon, A. G. (1971). Absorption spectra of ethylene diffusion flames. *Combust. and Flame* **16**, 55–59.
36. Chakraborty, B. B. and Long, R. (1967). Gas chromotographic analysis of polycyclic aromatic hydrocarbons in soot samples. *Environ. Sci. and Technol.* **1**, 828–834.
37. Cotton, D. H., Friswell, N. J. and Jenkins, D. R. (1971). The suppression of soot emission from flames by metal additives. *Combust. and Flame* **17**, 87–98.
38. Boyer, K. W. and Laitinen, H. A. (1974). Lead halide aerosols. Some properties of environmental significance. *Environ. Sci. and Technol.* **8**, 1093–1096.
39. Lee, R. E., Patterson, R. K., Crider, W. L. and Wagman, J. (1971) Concentration and particle size determination of particulate emission in automobile exhaust. *Atmos. Environ.* **5**, 225–237.
40. World Metal Statistics, Vol. 29, pp. 72, 73 and 78. Published by World Bureau of Metal Statistics (1976).
41. Patterson, C. (1965). Contaminated and natural lead environments in man. *Arch. Environ. Health*, **11**, 344–360.
42. Egorov, V. V., Zhigalovskaya, T. N. and Malakhov, S. G. (1970). Microelement content of surface air above the continent and the ocean. *J. Geophys. Res.* **75**, 3650–3656.
43. Chow, T. J., Earl, J. L. and Bennett, C. F. (1969). Lead aerosols in marine atmosphere. *Environ. Sci. and Technol.* **3**, 737–740.
44. Habibi, K. (1970). Characterization of particulate lead in vehicle exhaust. Experimental techniques. *Environ. Sci. and Technol.* **4**, 239–248.
45. Gillette, D. A. (1972). A study of ageing lead aerosols, II. *Atmos. Environ.* **6**, 451–462.
46. Ter Haar, G. L. and Bayard, M. A. (1971). Composition of airborne particles. *Nature* **232**, 553–554.
47. Francis, C. W., Chesters, G. and Haskin, L. A. Determination of ^{210}Pb mean residence time in the atmosphere. *Environ. Sci. and Technol.* **4**, 586–589.
48. Drake, P. F. and Hubbard, E. H. (1966). Combustion system aerodynamics and their effect on the burning of heavy oil. *J. Instit. Fuel* **39**, 98–109.

49. Gills, B. G. (1973). Production and emission of solids, SO_x and NO_x from liquid fuel flames. *J. Instit. Fuel* **46**, 71–76.
50. von Lehmden, D. J., Hangebrauck, R. P. and Meeker, J. E. (1965). Polynuclear hydrocarbon emissions from selected industrial processes. *J. Air Pollut. Contr. Assoc.* **15**, 306–312.
51. Hangebrauck, R. P., von Lehmden, D. J. and Meeker, J. E. (1964). Emission of polynuclear hydrocarbons and other pollutants from heat-generation and incineration processes. *J. Air Pollut. Contr. Assoc.* **14**, 267–278.
52. Commins, B. T. (1969). Formation of polycyclic aromatic hydrocarbons during pyrolysis and combustion of hydrocarbons. *Atmos. Environ.* **3**, 565–572.
53. Davies, I. W., Harrison, R. M., Perry, R., Ratnayaka, D. and Wellings, R. A. (1976). Municipal incinerator as source of polynuclear aromatic hydrocarbons in environment. *Environ. Sci. and Technol.* **10**, 451–453.
54. Palmer, T. Y., Combustion sources of atmospheric chlorine. (1976). *Nature* **263**, 44–46.
55. Galbally, I. E. (1976). Man-made carbon tetrachloride in the atmosphere. *Science* **193**, 573–576.
56. Molina, M. J. and Rowland, F. S. (1974). Predicted present stratospheric abundances of chlorine species from photochemical dissociation of CCl_4. *Geophys. Res. Lett.* **1**, 309–312.
57. Yung, Y. L., McElroy, M. B. and Wofsy, S. C. (1975). Atmospheric halocarbons: A discussion with emphasis on chloroform. *Geophys. Res. Lett.*, **2**, 397–399.
58. Morris, J. C. (1975). "Formation of halogenated organics by chlorination of water supplies: A review." Draft report to Office of Research and Development, EPA(RD683) Washington DC.
59. Wofsy, S. C., McElroy, M. B. and Yung, Y. L. (1975). The chemistry of atmospheric bromine. *Geophys. Res. Lett.* **2**, 215–218.
60. Molina, M. J. and Rowland, F. S. (1974). Stratospheric sink for chlorofluoromethanes: Chlorine atom catalyzed destruction of ozone. *Nature*, **249**, 810–812.
61. Woodwell, G. W., Craig, P. P. and Johnson, H. A. (1971). DDT in the biosphere: Where does it go? *Science*, **174**, 1101–1107.
62. Cawse, P. A. (1976). "A survey of atmospheric trace elements in the U.K.: Results for 1975." U.K.A.E.A. Harwell, AERE-R8398, H.M.S.O.
63. Merritt, W. F. (1976). "Trace element content of precipitation in remote areas," pp 75–86. Internat. Atomic Energy Agency, Vienna.
64. Dufour, L. (1969). "The atmospheric aerosol." Atmospheric Sciences Laboratory, White Sands Missile Range, New Mexico, DDC. No. AD 685851.
65. Went, F. W. (1964). The nature of Aitken condensation nuclei in the atmosphere. *Proc. U.S. Nat. Acad. Sci.* **51**, 1259–1267.
66. Greenfield, S. M. (1957). Rain scavenging of radioactive particulate matter from the atmosphere. *J. Meteorol.* **15**, 115–125.
67. Junge, C. E. (1963). "Air Chemistry and Radioactivity". Academic Press, New York.
68. Murozumi, M., Chow, T. J. and Patterson, C. (1969). Chemical concentrations of pollutant lead aerosols, terrestrial dusts, and sea salts in Greenland and Antarctic snow strata. *Geochem. Cosmochem. Acta.* **33**, 1247–1294.

69. Junge, C. E. and Abel, N. (1965). "Modification of aerosol size distribution in the atmosphere and development of an ion counter of high sensitivity." University of Gotenburg., Mainz, Germany. Final Technical Report, Contract No. Da 91-591-EUC-3483. DDC No. AD 469376.
70. Lamb, H. H. (1970). Volcanic dust in the atmosphere; with a chronology and assessment of its meteorological significance. *Philos. Transact. Roy. Soc. (London) A* **266**, 425–533.
71. Chamberlain, A. C. (1967). Radio-active aerosols and vapours. *Contemp. Phys.* **8**, 561–581.
72. "Reducing pollution from selected energy transformation sources." Chem. Systems International Ltd. Published by: Graham and Trotman Ltd. for the EEC (1976).
73. Crawford, K. C. and Lindsay, R. (1970). "Future of the Gasoline Engine—Environmental Conservation." Shell Publication MOR 558F, April 1970.
74. Shelef, M. and Gandhi, H. S. (1972). Ammonia formation in catalytic reduction of nitric oxide by molecular hydrogen. I. Base-metal oxide catalysts. *Ind. Eng. Chem., Product. Res. and Develop.* **11**, 2–11.
75. Shelef, M. and Gandhi, H. S. (1972). Ammonia formation in catalytic reduction of nitric oxide by molecular hydrogen. II. Noble metal catalysts. *Ind. Eng. Chem., Product Res. and Develop.* **11**, 393–396.
76. Shelef, M. and Gandhi, H. S. (1974). Ammonia formation in the catalytic reduction of nitric oxide. III. The role of water gas shift, reduction by hydrocarbons and steam reforming. *Ind. Eng. Chem., Product Res. and Develop.* **13**, 80–85.
77. Vartuli, J. C. and Gonzalez, R. D. (1973). Ammonia formation in the nitric oxide-methane reaction. *Ind. Eng. Chem., Product Res. and Develop.* **12**, 171–175.
78. Taylor, K. C. and Klimisch, R. L. (1973). The catalytic reduction of nitric oxide over supported ruthenium catalysts. *J. Catalysis* **30**, 478–484.
79. Taylor, K. C. and Klimisch, R. L. (1973). Ammonia intermediacy as a basis of catalyst selection for nitric oxide reduction. *Environ. Sci. and Technol.* **7**, 127–131.
80. Unland, M. L. (1973). Isocynate intermediates in the reaction NO+CO over a Pt/Al_2O_3 catalyst. *J. Phys. Chem.* **77**, 1952–1956.
81. Unland, M. L. (1973). Isocyanate intermediates in the reaction of NO and CO over noble metal catalysts. *J. Catalysis* **31**, 459–465.
82. Butler, J. D. and Davis, D. R. (1976). Kinetics of reaction between nitrogen mono-oxide and carbon mono-oxide over palladium-alumina and ruthenium-alumina catalysts. *J. Chem. Soc. (Dalton)*, 2249–2253.
83. Transport and Road Research Laboratory Report 662 (1974). "The Assessment of a Lead Trap for Motor Vehicles."
84. Cantwell, E. N., Jacobs, E. S., Kunz, W. G. and Liberi, V. E. (1972). Control of particulate lead emissions from automobiles. National Society of Automotive Engineers, May, Detroit, Michigan, paper. no. 720672.

3

Air Sampling and Collection

3.1 Sampling time

Central and local governments are generally interested in long-term pollution trends and favour monitoring programmes designed to supply this sort of information. At the same time, any assistance that these programmes may provide, in helping to prevent or predict the type of disasters discussed in Chapter 1, is welcomed. Such surveys may also help to establish any long-term effects on health, especially when integrated with epidemiological studies. On the other hand, shorter sampling time periods and more detailed information may be required to assess a health risk in selected instances. People exposed to the discharge from a factory, or operating and working in the presence of known toxic substances, living in close proximity to heavy vehicular traffic, etc. are cases in point. These factors impose their own conditions on any monitoring exercise. They must be matched by the ability of the sample collection and analytical procedures adopted to estimate pollution concentrations accurately. To add to the difficulty there is often a paucity of data between dose and effect relationships for long-term low-levels of exposure to a pollutant.

3.1.1 Biological half-lives

One of the guidelines laid down to judge the toxic qualities of a pollutant is called the biological half-life. This is defined as the time taken for the amount of the substance present in the body to be halved. Some examples of the biological half-lives of airborne pollutants are given in Table 3.1.

Roach [1] has considered the effect of the exposure to an air pollutant in terms of the body burden. For the case in which the total concentration of pollutant remains constant, and assuming that:

(1) the absorption of pollutant after inhalation is directly proportional to its concentration;

(2) the greater the amount of pollutant present in the body, the greater the tendency for it to be removed—

an equation may be derived from which the body burden x may be calculated.

TABLE 3.1 *Biological half-lives of air pollutants in relation to the human body or the most sensitive organ of the body* [1]

Pollutant	Sampling period	Biological half-life
SO_2	4 min	<20 min
H_2S	2 min	<20 min
Cl_2	2 min	<20 min
NO_2	5 min	~60 min
CO	10 min	2 h
Fe_2O_3(particulate)	60 min	12 h
Hg	16 h	5 weeks
Pb	80 h	6 months

Let C = concentration of pollutant (in mg m^{-3}) contained in the inspired air, then the rate at which the substance is taken into the body from the air = kC mg h^{-1}, where k is a constant characteristic of the substance. At the same time the body is attempting to eliminate the substance and after time t from the start of exposure, x mg h^{-1} is the body burden, then $k'x$ will be the rate of elimination, where k' is another characteristic constant. Thus, at any instant the rate of change of body burden with time will be given by:

$$\frac{dx}{dt} = kC - k'x \tag{3.1}$$

which upon integration gives

$$x = \frac{kC}{k'} + \left(x_0 - \frac{kC}{k'}\right) e^{-k't} \tag{3.2}$$

where x_0 is the original body burden at time $t = t_0$. When the person comes out of the contaminated environment and breathes unpolluted air, C may be put equal to zero, hence

$$x = x_0 e^{-k't} \tag{3.3}$$

The time for x to become $x_0/2$ is $t_{1/2}$, where $t_{1/2}$ defines the biological half-life time and

$$k' = \frac{\ln 2}{t_{1/2}} = \frac{0{\cdot}693}{t_{1/2}} \tag{3.4}$$

It will be seen from equation (3.2) that when the exposure t is long, $e^{-k't}$ will approach zero, and under these circumstances the equation reduces to

$$x \sim \frac{k}{k'}C \tag{3.5}$$

In other words, an equilibrium condition exists in which the rate of intake equals the rate of elimination. This situation also arises when the half-life time $t_{1/2}$ is very small, because from equation (3.4), k' will then be large and once more $x \sim kC/k'$, i.e. the body burden will be directly proportional to the concentration of the pollutant.

The accumulation of body burden with exposure to a substance which has a very long half-life time can best be appreciated by considering the case when the initial body burden is zero, viz. $x_0 = 0$ in equation (3.2). Then,

$$x = \frac{kC}{k'}(1 - e^{-k't}) \tag{3.6}$$

or expanding,

$$x = kC\left(t - \frac{k't^2}{2!} + \frac{k'^2t^3}{3!} \cdots\right) \tag{3.7}$$

Since the half-life time is very long k' is nearly zero, so that

$$x \sim kCt \tag{3.8}$$

This means that for very long half-life times the body burden is directly proportional to the concentration as well as the total time of exposure. This is the total dose of the substance inhaled. In any well planned monitoring exercise it is evident that biological half-lives applicable to the substance of interest must be taken into consideration and appropriate sampling times relevant to the substance in question chosen. Roach [1] recommends that the duration of individual samples should be no more than one-tenth of the biological half-life time.

In statistical terms the number of samples taken, n, is usually small, often about 10–12. In these circumstances the standard deviation SD for the dispersion or range of values is given by

$$\text{SD} = \frac{\text{Difference between highest and lowest pollution levels}}{(n-1)^{1/2}} \tag{3.9}$$

If this value when added to the average value, viz. average + SD is less than the threshold limit value, the TLV, then the pollutant concentration may be deemed acceptable:

$$\text{average} + \text{SD} < \text{TLV} \quad \text{acceptable}$$

$$\text{average} + \text{SD} > \text{TLV} \quad \text{unacceptable}$$

3.1.2 Averaging times

Figure 3.1 shows a continuous record of a pollutant concentration in air over an 8 h period. The results reveal that the maximum concentration recorded was 200 mg m^{-3}, but over a 30 min period this would have fallen to around 160 mg m^{-3} and for a 2 h period to about 120 mg m^{-3}. The fluctuations themselves depend upon the rate of emisson from the source and the proximity of source from detector; also meteorological parameters may influence dispersion and dilution in the atmosphere. Given a sufficient time span, diurnal-nocturnal along with seasonal fluctuations are often discernible in the data collected.

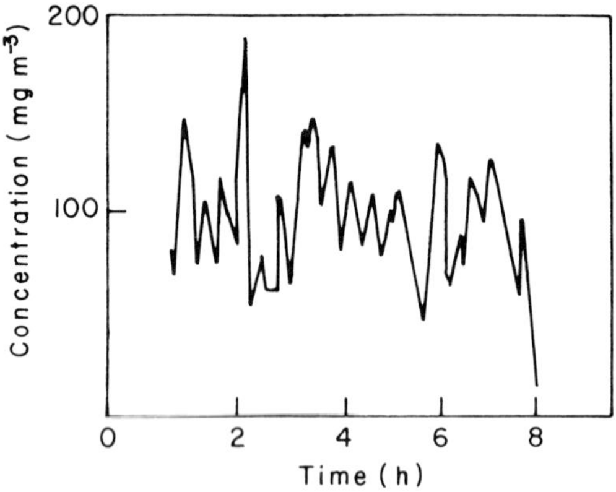

FIG. 3.1 Variation in pollutant concentration measured continuously with time

McGuire and Noll [2] have used empirical equations suggested by Larsen [3] to show the effect of averaging the concentrations of a selected pollutant over different time periods. For example, the average hourly maximum concentration C_h can be related to the average maximum concentration C_x, averaged over a time period t by the equation:

$$C_x = C_h t^b \tag{3.10}$$

where b is a constant characteristic of the pollutant. Taking logarithms this becomes

$$\log C_x = \log C_h + b \log t \tag{3.11}$$

and a plot of log C_x *vs* log t will give a straight line of slope b. Figure 3.2 shows this plot for the maximum carbon monoxide concentration (parts 10^{-6}) for different averaging times taken from observational data at a site in Los Angeles in 1967. Similar plots for other pollutants, nitrogen oxides,

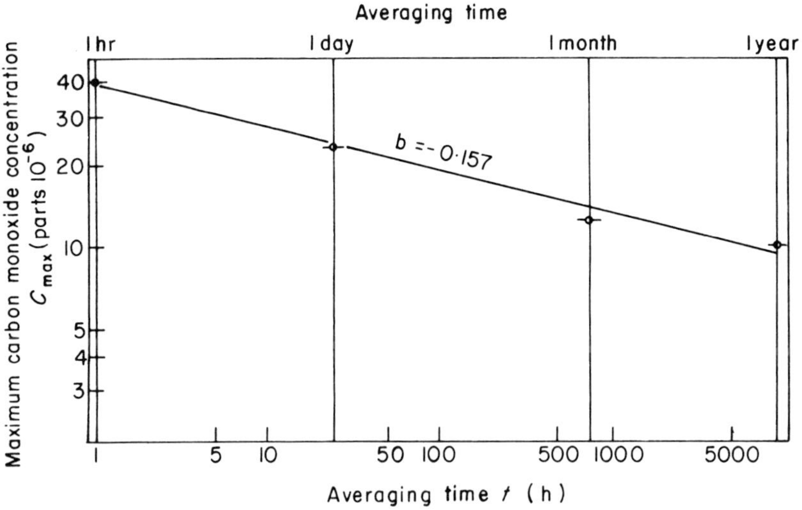

FIG. 3.2 Maximum concentration *vs* averaging time for carbon monoxide, downtown Los Angeles 1967 [2]

sulphur dioxide and oxidants also give straight lines with characteristic slopes. The constancy of the slope for a particular pollutant is illustrated by the data, also from Los Angeles for the years 1963–67, where the variation in the b value for oxidant concentration was $-0{\cdot}276 \pm 0{\cdot}02$. Once the value of the exponent b has been evaluated then it is possible to predict an average value for another time period and compare the observed and calculated values. Table 3.2 shows that an error of about 10% may be expected for some typical gaseous pollutants calculated by this method.

3.1.3 Changes in pollution patterns

Changes in pollution patterns that follow seasonal variations in climate immediately introduce the idea that meteorology affects pollution concentrations. The production of ozone and photochemical smog during summer may be quoted as obvious examples of such behaviour. More ambiguous examples are the elevated smoke and particulate loadings of the atmospheric aerosol during winter. Here, increased combustion of fossil fuel

TABLE 3.2 *Comparison of calculated maximum concentrations with observed maximum concentrations calculated from line of best fit downtown Los Angeles station 1967* [2]

		Maximum hourly average			Maximum daily average			Maximum monthly average			Annual average		
Pollutant	b	Obs. (parts 10^{-8})	Calc. (parts 10^{-8})	Error (%)	Obs. (parts 10^{-8})	Calc. (parts 10^{-8})	Error (%)	Obs. (parts 10^{-8})	Calc. (parts 10^{-8})	Error (%)	Obs. (parts 10^{-8})	Calc. (parts 10^{-8})	Error (%)
Carbon monoxide	−0·157	4000	3850	−4	2300	2340	+2	1230	1360	+11	1000	930	−7
Nitrogen dioxide	−0·249	54	59	+9	29·3	26·8	−9	10·7	11·4	+7	6·7	6·2	−7
Oxides of nitrogen	−0·240	134	154	+15	84·2	71·8	−15	33·3	31·6	−5	16·5	17·4	+5
Oxidant	−0·260	36	32	−11	11·7	14·0	+20	5·4	5·8	+7	3·3	3·0	−9
Sulphur dioxide	−0·197	13	11	−15	5·0	5·9	+18	2·7	3·0	+11	2·0	1·8	−10

from both domestic and power generation sources are at least partly to blame. The simultaneous occurrence of a temperature inversion can lead to atmospheric stagnation in the boundary layer of air lying close to the earth's surface. This aspect is discussed in Chapter 6. In cases of increased emissions coupled with temperature inversions severe pollution episodes can develop.

In urban and suburban environments pollution fluctuations are often coincident with days of the week. Figure 3.3 traces the changes in the lead aerosol concentration on a daily basis at two sites in a residential district of Birmingham, UK. Clearly, the lowest pollution levels occur on Saturday and Sunday. Nevertheless, superimposed on this pattern, peak concentrations which can only be attributed to meteorological influences are also apparent.

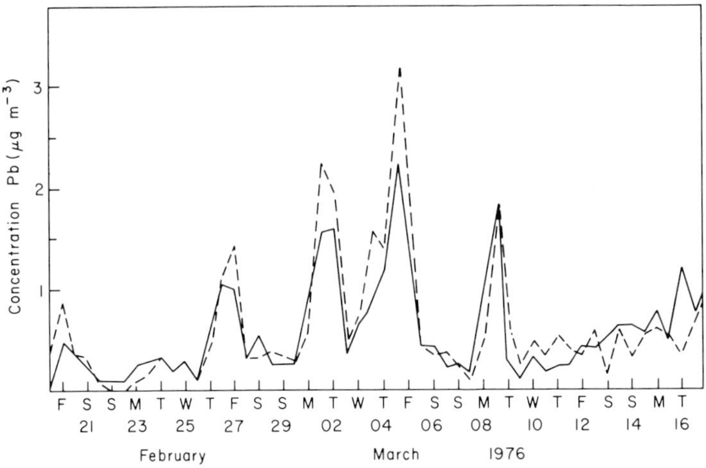

FIG. 3.3 Atmospheric particulate lead concentrations monitored at two suburban sites 1·6 km apart. Source: Biggins, P. D. E., "Studies in atmospheric pollution by lead in an urban environment." M.Phil. thesis, 1976, Department of Chemistry, University of Aston, Birmingham.

Most of the methods of chemical analysis reviewed in this chapter—those employing wet chemical analysis—generally need collection periods from a few minutes to several hours before sufficient pollutant is available to affect the reagent. In contrast, instrumental methods of sampling and analysis discussed in Chapter 4 normally provide a continuous record on chart paper, or print out values at frequent time intervals.

3.2 Air sampling principles

3.2.1 Representative samples and equipment requirements

The objective in any sampling exercise is to obtain a genuine and representative sample. This may not always be easy or possible. Frequently, the very act of sampling causes disturbances within the system that create differences between the actual sample and the collected sample. Isokinetic sampling can be defined as a process in which the system to be sampled enters the probe at the same linear velocity as that of the bulk suspending medium. When either calm or turbulent air is being sampled this is impossible to achieve. The passage of sample through sample lines or into sample chambers can cause aerosol modification by precipitation of the larger particles, diffusion of smaller particles to the tube walls and impaction on the walls. In order to avoid these errors tubes should be kept as short as possible, free from bends and be chemically inert. The problem has been considered by Davies [4] who suggested the permissible tube radii for aerosol sampling given in Table 3.3.

TABLE 3.3 *Permissible tube radii (cm) for aerosol sampling* [4]

Particle diameter	Flowrate ($cm^3 s^{-1}$)					
μm	1	10	10^2	10^3	10^4	10^5
1	0·033–1·9	0·071–6·0	0·15–19	0·33–60	0·1–190	1·5–600
2	0·051–1·0	0·11–3·2	0·23–10	0·51–32	1·1–100	2·3–320
5	0·093–0·41	0·20–1·3	0·43–4·1	0·93–13	2·0–41	4·3–130
10	0·15–0·21	0·31–0·65	0·68–2·1	1·5–6·5	3·1–21	6·8–65

Since pollutants are reported on a concentration basis, either parts 10^{-6} or $\mu g\ m^{-3}$ an accurate assessment of the volume or weight of pollutant in a stated volume must be known. When sampling is carried out, therefore, two requirements are essential:

(1) the volume of air sampled must be measured, and

(2) the efficiency of the trapping system employed must be verified.

Figure 3.4 shows a general layout of air sampling equipment designed with these points in mind. Whether the flowmeter is positioned before or after the sample collection stage depends on the pressure drop across the collector. If the pressure drop is negligible then either position may be used, but when this is not the case the flowmeter must be placed in front of the sampler. This is because most flowmeters are designed to operate at and are calibrated at atmospheric pressure. Serious error can result when instruments calibrated at atmospheric pressure are operated at reduced pressure. It is also essential to employ a reliable pump which gives a constant flow of air and sampling should not continue for such long time periods that the collector or filter become clogged, thereby restricting gas flow. Finally, the collector must be efficient.

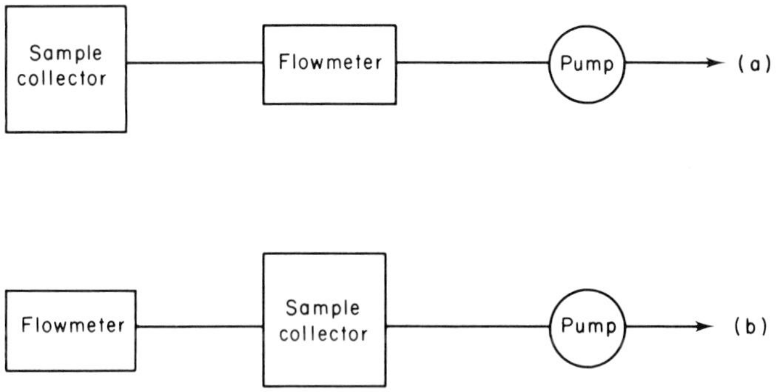

FIG. 3.4 Typical component arrangement required for atmospheric sampling

The choice of collector is governed by the nature of the pollutant to be measured and the method of analysis to be employed. For particulates, the air sample is drawn through a filter or series of filters to separate the suspended matter from the air. When this method is used filter efficiency must always be examined. Usually, when filter units in series are employed, graded Millipore or Nucleopore filter media of precise pore size characteristics enable some sort of check to be made of the passage of particles through the system. The last filter in the line ultimately carries the responsibility of being the "absolute" filter through which not even the smallest particles can penetrate. Some of the problems encountered with filters are discussed elsewhere; for filter efficiencies section 3.4.2 and for sublimation losses section 4.3.1 should be consulted.

3.3 Collection of gaseous pollutants

For some common pollutant gases (SO_2, SO_3, NO_x, O_3, H_2S, CO, HF, HCl, and low molecular weight hydrocarbons), special equipment relying on some physico-chemical analytical procedure is nowadays widely used. These instruments incorporate suitable sampling devices in their design. A description of these features is included in the sections where these instruments are explained in Chapter 4.

The choice of collector system which utilizes a wet chemical analytical approach normally falls within the following categories:

(1) Air is drawn through a chemical solution which reacts with gaseous pollutant.
(2) Air is passed through a suitable absorbing medium, e.g. charcoal, from which the pollutant can subsequently be recovered.
(3) Air is allowed to pass into and fill an inert container by either liquid or air displacement.
(4) Air is allowed to inflate a plastic bag: "grab-sampling".
(5) Air is passed through a series of collectors cooled to selectively lower temperatures and pollutants frozen out in either liquid or solid form.

3.3.1 Absorption in liquids

When air is bubbled through an absorbing solution good contact between gas and liquid is essential. This can be achieved by impinging the gas stream onto a flat surface, which can be the bottom of the absorbing vessel, or by passing the gas through a sintered glass porous frit (Fig. 3.5). The latter method gives good distribution of gas through the absorbing solution but the glass frit can become blocked and is often difficult to clean after

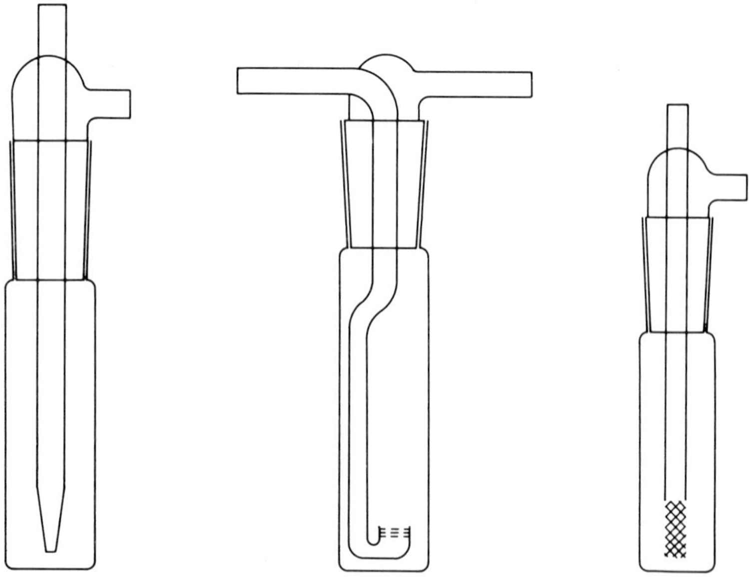

FIG. 3.5 Different designs of impingers, bubblers and glass-frit assemblies commonly used in air sampling

use. Some examples of analytical procedures which depend upon this method of collection are given in Table 3.4.

3.3.2 Gas adsorption on solids

In the case of organic vapours, solids such as activated charcoal or chromatographic materials like Tenax* and Porapak N have been used [27]. The adsorbent is packed into a tube about 7–10 cm long by 0·5 cm in diameter, through which air containing pollutant is drawn. The pollutant is retained by the packing and thereby concentrated. Analysis can be carried out after the pollutant has been recovered by solvent extraction or volatilization.

As an illustration of recovery by solvent extraction, the collection of vinyl chloride monomer on charcoal may be cited [28]. After collection of the vinyl chloride monomer the charcoal packing is emptied into a 2 ml glass vial and sealed with a rubber septum. The vial is cooled to −78°C for several minutes before 0·5 ml of CS_2 is injected through the septum. After removal from the coolant the vial is allowed to stand for 5 min before 5 μl of solution are removed and injected into a gas chromatograph. 100 mg of activated charcoal has a capacity for retaining at least 65 μg of vinyl chloride.

Alternatively, the collection and analysis of lead alkyls in air has been accomplished by drawing air through absorbent packings followed by flash heating to recover the volatile lead compounds [29]. Two principles—frontal chromatography and elution chromatography—are involved at the collection stage when this method is employed. In frontal chromatography the sample is continuously injected onto the column and a "front" of each component travels through the column. Normally, elution chromatography is used in analysis where only a small "plug" of each component moves through the column.

The relationship between frontal and elution chromatographic techniques is shown in Fig. 3.6. This indicates how the mass of a particular component leaving the column, per unit time, varies with time. Hence, in order to use frontal chromatography to collect a sample, the sample must be continuously injected onto the column until the amount of all components leaving the column in unit time is constant. This occurs when the amount of any component entering the column equals the amount leaving the column. This condition is usually attained when the sampling time is about twice the retention time of the most strongly retained component. Following Novák *et al.* [30] for a system of k components at

* Polymer of 2,6-diphenyl-*p*-phenylene oxide having good high-temperature stability (375°C).

TABLE 3.4 *A selection of procedures for the analysis of gaseous pollutants by wet chemical methods*

Pollutant	Reagent solution	Procedure	Sensitivity	Interference	Reference
CO	SiO_2 impregnated with ammonium molybdate/Pb in sulphuric acid packed in an indicator tube.	Draw known volume of air through indicator tube. Blue colour develops intensity proportional to CO present.	12·5 mg m^{-3} (10 parts 10^{-6}). Can be adapted down to 1–2 parts 10^{-6}.	Organics capable of reducing Mo.	5
	Reduction of CO to CH_4 with H_2 and Ni catalyst supported on firebrick at 250°C.	Determine CH_4 gas chromatographically with fid.	3·7 mg m^{-3} (3 parts 10^{-6}).		6, 7
Cl_2	Chlorine gives a yellow colour with *o*-toluidine.	Draw known volume of air through NaOH solution. After neutralization with H_2SO_4 treat with *o*-toluidine in HCl.	1 μg in volume analysed.	NO_x, Fe(III), (Mn(III)).	8
HCl	$AgNO_3$ solution.	2 m^3 of air are passed through chloride-free distilled water filtered and filtrate treated with $AgNO_3$ in HNO_3. AgCl is determined either nephelometrically, spectrophotometrically at 560 nm or titrometrically using dichlorofluorescein as indicator.	1 μg Cl-ion in 2 ml of solution.	Other acids, halogen acids and HCN.	9
Fluorides	Alizarin-thorium or alizarin-complexone-lanthanum lake.	Draw known volume of air through mixture of $NaOH/H_2O_2$, steam distil/$HClO_4$ measure bleaching action on lake reagent spectrophotometrically.	0·6 μg m^{-3}.		10
HCN	Potassium bis(5-sulphoxino)-Pd(11) reacts with HCN to give 8-hydroxyquinoline-5-sulphonic acid which coordinates with Mg to give a fluorescent compound.	Draw known volume of air through KOH solution of reagent glycine and $MgCl_2$ added.	0·02 μg ml^{-1} of test solution.	Fluorides and HF quench fluorescence.	11

H_2S	Sulphide ion reacts with *p*-aminodimethyl-aniline and Fe(III) chloride to give methylene blue.	Draw known volume of air through a suspension of $Cd(OH)_2$. A solution of *N*,*N*-dimethyl-*p*-phenylenediamine in H_2SO_4 and $FeCl_3$ is added. Measure intensity of methylene blue spectrophotometrically at 670 nm.	0·05 $\mu g\ ml^{-1}$ of test solution.	NO_x, Cl_2, O_3.	12
NH_3	Nessler's solution ($HgI_2 \cdot 2KI$) with ammonia gives a brown colour.	Draw a known volume of air through dilute H_2SO_4. Treat with reagent solution and measure colour intensity spectrophotometrically.	1 μg (3 ml)$^{-1}$.	Ammonium salts, H_2S, aliphatic amines.	13
Oxidants O_3, Cl_2, H_2O_2 and organic peroxide	Oxidants liberate iodine from KI solution.	Draw known volume of air through impinger or bubbler containing KI in a buffer solution of KH_2PO_4/Na_2HPO_4 pH 6·8. Measure absorbance of solution at 352 nm spectrophotometrically.	0·01 parts 10^{-6} expressed as O_3.	SO_2, H_2S and reducing gases.	14
	Oxidants react with alkaline KI to give hypoiodite from which iodine is liberated by sulphamic and phosphoric acids.	Draw known volume of air through impinger or bubbler containing KI/NaOH solution. Add solution of sulphamic and phosphoric acids and determine liberated iodine at 352 nm.	0·02 parts 10^{-6} as O_3.	Reducing gases.	14
NO_2	NO_2 reacts with *N*-(1-naphthyl)-ethylene-diamine dihydrochloride to give a red-violet colour. Modified Griess–Ilosvay method.	Draw known volume of air through impinger or bubbler containing reagent solution and sulphamic acid in glacial acetic acid. Measure colour intensity spectrophotometrically at 550 nm.	1 $\mu g\ ml^{-1}$ of test solution equivalent to 0·005–5 parts 10^{-6} or 0·01–10·0 $mg\ m^{-3}$.		15
NO	NO oxidized to NO_2 by filter paper impregnated with $K_2Cr_2O_7/H_2SO_4$.				16

TABLE 3.4—*continued*

Pollutant	Reagent solution	Procedure	Sensitivity	Interference	Reference
Nitrates	2,4-xylenol in glacial acetic acid.	For nitrates in suspended particulate matter. Use Hi-Vol air sampler and GFA filters to filter 20 m^3 of air. Reflux filter in distilled water for 90 min and filter. Acidify filtrate with H_2SO_4 and 1 ml of reagent solution, mix at 60°C for 30 min. Cool. Extract organic with toluene and then NaOH solution. Measure absorbance of aqueous layer at 435 nm spectrophotometrically.	0·1 $\mu g\ m^{-3}$.	Chloride, nitrites and oxidizing agents.	17
SO_2	SO_2 reacts with sodium tetrachloromercurate to yield dichlorosulphitomercurate which gives a red-purple colour with acid bleached *p*-rosaniline and formaldehyde. West and Gaeke method.	Aspirate air through reagent solution in bubbler or impinger and measure colour intensity after addition of *p*-rosaniline HCl in HCl and formaldehyde, spectrophotometrically at 560 nm.	0·015–0·6 mg m^{-3} (0·005–2·2 parts 10^{-6}).	NO_2, O_3 sulphides, nitrites, iron and heavy metal salts.	18 19 20 21
	Decolorized fuchsin-formaldehyde solution.	Aspirate air for 30 min at 5 l min^{-1} through alkaline glycerol solution. Add reagent solution and measure intensity of red-violet solution spectrophotometrically.	0·024 mg m^{-3} (0·008 parts 10^{-6}).	Thiols, thiosulphates. sulphides, NO_x.	22
H_2SO_4 mist		Aspirate air through Whatman No 1 filter paper. After exposure paper is washed with de-ionized water and the acid content determined by titration with standard sodium tetraborate solution to pH 7·0.	1 $\mu g\ m^{-3}$ for 10 m^3 air collected at 20 l min^{-1}.	Soluble bases and NH_3.	23

Aldehydes	3-methyl-2-benzothiazolone hydrozone (MBTH)	Aspirate air through reagent solution and develop blue colour by treating with sulphamic acid and iron (III) chloride. Measure intensity spectrophotometrically at 628 nm.	$2{\cdot}7\ \mu g\ m^{-3}$.	Aromatic amines and carbazole.	24 25
Acrolein	4-hexylresorcinol, mercury(11)chloride and trichloracetic acid.	Aspirate air through reagent solution contained in impingers in series. Stand solution in water bath at 60°C to develop blue-purple colour. Measure intensity spectrophotometrically at 605 nm.			26

constant pressure and temperature, the Gibbs–Duhem equation describing the relation between changes in chemical potential with change in composition may be written:

$$\sum_{i=1}^{k} N_i \frac{\partial \mu_i}{\partial N_j} = 0 \tag{3.12}$$

where

N_i = number of moles of component i,
μ_i = chemical potential of the ith component,
j = particular component of interest,
N_j = number of moles of this component.

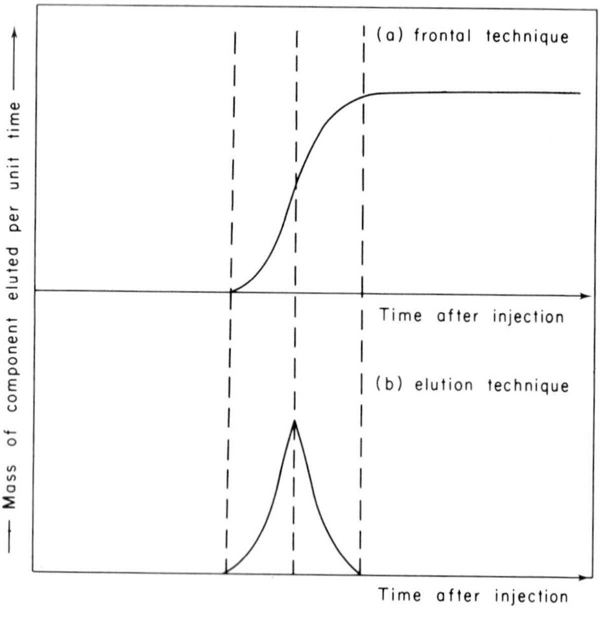

FIG. 3.6 Relationship between frontal and elution chromatography

In the chromatographic column there will be N_z moles of liquid stationary phase, z, having chemical potential μ_z. With elution chromatography only two components need to be considered: the liquid stationary phase z and the component of interest j. The Gibbs–Duhem equation then becomes,

$$N_z \frac{\partial \mu_z}{\partial N_j} + N_j \frac{\partial \mu_j}{\partial N_j} = 0 \tag{3.13}$$

or rearranging,

$$\left(\frac{\partial \mu_j}{\partial N_j}\right)_e = -\frac{N_z}{N_j}\frac{\partial \mu_z}{\partial N_j} \tag{3.14}$$

where $(\partial \mu_j/\partial N_j)_e$ is the variation in chemical potential of component j with the number of moles of j, in elution chromatography (elution denoted by subscript e).

In frontal chromatography all components must be considered and the Gibbs–Duhem equation becomes:

$$N_z\frac{\partial \mu_z}{\partial N_j} + N_j\frac{\partial \mu_j}{\partial N_j} + \sum_{i=1}^{k} N_i\frac{\partial \mu_i}{\partial N_j} = 0 \tag{3.15}$$

and then

$$\left(\frac{\partial \mu_j}{\partial N_j}\right)_f = -\left[\frac{N_z}{N_j}\frac{\partial \mu_z}{\partial N_j} + \frac{\sum_{i=1}^{k} N_i}{N_j}\cdot\frac{\partial \mu_i}{\partial N_j}\right] \tag{3.16}$$

where $(\partial u_j/\partial N_j)_f$ is the variation in chemical potential of component j with the number of moles of j, in frontal chromatography (frontal chromatography denoted by subscript f).

If

$$\frac{N_z}{N_j}\frac{\partial \mu_z}{\partial N_j} \gg \frac{\sum_{i=1}^{k} N_i}{N_j}\cdot\frac{\partial \mu_i}{\partial N_j}$$

Then

$$\left(\frac{\partial \mu_j}{\partial N_j}\right)_f \sim \left(\frac{\partial \mu_i}{\partial N_j}\right)_e \tag{3.17}$$

Hence, the properties of the component of interest j in the liquid stationary phase are almost uninfluenced by other dissolved components, provided the amounts of these components are small compared to the amount of liquid stationary phase. This condition will be obeyed in trace analysis of atmospheres. Under these conditions the partition coefficient appropriate to elution chromatography also applies to frontal chromatography. Then the partition coefficient is

$$K = \frac{C_L}{C_G} \tag{3.18}$$

where C_L is the concentration of component j in the liquid stationary phase and C_G is the concentration of component j in the gas phase. This partition coefficient is related to the specific retention volume of the liquid stationary phase by the equation

$$K = \frac{V_g^0 \rho_L T}{273} \tag{3.19}$$

where V_g^0 is the specific retention volume of the liquid stationary phase at absolute temperature T, and ρ_L is the density of the liquid stationary phase. Combining equations (3.18) and (3.19),

$$C_G = \frac{273 C_L}{V_g^0 \rho_L T} \tag{3.20}$$

In this equation

$$C_L = \frac{\text{mass of trapped component } (m)}{\text{volume of liquid stationary phase}} \tag{3.21}$$

and

$$\rho_L = \frac{\text{mass of liquid stationary phase in the column } (W_L)}{\text{volume of liquid stationary phase}} \tag{3.22}$$

so that

$$C_L = \frac{m \rho_L}{W_L} \tag{3.23}$$

and C_G becomes

$$C_G = \frac{273\ \mathrm{m}}{V_g^0 W_L T} \tag{3.24}$$

The equation enables C_G (the concentration of lead alkyl in this case) to be determined, if the specific retention volume, the weight of liquid stationary phase, the temperature and the mass of lead alkyl (m) collected are known. The only unknown, m, can be found by flash heating the sample tube and injecting the desorbed material on to a suitable gas chromatographic column for separation and analysis. The use of an electron capture detector (ecd) for lead alkyls separated by gas chromatography was first suggested by Lovelock and Zlatkis [31]. These detectors have improved sensitivity compared with flame ionization detectors (fid) for these compounds. When used for the analysis of petrol or petroleum vapours, however [32, 33], complications arise because of ethylene dichloride and ethylene dibromide added as scavenging agents in the fuel. These compounds have a large electron capture cross-section and this interferes with the performance of the ecd. One way of overcoming this interference is to insert a pre-column scrubber packed with porous support saturated with silver nitrate. Figure

3.7 shows the gas-chromatographic separation of lead alkyls in a petrol sample, in which this scrubbing procedure has been used.

Instead of collecting lead alkyls from the atmosphere by aspiration through porous packings at ambient temperatures an alternative method

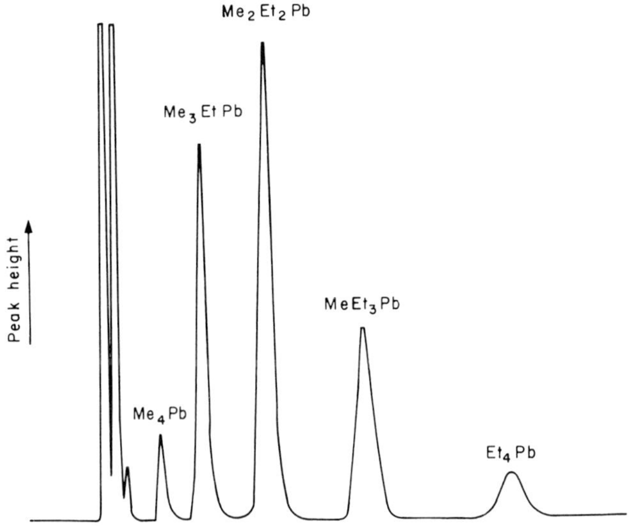

FIG. 3.7 Chromatographic separation with ecd of lead alkyls on a 3 m × 3·2 mm stainless steel column of 10% 1,2,3-tris(2-cyanoethoxy)propane (TCEP) on HMDS treated 80/100 Chromosorb W. The ecd was protected from halogen scavengers, using a scrubber section of 15 cm × 3·2 mm stainless steel packed with 20% Carbowax 400 (saturated $AgNO_3$) on 30/60 Chromosorb W precoated with 8% KOH, positioned between the analytical column and the ecd [32].

by Harrison *et al.* [34] employs liquid nitrogen temperatures. After collecting the sample over a 30–40 min period in a stainless steel absorbing tube containing 10% Apiezon L on a porous support, the tube is sealed and heated to 800°C in an oven. The tube contents are purged with nitrogen into the flame of an atomic absorption spectrophotometer. A sensitivity of 0·003 $\mu g\ m^{-3}$ from a 0·06 m^3 air sample has been claimed. Table 3.5 summarizes some of the procedures for the estimation of lead alkyls that appear in the literature.

3.3.3 "Grab-sampling"

Air sampling in which samples are collected by filling evacuated containers or inflatable bags has been called "grab-sampling". The technique has

TABLE 3.5 *Summary of methods for the collection and determination of lead alkyls*

Collector	Reagent	Procedure	Sensitivity	Refs.
Impinger or porous glass-frit scrubber.	Iodine monochloride solution prepared *in situ* $KIO_3 + 2KI + 6HCl = 3ICl + 3KCl + 3H_2O$. Followed by treatment with dithizone reagent (ammonium pyrrolidinedithiocarbamate).	Filtered air is passed through ICl reagent in scrubber. Dithiozone is added to give complex. Extract with methylisobutyl ketone and measure colour intensity spectrophotometrically at 520 nm or determine Pb by atomic absorption.	0·3 mg $(10\ m^3)^{-1}$.	35 36
Iodine crystals packed into a tube.	Dithizone.	Air is passed through the scrubber tube packed with iodine crystals. Pb recovered by addition of KI solution followed by dithizone reagent. H_2S interferes.	0–4 μg Pb ±12%.	37
10% Apiezon L on porous support at liquid N_2 temperature.		Draw air through collector cooled in liquid N_2 for 30–40 min. Heat tube to 800°C and purge with N_2 into atomic absorption spectrophotometer.	0·003 μg m^{-3} from 0·06 m^3 of air.	34
	Dithizone.	Inject 100 μl of gasoline onto a 0·3 m column packed with 40% nujol on chromosorb at 70°C. Collect vented gas into iodine scrubber followed by dithizone reagent.	±2%	38 39

been widely used because of its simplicity. The volume and pressure of air sampled is easily determined and in some cases the analysis can be carried out *in situ* by placing reagent in the bag or sample vessel. Precautions that should be taken include the following:

(1) Containers should be chemically inert; Milar bags have generally been found to be satisfactory for samples taken into inflatable containers. Walls of the container must not catalyse reaction between the compounds in the mixture collected [40].

(2) Analyse sample as quickly as possible after collection. Complex mixtures obtained from engine emissions, for example, can react chemically or be removed from the gas phase by adsorption on the walls of the container or by absorption in solution of condensed water vapour [41]. Stability can be improved by diluting the sample with air.

(3) Avoid exposing collected sample to sunlight, since this may promote chemical reaction between components in the aerosol [42].

(4) When collecting by air displacement, as is done with a syringe, pump the plunger for at least 1 min to ensure that the sample collected is as representative as possible.

3.4 Collection of particulate pollutants

3.4.1 Sedimentation

Stokes' law, as previously mentioned in section 2.4.1, may be applied to the collection of particles that are settling under gravity in calm air. The equation

$$6\pi\eta r v = \tfrac{4}{3}\pi r^3(\rho_p - \rho_{\text{air}})g \tag{3.26}$$

holds, where η is the viscosity of air, r is the radius of the particle, g is the acceleration due to gravity, v is the terminal settling velocity, ρ_p and ρ_{air} are the densities of the particles and the air, respectively. When $\rho_p \gg \rho_{\text{air}}$, the equation becomes

$$v = \frac{2g\rho_p r^2}{9\eta} \tag{3.27}$$

This equation is strictly only valid at very small values of the Reynolds number, *Re*, defined by

$$Re = \frac{2rv\rho_{\text{air}}}{\eta} \tag{3.28}$$

This natural settling of dust from the atmosphere provides, in principle, the simplest method for collection of atmospheric particulate matter. Particles larger than about 10 μm may be collected in suitably shaped receptacles left at selected sites. The method is unreliable for particles smaller than 1 μm in diameter. These sites should be free from overhead obstructions and as far as possible remote from interference from local sources such as chimneys. The sample collector should be positioned about 3 m above ground level and not nearer to a vertical wall than that subtended by an angle of 30° from the top of the wall to the sampler. The collector, which is often tapered and cylindrical in design having approximate dimensions $30 \times 12 \times 9\ cm^3$, may be constructed of any suitable inert material, e.g. glass, polyethylene or stainless steel.

Sample periods of about 1 month are recommended for equipment of this type. After this time the receptacle contents are transferred quantitatively to an evaporating basin of known weight, the contents evaporated to dryness at a maximum temperature of 105°C, and the final weight obtained. The results are expressed as the weight of solid deposited (unit area)$^{-1}$ month^{-1}. If desired, chemical analysis can be performed on the residue. The results can then be conveniently expressed in terms of total sulphate, total chloride, total nitrate, benzene soluble fraction, combustible material, and metal content.

3.4.2 Filtration

The mechanism and hence the theory of filtration depends upon the filter structure, the material of construction, and the method of manufacture. Analytical filters made from cellulose esters or polymers containing capillaries with well defined pore diameters and lengths are extensively used in aerosol science. Since 1965 a technique has been developed which allows porous filters to be made from sheets of polycarbonate [43].

A thin sheet of polycarbonate, 10 μm thick, is placed in contact with sheets of uranium and put into a reactor. The neutron flux generated causes fission of uranium-235 and the resulting fragments bombard the sheet, producing trails of damage. The polycarbonate sheets are etched to develop this damage. The extent of enlargement of the pores is determined by the type of bath, the reagents employed, and the duration and temperature of treatment. The resulting material called Nucleopore can be produced with reasonably graded pores of diameters from 0·5 μm up to 8 μm, which run parallel to each other in a matrix of density $0{\cdot}95\ g\ cm^{-3}$. Compared with the Millipore membrane type of filter which has three separate structures, viz. an upper surface, an interior and a lower surface, the Nucleopore has a simple uniform structure throughout. The upper surface is smooth, the pores are circular in cross-section and the matrix is mechanically strong. Since flow-rate pressure-drop characteristics are similar for these materials the theoretical treatment of collection efficiencies discussed by Pich [44], and Spurny and Pich [45] can be applied.

During filtration, particles are retained by the filter media by

(1) inertial impaction,
(2) diffusional deposition,
(3) direct interception,
(4) electrostatic deposition,
(5) sedimentation within the pore.

If r is the radius of the particle and R_0 is the radius of the filter pore then when $r > R_0$ the particle cannot penetrate the filter and the efficiency E will be unity. Cases of interest in aerosol filtration problems arise when $r < R_0$ and when $r \ll R_0$. Inertial impaction around a circular pore is represented by

$$e_i = \frac{2e_i'}{(1+Q)} - \frac{e_i'^2}{(1+Q)^2} \tag{3.29}$$

where

$$e_i' = 2\,\mathrm{Stk}\,Q^{1/2} + 2\,\mathrm{Stk}^2\,Q\exp(-1/\mathrm{Stk}\,Q^{1/2}) - 2\,\mathrm{Stk}^2\,Q \tag{3.30}$$

with $Q = P^{1/2}/(1-P^{1/2})$, P is the porosity. The Stk is defined by

$$mq/6\pi\eta rR_0 \tag{3.31}$$

with

$$m = \tfrac{4}{3}\pi r^3 s$$

where

q = gas velocity at the filter face (cm s^{-1})
η = viscosity of suspending medium (g $\mathrm{cm}^{-1}\,\mathrm{s}^{-1}$—$181\times10^{-6}$ for air)
s = aerosol density (g cm^{-3})

Diffusional deposition within the pore is given by

$$e_D = 1 - n/n_0 \tag{3.34}$$

where

$$n/n_0 = 0{\cdot}81904\exp(-3{\cdot}6568\,N_D) + 0{\cdot}09752\exp(-22{\cdot}3045\,N_D) + 0{\cdot}03248\exp(-56{\cdot}95\,N_D) + 0{\cdot}0157\exp(-107{\cdot}6\,N_D)\ldots \tag{3.35}$$

and

$$N_D = \frac{LDP}{R_0^2 q} \tag{3.36}$$

with

L = filter thickness or pore length (cm)
D = aerosol particle diffusivity for Brownian motion ($cm^2\,s^{-1}$)

$$= \frac{CkT}{6\pi\eta r} \tag{3.37}$$

and

$$C = 1 + 1.246\lambda/r + (0{\cdot}42\lambda/r)\exp(-0{\cdot}87r/\lambda) \tag{3.38}$$

λ = the mean free path of the suspending gas molecules, k and T are Boltzmann's constant and temperature °K, respectively. Some calculated values of the functions "Stk" and N_D for different pore sizes R and aerosol size r at face velocities between 0·1 and 40 cm s^{-1} have been given by Spurny and Lodge [46]. A limited selection of these values are included in Tables 3.6 and 3.7, so that the effect of gas face velocity and pore radius on these parameters can be seen.

TABLE 3.6 *Computed values of the "Stk" function for different pore sizes* $R_0(\mu m)$, *for aerosol particles having* $r = 0{\cdot}1\times10^{-4}$ *cm, density* $s = 2{\cdot}2$ *g* cm^{-3} *at face velocities between 0·1 and 40 cm* s^{-1}. $\eta = 181\times10^{-6}$ *poise* [46]

	$R_0(\mu m)$		
q cm s^{-1}	0·5	2·0	4·0
0·1	0·000540	0·000135	0·000068
1·0	0·005402	0·001351	0·000675
5·0	0·027010	0·006755	0·003375
25·0	0·135052	0·033775	0·016875
40·0	0·216083	0·054040	0·027000

TABLE 3.7 *Computed values of* N_D *for different pore sizes* R_0 (μm), *for pore length* $L = 2\times10^{-2}$ *cm and porosity* $P = 0{\cdot}76$, $r = 0{\cdot}1\times10^{-4}$ *cm*, $D = 2{\cdot}21\times10^{-6}$ $cm^2\,s^{-1}$ *at face velocities between 0·1 and 40 cm* s^{-1} [46]

	R_0 (μm)		
q cm s^{-1}	0·5	2·0	4·0
0·1	134·3680	8·3980	2·0995
1·0	13·4368	0·8398	0·2099
5·0	2·6874	0·1680	0·0420
25·0	0·5375	0·0336	0·0084
40·0	0·3359	0·0210	0·0052

Direct interception is given by

$$e_R = \frac{r}{R_0}\left(2 - \frac{r}{R_0}\right) \quad \text{where } r/R_0 \leqslant 1 \tag{3.38}$$

This equation takes into account the possibility of a particle which enters a pore being retained on the pore wall when it approaches the pore wall to a distance $\mathrm{d}x \leqslant r$.

The most important mechanisms are inertial impaction and diffusional deposition. The total efficiency for these processes E_1 is given by

$$\begin{aligned} E_1 &= e_i + e_D(1 - e_i) \\ &= e_i + e_D - e_i e_D \end{aligned} \tag{3.39}$$

When direct interception is included it has been estimated that this could account for about 15% where $r \ll R_0$, then

$$E_2 = E_1 + \delta e_R(1 - e_i) \quad \text{with } \delta = 0{\cdot}15 \tag{3.40}$$

when $r \leqslant R_0$ the influence of interception can be greater than that suggested by the above equation and accordingly,

$$E_3 = E_1 + 0{\cdot}63^{(1 - r/R_0)} e_R(1 - e_i) \tag{3.41}$$

Spurny and Lodge [47, 48] have also considered the pressure drop across membrane filters. This depends upon the Knudsen number K_n defined by λ/R_0. In the viscous flow region where $K_n \ll 1$, the pressure drop may be represented by

$$\Delta p = P_1 - P_1\left[1 - 5{\cdot}093\left(\frac{\eta L}{P_1}\right)\left(\frac{q}{R_0^4 N_p}\right)\right]^{1/2}, \tag{3.42}$$

where

P_1 = gas pressure in front of the filter (dyn cm^{-2})
Δp = pressure drop across the filter (dyn cm^{-2})
N_p = number of pores per square centimetre of filter surface (cm^{-2}).

At slightly larger values of K_n, $K_n < 1$, a correction factor must be introduced for gas slippage so that

$$\Delta p = P_1 - P_1\left[1 - 5{\cdot}093\left(\frac{\eta L}{P_1}\right)\left(\frac{q}{R_0^4 N_p(1 + 5{\cdot}5\lambda/R_0)}\right)\right]^{1/2} \tag{3.43}$$

When K_n is near unity or larger

$$\Delta p = P_1 + 3{\cdot}42\left(\frac{0{\cdot}75\eta}{R_0}\right)\left(\frac{2\pi RT}{M}\right)^{1/2}$$

$$-\left\{\left[P_1 + 3{\cdot}42\left(\frac{0{\cdot}75\eta}{R_0}\right)\left(\frac{2\pi RT}{M}\right)^{1/2}\right]^2 - 5{\cdot}093(\eta L P_1)\left(\frac{q}{R_0^4 N_p}\right)\right\}^{1/2} \quad (3.44)$$

where R is the gas constant $8{\cdot}31 \times 10^7$ g cm^2 s^{-2} K^{-1}, T is the temperature (°K), and M is the molecular weight. Some computed pressure drop changes with filtration time for various solid and liquid aerosols are shown in Fig. 3.8.

The filter efficiency and the pressure drop across a filter so far considered do not take into account the effect that deposition of material within the filter may have on these parameters. As material builds up in the filter pores correction factors must be applied if realistic estimates are to be made. Provided the diffusional efficiency is effective or alternatively direct interception of aerosol particles is unimportant, R_0 can be replaced by R_t, where R_t represents the restricted radius after a duration of use, t, seconds. In the case of a solid aerosol, deposition is given by

$$R_t = \left[R_0^2 - 2\left(\frac{R_0}{R_0 + ar}\right)r^3 nt\left(\frac{q}{LN_p}\right)\right]^{1/2} \quad (3.45)$$

where a is a coefficient for the pore-clogging process defined by

$$a = \left(\frac{R_0}{r}\right)^{0{\cdot}57} \quad (3.46)$$

and for liquid aerosols:

$$R_t = \left[R_0^2 - 1{\cdot}33 r^3 nt\left(\frac{q}{LN_p}\right)\right]^{1/2} \quad (3.47)$$

and

$$n \text{ is the aerosol particle concentration (cm}^{-3}) \quad (3.48)$$

R_t is now the effective radius of the filter pore and strictly should replace R_0 in the equations given for estimation of collection efficiency.

3.4.3 Particle size measurement

In view of the health aspects associated with air pollution much effort has been directed to investigations which enquire into the nature, the properties and the behaviour of aerosol particles. In this regard, information on particle size is important because this property can affect aerosol stability, meteorology, respiration, and the extent of lung penetration and retention. In this section the discussion is limited to particle size measurements mainly of interest to chemists. Optical measurements such as light-scattering techniques are not considered.

Owing to the heterogeneous nature of suspended particles, they have an infinite variety of different shapes and sizes. It is necessary to take this

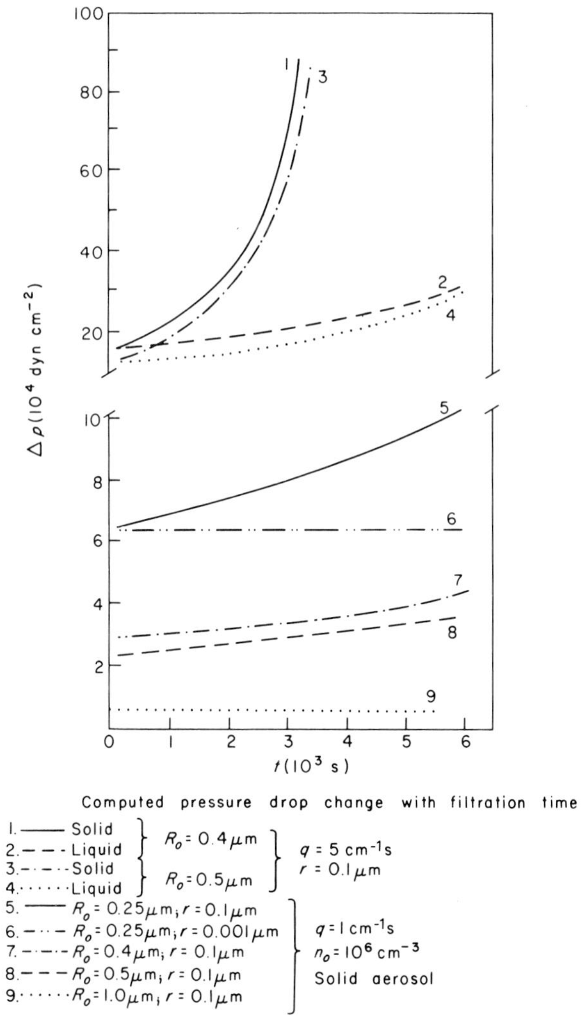

FIG. 3.8 Pressure drop characteristics of filters [48]

feature into account in any treatment of particle size. One way in which this can be done is to define the "aerodynamic diameter" as the diameter of an equivalent spherical particle of unit density that behaves in the same manner as the actual aerosol particle. Particle sizing on this basis can be accomplished with equipment which collects the particulate component of the aerosol at a surface by impaction. Particle size discrimination can be controlled at the collection stage. Subsequent chemical analysis of each size fraction collected is usually possible when sampling is performed with instruments of this type.

3.4.3.1 The Andersen cascade impactor

When the velocity imparted to a particle is sufficiently great, it will have enough inertia to escape from a stream of gas being reflected from a surface upon which it impinges (Fig. 3.9). Hence, the incident and reflected aerosol streams will have a different composition. The reflected stream of gas will contain fewer larger particles compared with the incident stream.

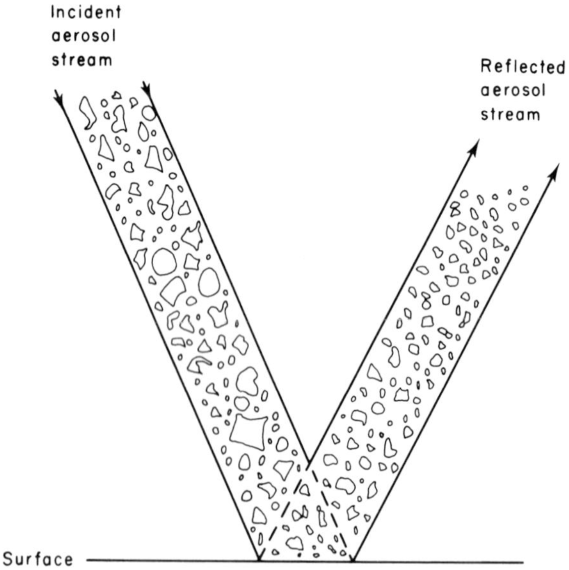

FIG. 3.9 Illustration of the principle used to separate particles of different sizes by impaction at a surface. Retention of the larger particles at the surface impacted will result in different compositions of the incident and reflected air streams

The process can be repeated and if, in addition, the gas velocity is increased, particles in another smaller size range can be selectively removed from the incident stream. The Andersen particle size fractionator (Fig. 3.10) operates on this principle.

The theory of the process has been treated by Ranz and Wong [49]. For the impaction of a round aerosol jet on a flat plate of infinite extent they have derived the following equation for the efficiency, E:

$$E = \left[\frac{S_2 - S_1}{S_2 \exp S_1 h - S_1 \exp S_2 h}\right]^2 \tag{3.49}$$

where

$$S_{1,2} = (-1/4K) \pm [(1/4K)^2 + 1/2K]^{1/2} \tag{3.50}$$

$$h = \frac{1}{q} \tan^{-1}\left[\frac{(1-4K)8Kq}{(8K-1)+(4Kq)^2}\right] \tag{3.51}$$

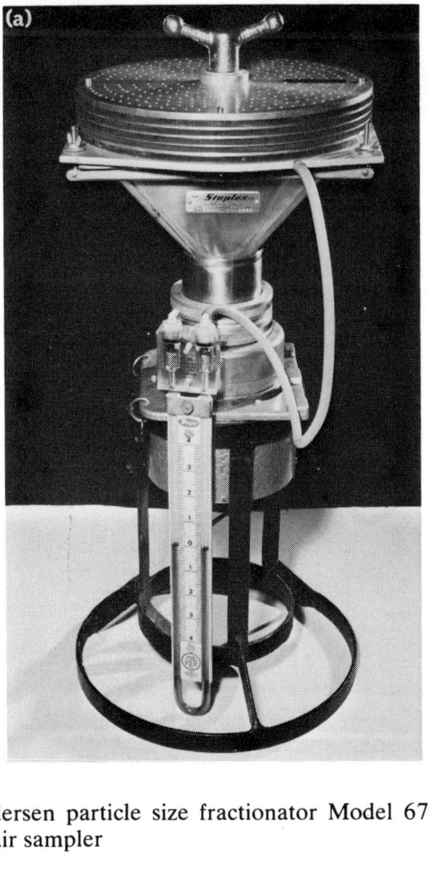

FIG. 3.10(a) Andersen particle size fractionator Model 67-000 mounted on a Staplex "Hi-Vol" air sampler

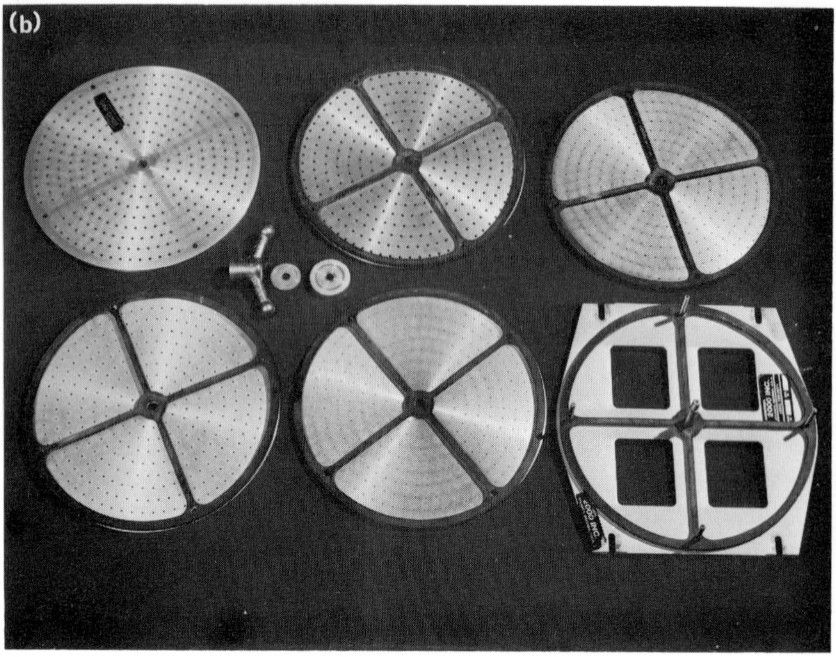

FIG. 3.10(b) Exploded view showing the different stages and the arrangement of holes and gaskets

and

$$q = [(1/K) - (1/4K)^2]^{1/2} \tag{3.52}$$

Provided the particles are large compared with the mean free path of the suspending gas molecules, K is defined by

$$K = \frac{\rho v d^2}{18 \eta D} \tag{3.53}$$

where

ρ = particle density (g cm^{-3})
v = velocity of aerosol stream (cm s^{-1})
d = diameter of aerosol particle (cm)
D = diameter of aerosol jet (cm)

and

η = gas viscosity (poise).

Some calculated values of K, q, h, S_1, S_2 and E for various values of d in the range 2·4–4·0 μm are shown in Table 3.8. The values given in the table illustrate the way in which the different parameters move and how the efficiency changes with the particle size of the aerosol. Under the conditions stipulated in the table the theoretical calculation indicates that particles larger than 4 μm in size will be retained. The impaction efficiency, however, decreases for smaller particles until eventually for particles rather smaller than 2·4 μm the efficiency falls to zero.

TABLE 3.8 *Computed values of* E *and* $K^{1/2}$ *for selected values of* $d(\mu m)$, *when* $\rho_p = 1{\cdot}0$; $v = 760\ cm\ sec^{-1}$; $\eta = 180 \times 10^{-6}$ *poise and* $D = 0{\cdot}158\ cm$

d μm	K	q	h	S_1	S_2	E	$K^{1/2}$
2·40	0·085	1·77	0·85	0·87	−6·73	0·29	0·29
2·45	0·089	1·83	0·77	0·87	−6·49	0·34	0·30
2·50	0·093	1·87	0·70	0·86	−6·26	0·38	0·30
2·75	0·112	1·99	0·48	0·84	−5·30	0·58	0·33
3·0	0·133	2·00	0·35	0·82	−4·57	0·74	0·36
3·5	0·181	1·90	0·16	0·78	−3·53	0·94	0·43
4·0	0·235	1·76	0·03	0·74	−2·85	1·00	0·49

In practice, the design shown in Fig. 3.10 comprises a series of plates that have holes drilled through them to allow passage of the aerosol. Each stage or plate has smaller diameter holes than the preceding plate, so that aerosol velocity is increased from one stage to the next. Thus, each stage collects particles within a specific size range. Ideally, each stage collects all particles larger than a certain size and none smaller. This cannot in fact be achieved, so that, as shown in Fig. 3.11, the collection efficiency varies from 0% to 100% within the size range. The diameter chosen by Mercer [50] to represent the average aerosol particle sampled at a stage, or the "cut-off" point, is that diameter for which the collection efficiency is 50%. The effective cut-off diameters (ECDs) of the different stages of a cascade impactor are specified by the manufacturer for spherical particles of unit density. Ambient aerosols do not have such unique characteristics and this leads to uncertainty about the size fraction actually collected.

Other features which cause distortion of the particle size distributions are losses on the walls of the impactor and the re-entrainment of particles. Wall losses are believed to be small with the Andersen impactor but re-entrainment problems always occur with any design of this type. Impaction surfaces used in the earlier models for particle size measurements on micro-organisms were petri dishes [51]. Later models (Fig. 3.10) employ filter papers as collection surfaces. In either case, the amount of material collected on each stage can be determined by weighing and, therefore, the weight composition of the aerosol particle size distribution can be estimated. Furthermore, because each stage consists of plates containing between 200 and 400 holes, high flowrates of up to 500 or $600 \, l \, min^{-1}$ are possible and so relatively large amounts of particulate matter can be collected per stage. This enables chemical analysis of metals or polynuclear hydrocarbons, for example, to be carried out on each stage and the particle size distributions of these compounds can be ascertained.

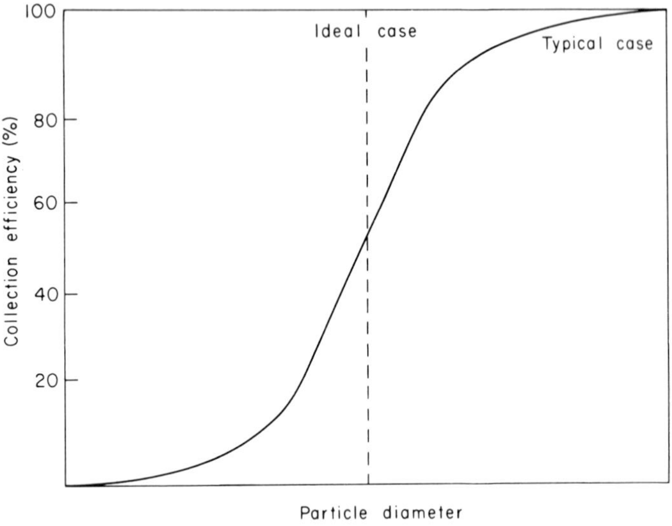

FIG. 3.11 Collection efficiencies of an impactor

The problems of particle re-entrainment mentioned above has received theoretical attention by Dahneke [52], who considered the collision between a surface and a particle in terms of the energy of the system. Let a particle moving towards a surface have an incident normal velocity V_{ni} and kinetic energy KE_{ni}. Upon nearing the surface the particle will fall into the particle-surface potential energy well of depth E. If the particle does not have sufficient energy on rebound it will not escape this potential well and will be captured by the surface. In fact perturbations produced at the surface by the act of collision can alter the depth of this well. To account for this feature two potential well depths will be defined. E is the depth seen by the incident particle and E_r is the depth seen by the reflected particle.

In order to calculate the kinetic energy of the particle at the instant of rebound a coefficient of restitution e has been defined as the ratio of the normal particle velocity at the instant of rebound over the normal particle velocity at the instant of contact. Hence, the kinetic energy of the particle at the instant of rebound is given by,

$$KE_{\text{at rebound}} = (KE_{ni} + E_i)e^2 \tag{3.54}$$

where the sum of KE_{ni} and E_i is the kinetic energy of the particle at the instant of contact and e^2 is the fraction of this energy recovered by the particle. Upon rebound, the particle must exchange kinetic energy for potential energy as it climbs out of the well. The final kinetic energy of the reflected particle beyond the influence of the surface, is KE_{nr}, where

$$KE_{nr} = (KE_{ni} + E_i)e^2 - E_r \tag{3.55}$$

Equation (3.55) can be rewritten in terms of particle velocity V_n and mass m by

$$\frac{V_{nr}}{V_{ni}} = \left[e^2 - \frac{2(E_r - e^2 E_i)}{mV_{ni}^2}\right]^{1/2} \tag{3.56}$$

In the limiting case, when all the particle rebound energy is required to lift it out of the potential well, $K_{nr} = 0$, and V_{ni}^* defines the particle capture limit. Then

$$V_{ni}^* = \left[\frac{2}{me^2}(E_r - e^2 E_i)\right]^{1/2} \tag{3.57}$$

This capture occurs when $V_{ni} < V_{ni}^*$ and rebounds when $V_{ni} > V_{ni}^*$. Two special cases can be distinguished:

(i) When $E_r = E_i = E$

$$V_{ni}^* = \left[\frac{2E}{m} \cdot \frac{1 - e^2}{e^2}\right]^{1/2} \tag{3.58}$$

(ii) When $E_i \ll E_r$

$$V_{ni}^* = \left(\frac{2E_r}{me^2}\right)^{1/2} \tag{3.59}$$

The depth of the potential well for two contacting spheres of diameter d_1 and d_2 is given by

$$E = \frac{Ad'}{12z_0} \tag{3.60}$$

where A is a constant, d' is the reduced mass $d_1d_2/(d_1+d_2)$ and z_0 is the equilibrium separation of the spheres. If $d_2=\infty$, equation (3.60) is applicable to a sphere of diameter d and density ρ with a plain surface, hence, combining equations (3.58) and (3.60) gives

$$V_{ni}^* = \left[\frac{A}{z_0\rho\pi e^2}\cdot 1-e^2\right]^{1/2} \Big/ d \tag{3.61}$$

Equation (3.61) enables the dependence of V_{ni}^* on particle size to be determined for a particle striking a flat rigid surface. Results of the calculation performed by Dahneke for collision between quartz spheres and polystyrene spheres on a quartz surface are shown in Fig. 3.12. The curves for silica particles are thought to be accurate, but owing to the neglect of elastic flattening in the theoretical treatment outlined, the curves for polystyrene particles are probably too low.

Since particles reach velocities of the order of $100\,\mathrm{m\,s^{-1}}$ in impactor devices, a 10- or 10^2-fold increase in V_{ni}^* is reasonable and theory predicts that replacing a hard rigid impaction surface with a thin flexible membrane of soft material will have little effect. The only way to capture solid aerosol particles by impaction is to coat the impaction surface with a viscous liquid. This can be done with thin smears of grease or oil such as Vaseline. Although such coats can be applied before sample collection of atmospheric aerosols it may not always be necessary. Smog aerosols, for instance, are usually sufficiently "sticky" to make such application unnecessary.

The other principal source of error which can arise, first discovered by Burton *et al.* [53], concerns the pH of the collecting media. When glass fibre collecting surfaces of pH 11·0 are used the total weight of material collected can be 30% greater than that collected by a standard Hi-Vol sampler. The cause of the discrepancy is the retention of acidic material, primarily sulphate, by the alkaline media. Lowering the pH of the glass fibre to 6·5 results in the total particulate collected in the fractionator being about 94% of that collected by the conventional sampler. The difference of around 6% is generally attributed to losses on the internal non-collecting surfaces of the fractionator.

Results from particle size fractionation studies are interpreted mathematically by a probability density function called a log-normal distribution law. This distribution law which is actually a variant of the Gaussian of normal distribution law is written [54]:

$$y = \frac{1}{\ln\sigma_g\sqrt{2\pi}}\exp\left[-\frac{(\ln x-\ln M)^2}{2\ln^2\sigma_g}\right] \tag{3.62}$$

where y is the probability density equal to $n/\Delta\ln x$ and n is the number of particles whose diameters, x, have their logarithms lying in the interval

$\Delta \ln x$: σ_g is the geometric standard deviation and M the geometric mass median equivalent diameter (MMD). The law can be verified graphically from a plot on log-probability paper. As an illustration, the weights of

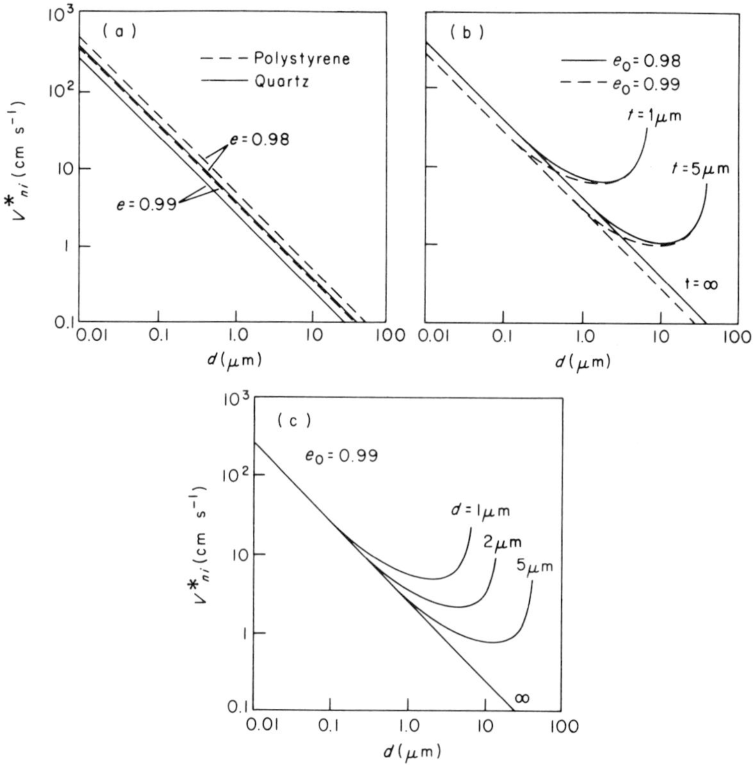

FIG. 3.12 Results of Dahneke's calculations [52] of capture limits of spheres impacting on surfaces under the conditions indicated. (a) Particle capture limits for polystyrene and silica (quartz) spheres colliding with a rigid quartz surface, i.e. the surface of a thick body. Particle capture occurs when V_{ni} is less than V^*_{ni}. (b) The capture limits for silica spheres colliding with quartz plates of various thickness t. Since thin plates have better ability to dissipate collision energy, they can capture particles striking them at higher velocities. (c) The capture limits for silica spheres striking quartz cylinders (fibres) of various diameters (d)

airborne dust which has been retained at each stage of the impactor are recorded and expressed in terms of particle size distributions. This is done by the procedure indicated in Table 3.9. The first column shows the stage on which the particulate has been collected. The second column gives the weights of particulates in milligrams. The third column is the cumulative weight which is derived from the second column by adding successively the individual weights of each stage together. These are expressed as a cumulative percentage in column four, then, 100 minus the cumulative percentage in column five, and the appropriate effective cut-off diameters of each stage in column six. The results are reported by plotting the values in the

TABLE 3.9 *Interpretation of data collected on an Andersen impactor to find the MMD and* σ_g

Birmingham, UK
Date 25-03-77 to 26-04-77

Column Number	1 Andersen stage	2 Nett weight mg	3 Cumulative weight mg	4 Cumulative weight %	5 100-Cumulative weight %	6 ECD μm
	1	216·6	216·6	18·25	81·75	7
	2	62·5	279·1	23·51	76·49	3·3
	3	162·0	441·1	37·16	62·84	2·0
	4	166·0	607·1	51·14	48·86	1·1
	5	580·0	1187·1	100·00		0·01

Air flowrate, 0·566 m^3 min^{-1}
Air volume, 26 039 m^3
Total particulate concentration, 45·6 $\mu g\ m^{-3}$
MMD, 1·15 μm
σ_g, 6·96

% Particulate $\leqslant$ 1 μm 48
% Particulate $\leqslant$ 2 μm 62
% Particulate $\leqslant$ 3 μm 69

fifth and sixth columns on log-probability paper. From these plots the mass median equivalent diameter (MMD) can be read off on the particle diameter axis which corresponds to 50% on the cumulative % mass $\leqslant$ particle diameter axis (Fig. 3.13). The geometric standard deviation σ_g may readily be calculated from the plot by using the equation,

$$\sigma_g = \frac{84\%\ \text{size}}{50\%\ \text{size}} = \frac{50\%\ \text{size}}{16\%\ \text{size}} \tag{3.63}$$

This treatment also enables the percentage which is less than a specified MMD to be read from the plot, e.g. % particulate $\leqslant$1 μm or $\leqslant$2 μm, etc. This is valuable because the amount of the total particulate in the aerosol which is respirable can now be assessed. Other parameters which are occasionally quoted such as the arithmetic mean diameter, d, is related to M and σ_g by

$$\ln d = \ln M + 1{\cdot}1513 \ln^2 \sigma_g \tag{3.64}$$

An equation which describes the particle size x_M about which will be clustered those particles with the greatest frequency of occurrence or for which the particle density $n/\Delta x$ is a maximum is given by [54]:

$$\ln x_M = \ln M - \ln^2 \sigma_g \tag{3.65}$$

Experience has shown that atmospheric aerosols very often exhibit bimodal distributions. These are most easily recognized by plotting (ECD) against $\Delta m/\Delta \ln (\text{ECD})_{\text{stage}}$ where Δm is the aerosol concentration retained

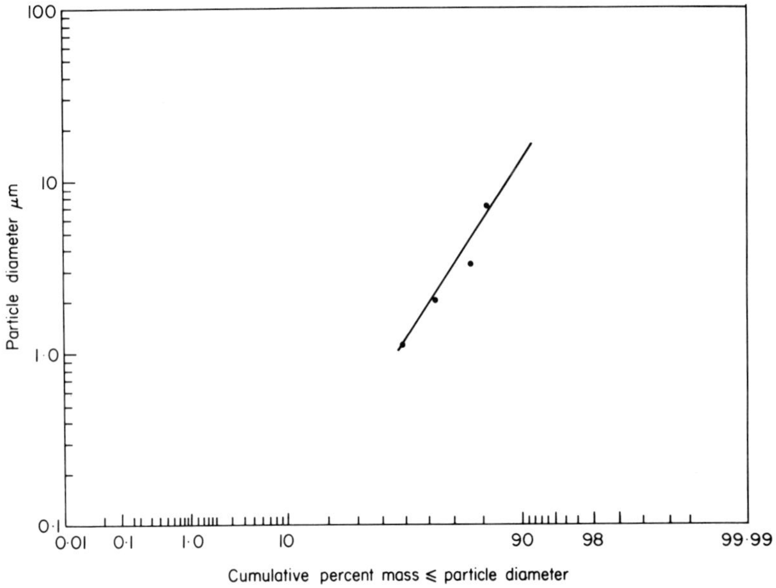

FIG. 3.13 Plot of data given in Table 3.9 on logarithm-probability paper

TABLE 3.10 *Treatment of data to reveal any tendency to bimodal characteristics*

Impactor stage	ECD (μm)	$\Delta \ln d$	Δm (μg m^{-3})	$\Delta m/\Delta \ln d$
1	7 (<30)	(1·45)	8·3	5·7
2	3·3	0·75	2·4	3·2
3	2·0	0·50	6·2	12·4
4	1·1	0·60	6·4	10·7
5	0·01	4·7	22·3	4·7

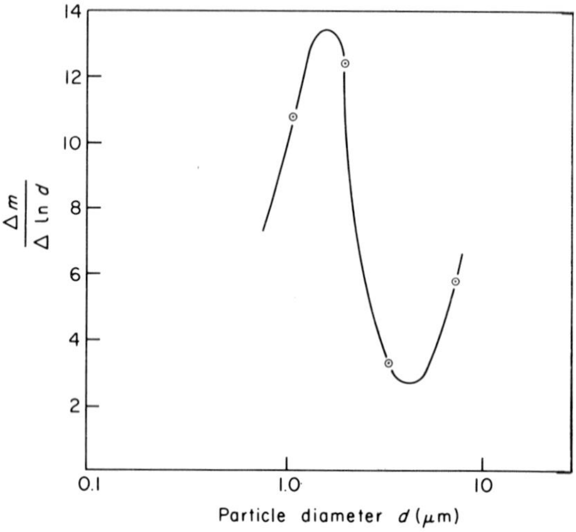

FIG. 3.14 Particle size distribution curve prepared by plotting the data given in Table 3.10

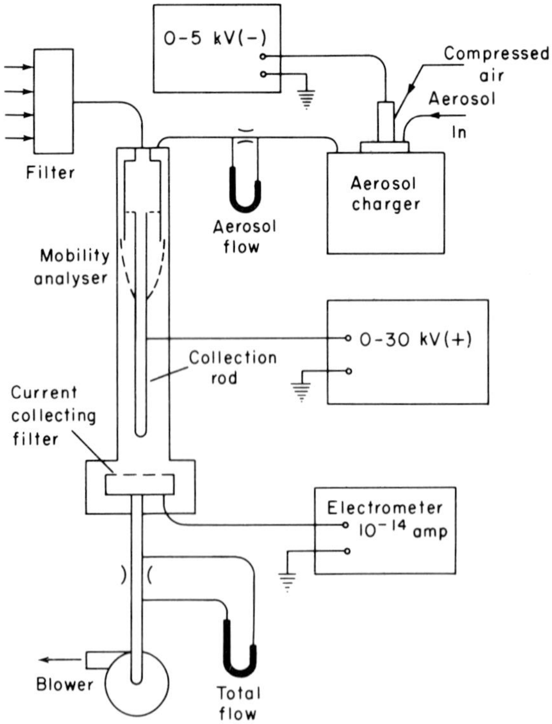

FIG. 3.15 Schematic representation of the electrical particle classifier [55]

by a stage in the impactor ($\mu g\ m^{-3}$) and $\Delta \ln (ECD)_{stage}$ is calculated from the equation,

$$\Delta \ln (ECD)_{stage} = \ln (ECD)_{previous\ stage} - \ln (ECD)_{stage} \qquad (3.66)$$

By way of example this has been done for the data given in Table 3.9 and is shown in Table 3.10 and Fig. 3.14. It is apparent from Fig. 3.14 that to construct the distribution curve, the greater the number of impactor stages, the more detail will be obtained in the plot.

3.4.3.2 Electrostatic precipitators

High voltage (5–20 kV) corona discharge induced between a central wire cathode and a concentric cylindrical anode causes ionization of gas molecules. These ionized molecules collide with particles present in the aerosol to produce electrostatically charged particles which can be collected by attraction to an electrode carrying a charge of opposite sign. The theory of electrostatic precipitators has been developed for industrial precipitators and is outside the scope of the present discussion. However, an application described by Flesch [55] will serve to show how the method can be used as a sub-micron aerosol classifier.

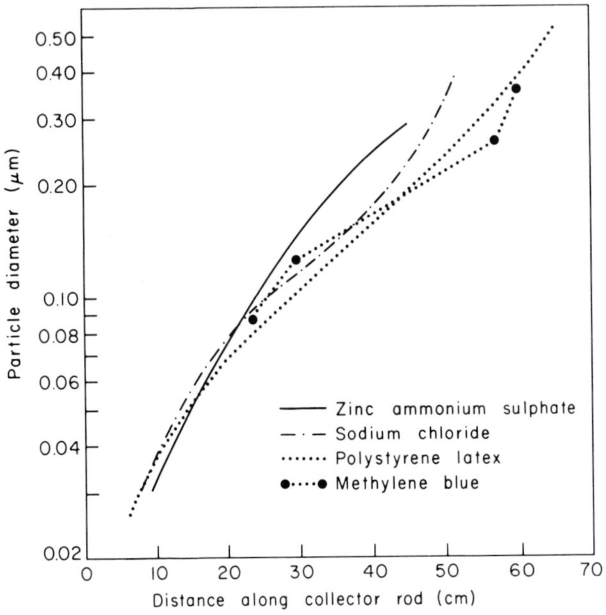

FIG. 3.16 Relationship of the arithmetic mean deposition distance *vs* particle diameter determined from electron micrographs for various aerosols [55]

Figure 3.15 shows a diagram of the equipment. The aerosol is charged and passes to the mobility analyser, where it is introduced as a thin annular cylinder around a core of clean air. The collector rod operates at constant potential so that the aerosol deposits along the rod according to decreasing mobility, i.e. increasing size. Particle size distributions along the rod for different aerosol preparations, estimated from electron micrographs of the deposits, are shown in Fig. 3.16. The equipment has been used to determine the lead particle size distribution in Cincinnati air. Figure 3.17 indicates the results obtained by Flesch using the electrical particle classifier with those determined by an Andersen cascade impactor. It must be remembered, of course, that particle diameters estimated with the Andersen are in terms of aerodynamically equivalent spheres of unit

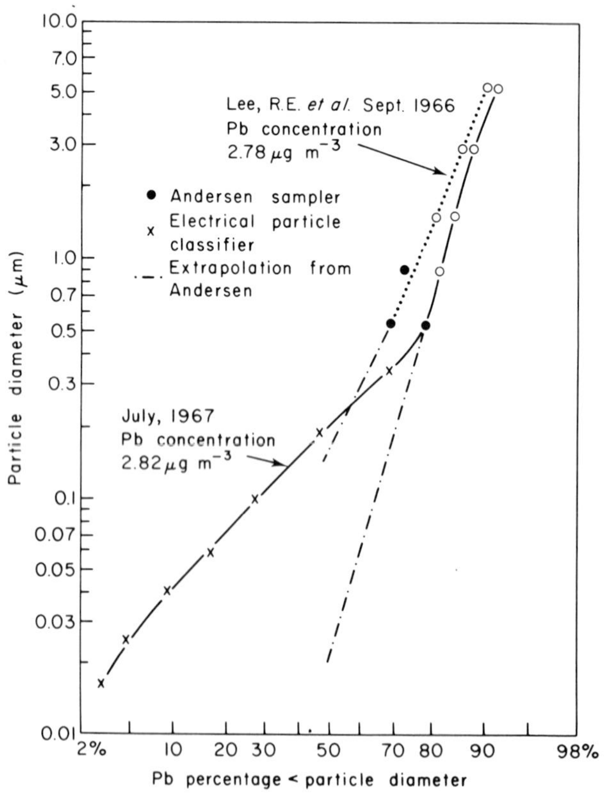

FIG. 3.17 Cumulative size distribution of lead in Cincinnati air. Comparison of results obtained using an Andersen cascade impactor with those from the electrical particle classifier [55]

density, whereas diameters reported by the electrical particle classifier represent equivalent projected area diameters independent of density.

References

1. Roach, S. A. (1966). A more rational basis for air sampling programmes. *Amer. Ind. Hyg. Assoc. J.* **27**, 1–12.
2. McGuire, T. and Noll, K. E. (1971). Relationship between concentrations of atmospheric pollutants and averaging time. *Atmos. Environ.* **5**, 291–298.
3. Larsen, R. I. (1969). A new mathematical model of air pollutant concentration averaging time and frequency. *J. Air Pollut. Contr. Assoc.* **19**, 24–30.
4. Davies, C. N. (1968). Zur Frage der Probenahme von Aerosolen. *Staub* **28**, 219–225.
5. Shepherd, M. (1947). Rapid determination of small amounts of carbon monoxide. *Anal. Chem.* **19**, 77–81.
6. Porter, K. and Volman, D. H. (1962). Flame ionization detection of carbon monoxide for gas chromatographic analysis. *Anal. Chem.* **34**, 748–749.
7. Dubois, L., Zdrojewski, A. and Monkman, J. L. (1966). Analysis of CO in urban air at the ppm level, and the normal CO value. *J. Air Pollut. Contr. Assoc.* **16**, 135–139.
8. Jacobs, M. B. (1960). "The chemical analysis of air pollutants." Chemical Analysis, Vol. 10, p. 195. Interscience Publishers, New York.
9. Jacobs, M. B. (1960). "The chemical analysis of air pollutants." Chemical Analysis, Vol. 10, p. 197. Interscience Publishers, New York.
10. Jeffery, P. G. and Williams, D. (1961). The determination of fluorine in deposit-gauge samples. *Analyst* **86**, 590–597.
11. Hanker, J. S., Gelberg, A. and Witten, B. (1958). Fluorometric and colorimetric estimation of cyanide and sulfide by demasking reactions of palladium chelates. *Anal. Chem.* **30**, 93–95.
12. Jacobs, M. B., Braverman, M. M. and Hochheiser, S. (1957). Ultramicrodetermination of sulfides in air. *Anal. Chem.* **29**, 1349–1351.
13. Jacobs, M. B. (1960). "The chemical analysis of air pollutants." Chemical Analysis, Vol. 10, p. 216. Interscience Publishers, New York.
14. Saltzman, B. E. and Wartburg, A. F. (1965). Absorption tube for removal of interfering sulfur dioxide in analysis of atmospheric oxidant. *Anal. Chem.* **37**, 779–782.
15. Saltzman, B. E. (1954). Colorimetric microdetermination of nitrogen dioxide in the atmosphere. *Anal. Chem.* **26**, 1949–1955.
16. Ripley, D. L., Clingenpeel, J. M. and Hurn, R. W. (1964). Continuous determination of nitrogen oxides in air and exhaust gases. *Int. J. Air Water Pollut.* **8**, 455–463.
17. Barnes, H. (1950). A modified 2,4-xylenol method for nitrate estimations. *Analyst* **75**, 388–392.
18. West, P. W. and Gaeke, G. C. (1956). Fixation of sulfur dioxide as sulfitomercurate (11) and subsequent colorimetric determination. *Anal. Chem.* **28**, 1816–1819.
19. Nauman, R. V., West, P. W., Tron, F. and Gaeke, G. C. (1960). A spectrophotometric study of the Schiff reaction as applied to the quantitative determination of sulfur dioxide. *Anal. Chem.* **32**, 1307–1311.

20. Pate, J. B., Lodge, J. P. and Wartburg, A. F. (1962). Effect of pararosaniline in the trace determination of sulfur dioxide. *Anal. Chem.* **34**, 1660–1662.
21. Terraglio, F. P. and Manganelli, R. M. (1962). Laboratory evaluation of sulfur dioxide methods and the influence of ozone-oxides of nitrogen mixtures. *Anal. Chem.* **34**, 675–677.
22. Urone, P. F. and Boggs, W. E. (1957). Acid bleached fuchsin in determination of sulfur dioxide in the atmosphere. *Anal. Chem.* **23**, 1517–1519.
23. Commins, B. T. (1963). Determination of particulate acid in town air. *Analyst* **88**, 364–367.
24. Hauser, T. R. and Cummins, R. L. (1964). Increasing sensitivity of 3-methyl-2-benzothiazolone hydrazone test for analysis of aliphatic aldehydes in air. *Anal. Chem.* **36**, 679–681.
25. Altshuller, A. P. and Leng, L. J. (1963). Application of the 3-methyl-2-benzothiazolone hydrazone method for atmospheric analysis of aliphatic aldehydes. *Anal. Chem.* **35**, 1541–1542.
26. Cohen, I. R. and Altshuller, A. P. (1961). A new spectrophotometric method for the determination of acrolein in combustion gases and in the atmosphere. *Anal. Chem.* **33**, 726–733.
27. Russell, J. W. (1975). Analysis of air pollutants using sampling tubes and gas chromatography. *Environ. Sci. and Technol.* **9**, 1175–1178.
28. Cuddeback, J. E., Burg, W. R. and Birch, S. R. (1975). Performance of charcoal tubes in determination of vinyl chloride. *Environ. Sci. and Technol.* **9**, 1168–1171.
29. Cantuti, V. J. and Cartoni, G. P. (1968). Gas chromatographic determination of TEL in air. *J. Chromat.* **32**, 641–647.
30. Novák, J., Vašák, V. and Janák, J. (1965). Chromatographic method for the concentration of trace impurities in the atmosphere and other gases. *Anal. Chem.* **37**, 660–666.
31. Lovelock, J. E. and Zlatkis, A. (1961). A new approach to lead alkyl analysis: Gas phase electron absorption for selective detection. *Anal. Chem.* **33**, 1958–1959.
32. Bonelli, E. J. and Hartmann, H. (1963) Determination of lead alkyls by gas chromatography with electron capture detector. *Anal. Chem.* **35**, 1980–1981.
33. Dawson, H. J. (1963) Determination of methyl-ethyl lead alkyls in gasoline by gas chromatography with an electron capture detector. *Anal. Chem.* **35**, 542–545.
34. Harrison, R. M., Perry, R. and Slater, D. H. (1974). An adsorption technique for the determination of organic lead in street air. *Atmos. Environ.* **8**, 1187–1194.
35. Moss, R. and Browett, E. V. (1966) Determination of tetra-alkyl lead vapour and inorganic lead dust in air. *Analyst* **91**, 428–438.
36. Purdue, L. J., Enrione, R. E., Thompson, R. J. and Bonfield, B. A. (1973). Determination of organic and total lead in the atmosphere by atomic absorption spectrophotometry. *Anal. Chem.* **45**, 527–530.
37. Snyder, L. J. and Henderson, S. R. (1961). Rapid spectrophotometric determination of triethyl lead, diethyl lead and inorganic lead ions, and application to the determination of tetraorgano lead compounds. *Anal. Chem.* **33**, 1172–1180.
38. Parker, W. W., Smith, G. Z. and Hudson, R. L. (1961). Determination of mixed lead alkyls in gasoline by combined gas chromatographic and spectrophotometric techniques. *Anal. Chem.* **33**, 1170–1171.

39. Parker, W. W., and Hudson, R. L. (1963). A simplified chromatographic method for separation and identification of mixed lead alkyls in gasoline. *Anal. Chem.* **35**, 1334–1335.
40. Clemons, C. A. and Altshuller, A. P. (1964). Plastic containers for sampling and storage of atmospheric hydrocarbons prior to gas chromatographic analysis. *J. Air Pollut. Contr. Assoc.* **14**, 407–408.
41. Butler, J. D. and Thorne, D. M. (1973). Estimation of nitric oxide emissions from an internal combustion engine. *J. Appl. Chem. Biotechnol.* **23**, 195–203.
42. Altshuller, A. P. and Cohen, I. R. (1964). Atmospheric photo-oxidation of ethylene-nitric oxide mixtures. *Int. J. Air Water Pollut.* **8**, 611–632.
43. Fleischer, R. L., Price, P. B. and Walker, R. M. (1965). Tracks of charged particles in solids. *Science* **149**, 383–393.
44. Pich. J. (1964). Impaction of aerosol particles in the neighbourhood of a circular hole. *Coll. Czech. Chem., Commun.* **29**, 2223–2226.
45. Spurny, K. and Pich, J. (1965). Analytical methods for determination of aerosols by means of membrane ultrafilters, VII. *Coll. Czech. Chem. Commun.* **30**, 2276–2286.
46. Spurny, K. and Lodge, J. P. (1968). Analytical methods for the determination of aerosols by means of membrane filters, XII. *Coll. Czech. Chem. Commun.* **33**, 3931–3938.
47. Spurny, K. and Lodge, L. P. (1968). Analytical methods for determination of aerosols by means of membrane ultra filters, XI. *Coll. Czech. Chem. Commun.* **33**, 3679–3693.
48. Spurny, K., Lodge, J. P., Frank, E. R. and Sheesley, D. C. (1969). Aerosol filtration by means of nucleopore filters. *Environ. Sci. and Technol.* **3**, 453–468.
49. Ranz, W. E. and Wong, J. B. (1952) Impaction of dust and smoke particles. *Ind. Eng. Chem.* **44**, 1371–1380.
50. Mercer, T. E. (1963). On the calibration of cascade impactors. *Ann. Occup. Hyg.* **6**, 1–14.
51. Andersen, A. A. (1958). New sampler for the collection, sizing and enumeration of viable airborne particles. *J. Bact.* **76**, 471–484.
52. Dahneke, B. (1971). The capture of aerosol particles by surfaces. *J. Coll. and Interface Sci.* **37**, 342–353.
53. Burton, R. M., Howard, J. N., Penley, R. L., Ramsay, P. A. and Clarke, T. A. (1973). Field evaluation of the high-volume particle fractionating cascade impactor. *J. Air Pollut. Contr. Assoc.* **23**, 277–281.
54. Smith, J. E. and Jordan, M. L. (1964). Particle size distribution analysis. *J. Colloid Sci.* **19**, 549–559.
55. Flesch, J. P. (1969). Calibration studies of a new sub-micron aerosol classifier. *J. Coll. Interface Sci.* **29**, 502–509.

4

Analysis of Pollutants by Instrumental Methods

Ideally, analysis by instrumental means embodies the exploitation of some property of a molecule which can be used to deliver a signal that is proportional to and derivable from that pollutant molecule being analysed. This concept immediately focuses attention upon physico-chemical techniques involving visible, infrared, ultraviolet, emission, absorption, fluorescence, and chemiluminescence spectral phenomena, conductometric and gas ionization properties of molecules. Response times between the excitation or development of such a characteristic and the measurement or recording of it are normally short. Unlike the wet chemical analytical approach, therefore, in which time is required for the collection as well as time to carry out the analysis, instrumental methods are capable of giving continuous information on pollution concentrations.

4.1 Trace gas analysis

4.1.1 Conductometric determination of SO_2

Sometimes the wet chemical analytical procedure can be modified so that the final evaluation is done instrumentally. The estimation of atmospheric sulphur dioxide by conductometry may be quoted as an example. In this conductometric method [1], sulphur dioxide is absorbed in a dilute solution of hydrogen peroxide which is converted into sulphuric acid. The process is carried out continuously by arranging for a metered supply of air containing sulphur dioxide to pass up from the bottom of an absorber column down which a metered dilute hydrogen peroxide solution is flowing. The dilute sulphuric acid solution issuing from the base of the scrubber-absorber column is passed into a conductivity cell. Another identical cell is used to measure the conductivity of the reagents. The difference between the two readings is proportional to the sulphur dioxide which has been converted to sulphuric acid. The reagent feed lines, absorption column and conductivity cells must be enclosed in an insulated, thermostatically controlled cabinet.

The method is not specific for sulphur dioxide because any soluble gases that yield electrolytes in solution will affect the conductivity. All hydrogen halides present will interfere. Except near special sources, however, these gases are only present in insignificant amounts in comparison with sulphur dioxide. Weak acidic gases like hydrogen sulphide cause little interference and similarly, provided the water used is free from bases, the presence of carbon dioxide can be ignored. Alkaline gases such as ammonia or finely divided alkaline dust particles interfere by neutralizing the sulphuric acid and under these circumstances the method gives low results.

The effectiveness of the method compared with the West–Gaeke colorimetric technique mentioned in section 3.3.1, Table 3.4 has been discussed in some detail [2]; generally agreement is good.

4.1.2 Hydrocarbon analysis by flame ionization detector (fid)

This detector is the same in principle as that used in gas chromatography. When hydrogen and oxygen burn to produce steam there are relatively few ions generated in the flame. Introduction of trace amounts of hydrocarbons into the flame produces a complex ionization in which large numbers of ions are formed. A polarizing voltage is applied in the vicinity of the flame by making the burner jet positive with respect to a wire loop (Fig. 4.1). This

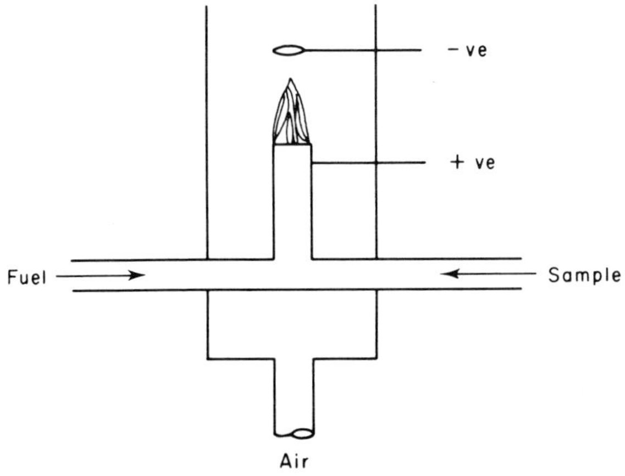

FIG. 4.1 Flame ionization detector

electric field induces ion migration such that positive ions are attracted to the wire loop collector electrode and negative ions are attracted to the burner. A small ionization current is thereby established between the electrodes which is proportional to the ion concentration in the flame. The current is amplified and displayed on an output meter or chart recorder. Since the hydrocarbon pollutants can be delivered by a pump into the flame a continuous recorded signal of changing hydrocarbon concentration sampled can be obtained.

One of the most important features of this type of detector is that the response is proportional to the number of carbon atoms present in the hydrocarbon being consumed in the flame. For example, a given volume concentration of a C_6-hydrocarbon will produce a signal twice as large as a C_3-hydrocarbon and six times that of methane. For this reason the instrument is usually calibrated against a standard mixture of methane in nitrogen and then the results are reported as a methane equivalent of the total hydrocarbons monitored.

Commercial instruments are available (Fig. 4.2) for estimation of hydrocarbons in the 0–3000 parts 10^{-6} range, although for ambient air

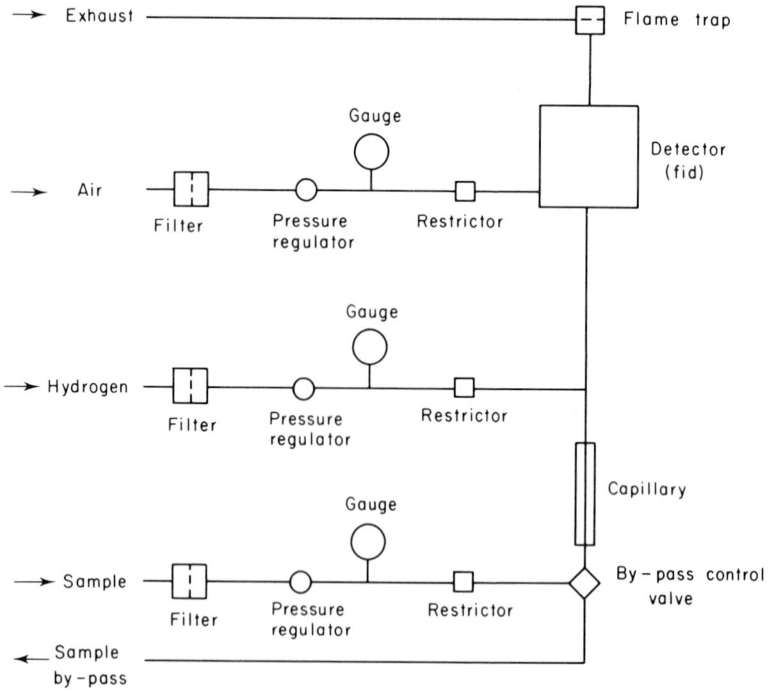

FIG. 4.2 Schematic diagram of a flame ionization total hydrocarbon analyser

monitoring the 0–10 or 0–25 parts 10^{-6} range factor switch positions are the most useful. Interferences are minimal, but there is evidence that organic halides and nitriles suppress signal response. The very nature of the detector, of course, ensures that there is no response from CO, CO_2, and water vapour.

Instruments of this type can be modified to measure, besides total hydrocarbons, methane and carbon monoxide. After first measuring total hydrocarbons the input supply of pollutant gas in hydrogen carrier gas is switched through a stripper column which removes water vapour and higher molecular weight hydrocarbons than methane. Methane and carbon monoxide from the stripper column are then separated gas chromatographically. Methane elutes first from the column and passes unchanged over a platinum reducing catalyst into the analyser where the methane response is measured. When carbon monoxide leaves the chromatographic column it is reduced by passage over the catalyst to methane. This is detected by the fid and recorded. Hence, the three signals from the instrument are due to total hydrocarbons, methane and carbon monoxide (as CH_4) [3].

Another simple modification of a fid system first discovered by Giuffrida and Karmen [4] enables compounds containing nitrogen, phosphorus and the halogens to be detected with enhanced sensitivity. In the presence of a wire, glass, or ceramic probe in the flame which has been coated with a sodium salt, these compounds increase the release of sodium vapour from the probe. The sodium vapour released into the flame ionizes and causes an increase in the electrical conductivity in the detector. Improvements in sensitivity of between 100 and 500 times that of a normal fid have been measured for organic nitrogen compounds in the presence of rubidium vapour [5]. Detectors of this type are called alkali flame ionization detectors (afid).

4.1.3 Total sulphur determination by flame photometric analysis

The flame photometric detector was developed from the flame ionization detector [6, 7]; the principle is illustrated in Fig. 4.3. When sulphur is burnt in a hydrogen–air flame light emission characteristic of sulphur occurs at 394 nm. The chemiluminescence arises from the "S_2" diatomic sulphur molecule which is produced when sulphur compounds are burnt in a hydrogen rich flame. By placing an optical filter, which transmits light of this wavelength between the flame and a photomultiplier tube, the light energy specific for sulphur can be converted to an electrical current output.

Stevens *et al.* [8] have used this detector coupled to a gas chromatographic column to separate and detect H_2S, SO_2, CH_3SH and $(CH_3)_2S$ in concentrations as low as 10 parts 10^{-9}. The separation of these sulphur compounds is achieved by using a Teflon column packed with powdered Teflon coated with polyphenyl ether (5-ring polymer). A small amount of orthophosphoric acid is mixed with the polyphenyl ether to prevent loss of acid sulphur gases and reduce tailing.

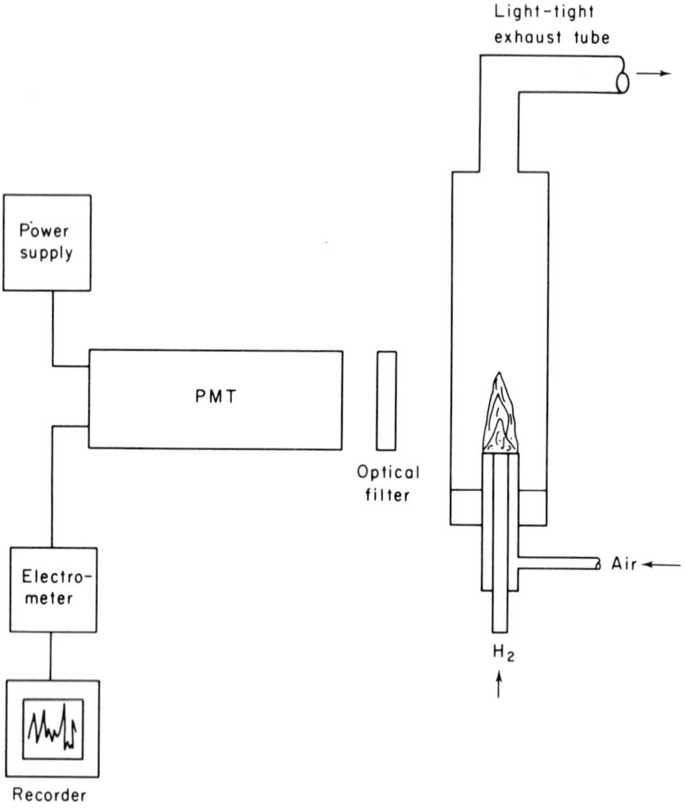

FIG. 4.3 Schematic diagram of a flame photometric detector

The instrument is capable of doing sulphur analyses of ambient atmospheres. In this capacity it has been shown that practically all the sulphur present is in the form of sulphur dioxide.

4.1.4 Phosphorus determination by flame photometric analysis

The flame photometric detector described in the previous section is equally applicable for the determination of volatile phosphorus compounds merely by changing the wavelength of the filter employed. In a cool, fuel-rich hydrogen–air flame, phosphorus forms POH radicals which emit a broad band spectrum with a peak at 526 nm. Interference occurs in the presence of sulphur and metals such as Mg and Ca which form stable compounds with phosphorus. The detector does not respond to any phosphorus that is in particulate form in the aerosol [9].

4.1.5 Halocarbons by electron capture detector

This detector was first described by Lovelock and Lipsky [10] in 1960. It consists of an ionization chamber which contains a source of ionizing radiation through which an inert gas stream of nitrogen is passing. Nickel-63 is a suitable convenient source of radiation. When a substance M capable of acquiring electrons enters the ionization chamber in the inert gas stream, an electron can be captured by the molecule M with the release of energy. The energy released is the electron affinity of the molecule M.

$$M + e \rightarrow M^- + E_{\text{electron affinity}} \tag{4.1}$$

Such negative ions M^- as indicated on the right-hand side of equation (4.1) are less mobile in an applied electric field than the electrons shown on the left-hand side of the equation. This means that a higher potential must be applied to the ionization chamber to collect these negative ions. When a relatively high potential is applied the negative ions may receive enough energy between collisions with inert gas molecules to dissociate. At sufficiently high potentials all the negative ions will be dissociated and a maximum current flow, or saturation current as it is called, will be observed. This effect is not only dependent upon the electron affinity of the compound in the ionization chamber but is also influenced by the shape and temperature of the chamber and the gas pressure.

The detector can be operated in two different modes. The potential across the ionization chamber can be applied continuously or as a pulse lasting for a few microseconds. The essential difference between the two operational modes is that under direct current conditions the electrons in the detector have a higher kinetic energy than the carrier gas molecules. Under pulse conditions, provided the pulse period is long compared with its width, the electrons are in thermal equilibrium with gas molecules in the ionization chamber. Anomalous responses encountered with a continuous electric field are mostly eliminated with pulse operation.

The detector when coupled to a gas chromatograph has proved very successful for analysis of organic halogen compounds. Practically all our knowledge of atmospheric concentrations of freons and pesticides, etc. has been gained through the employment of the ecd system. Under favourable conditions detection limits as low as 10^{-14} g are possible. Some specific examples in which ecd's have been used for air pollution analyses are quoted in Table 4.2 and Fig. 3.7.

With intensely electron-capturing substances, the ecd tends towards destructive detection in which a large portion of the substance entering the detector is ionized irreversibly. This feature of the detector can function as a gas phase coulometer [11]. Suppose all the solute molecules in a carrier gas stream entering the ecd are ionized, then the solute concentration will be proportional to the number of electrons absorbed. In cases where ionization efficiencies are less than 100%, the responses from two ecd's in series accounts for 100% ionization and hence the efficiency of the first detector can be calculated.

Let X_1 and X_2 be the signals in coulombs from two identical ecd's mounted in series, due to the passage of a plug of W solute molecules of compound M. If f is the fractional ionization of compound M at high ionization efficiencies and low solute concentration, then

$$X_1 = fQ \tag{4.2}$$

and

$$X_2 = f(Q - fQ) \tag{4.3}$$

and

$$f = 1 - \frac{X_2}{X_1} \tag{4.4}$$

where from Faraday's laws $Q = 96\,500\ W$. Hence, from equations (4.2) and (4.4)

$$W = \frac{X_1}{96\,500(1 - X_2/X_1)} \tag{4.5}$$

If V_D is the ecd volume and carrier gas is passing through the detector at a flowrate F, then assuming that electrons are captured by solute molecules with a pseudo-first-order rate law having reaction rate constant k, it can be shown that

$$\frac{1}{f} = 1 + \frac{F}{kV_D} \tag{4.6}$$

Plots of $1/f$ *vs* F should give a straight line passing through the origin. This has been verified for CCl_4, SF_6, CCl_3F, $CBrF_2 \cdot CBrF_2$, $CCl_2{=}CCl_2$, CH_3I and CCl_2F_2 [12].

4.1.6 *Infrared analysers for the determination of CO, CO_2, SO_2, acetylene and methane*

The principle of the method is shown in Fig. 4.4. Infrared radiation from two separate sources passes through two cells, one a reference cell and the other the sample cell. The reference cell is filled with a non-absorbing gas; the sample cell contains carbon monoxide molecules which absorb radiation at 4·6 μm. Thus, the radiation falling on the detector after passage through the sample cell will be less than that falling on the detector from the reference cell. The detector consists of two compartments separated by a thin metal diaphragm and filled with carbon monoxide. Since more radiation enters the left-hand compartment, gas contained therein expands and moves the diaphragm. The amount of movement is proportional to the

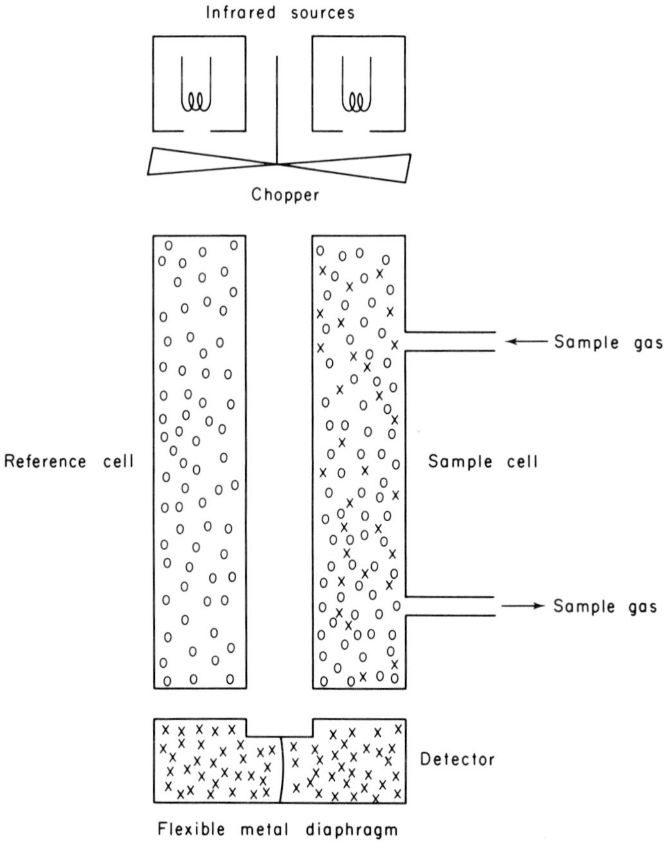

FIG. 4.4 Infrared gas analyser adapted from Bulletin 4129 Beckman Instrument Inc., Process Instruments Division, Fullerton, California, 92634

pressure difference between the two compartments; this in turn depends upon the amount of radiation absorbed by carbon monoxide molecules in the sample cell. The diaphragm can be made to pulsate by interrupting the infrared beams from their sources with a rotating chopper. The chopper alternately prevents and allows radiation from reaching the detector, causing the diaphragm either to be displaced or to return to the neutral position. By making the detector an electrical capacitance this mechanical movement can be converted into an electrical signal. The signal is amplified and displayed on an output meter or chart recorder in the usual way.

Although illustrated with reference to carbon monoxide the method is also applicable to carbon dioxide, water vapour, sulphur dioxide, acetylene and methane, merely by changing the gas in the detector.

4.1.7 Chemiluminescent methods

The use of chemiluminescent methods of analysis, especially when incorporated into detector systems of instruments designed to measure pollutants, has gained in popularity in the last few years. The reasons for this are the inherent sensitivity of the method, its specificity and its utility in the sense that support equipment such as compressed gas cylinders are frequently not required. Before discussing applications some remarks on the optimization of parameters of chemiluminescent detectors are relevant. Steffenson and Stedman [13] have shown that the detector signal from the photomultiplier tube is given by:

$$\text{Detector signal} = \frac{\text{reactor gas flow}}{\text{reactor pressure}} G\left[1 - \exp - \frac{F_{\text{react.}}}{F_{\text{poll.}}}\right] \tag{4.7}$$

where G is the geometry of photon collection of the reactor and $\exp(-F_{\text{react.}}/F_{\text{poll.}})$ is the fraction of pollutant molecules whose residence lifetime in the reactor ($F_{\text{react.}}$) is short compared with the reactive lifetime of ozone with pollutant. $F_{\text{react.}}$ is defined by

$$F_{\text{react.}} = \frac{A\,d\,P}{R}\ \text{seconds} \tag{4.8}$$

where A is the cross-sectional area of the reactor in cm^2, d is the reactor length in cm, P is the gas density in the reactor in molecules cm^{-3} and R is the flow in molecules s^{-1}.

The other notable feature of photomultiplier tubes that must be borne in mind, is that their signal-to-noise ratio is temperature dependent. The ratio is improved by reducing the temperature to below ambient. Operational temperatures of about $-20°C$ are frequently employed. Accurate and reliable thermostatic operation, therefore, is an essential feature of photomultiplier tube instrumentation.

4.1.7.1 Measurement of nitric oxide and oxides of nitrogen $NO + NO_2 = NO_x$ by chemiluminescence

When nitric oxide reacts with ozone nitrogen dioxide and oxygen are formed. About 7% of the nitrogen dioxide produced by this reaction is in an excited condition and reverts to the ground-state with emission of radiation:

$$NO + O_3 = NO_2^* + O_2 \tag{4.9}$$

$$NO_2^* = NO_2 + h\nu \tag{4.10}$$

The process was first investigated by Thrush and coworkers [14, 15], who established that the light intensity I emitted in the 600–875 nm spectral region is given by

$$I \propto \frac{[NO][O_3]}{[M]} \tag{4.11}$$

where $[M] = \text{air}$, and that the rate constant, k, of the reaction is defined by

$$\frac{-d}{dt}[O_3] = \frac{-d}{dt}[NO] = k[NO][O_3] \tag{4.12}$$

These findings form the basis of the development of the chemiluminescent determination of nitrogen oxides described by Fontijn *et al.* [16]. Figure 4.5 outlines the main components necessary for the construction of an NO_x analyser embodying these principles.

A light-tight reaction chamber maintained at a pressure of about 10 torr is screened from a photomultiplier tube by a red light filter to cut off radiation of $\lambda < 610$ nm. Ozonized air and polluted air containing nitric oxide are fed through nozzles into the reaction chamber. The gases mix, and light from the emitted radiation is detected by the PMT. The signal is amplified and displayed on an output meter or recorder. The instrument response is linear in the range $10–10^5$ parts 10^{-9} of NO. The lower limit of sensitivity is determined by the signal-to-noise ratio of the PMT. As already mentioned, commercial designs incorporate cooling devices to maintain the PMT at around −20°C. Another refinement includes a stainless steel coil held at 700°C, through which polluted gas can be by-passed before entering the reaction chamber. This coil ensures that any nitrogen dioxide present is decomposed to nitric oxide and oxygen before the gas is analysed. Hence, by first sampling gas directly for NO and then using the converter, the amount of NO_2 present in an NO_x mixture can be found by difference.

Interference by NO_2, CO_2, CO, C_2H_4, NH_3 and SO_2 does not occur at concentrations of these gases normally present in polluted ambient air. If desired the method can be used to measure ozone instead of NO_x by by-passing the ozonizer and substituting NO/N_2 mixture for ozonized air supply.

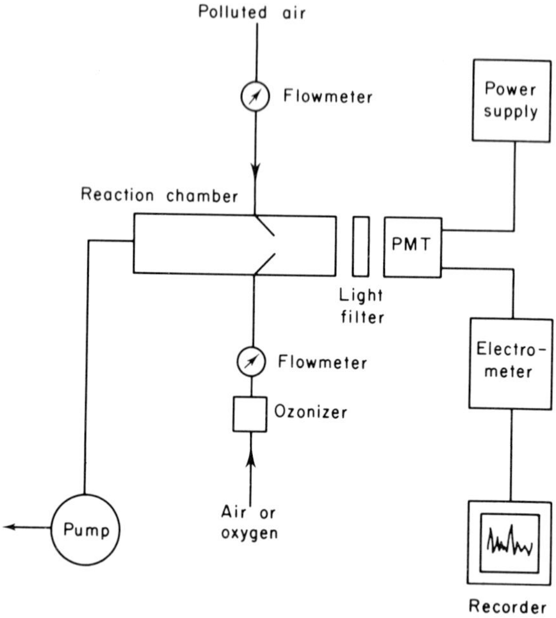

FIG. 4.5 The main components required for the determination of NO by chemiluminescence

4.1.7.2 Determination of N-nitrosamines

The *N*-nitrosamines were first recognized as carcinogens by Magee and Barnes in 1956 from experiments on rats [17]. These compounds are widely distributed: they probably occur in foods as a result of interaction of amines with nitrites; in tobacco smoke; in wheat kernels and flour, dairy products, fish, meat and mushrooms. Epidemiological studies have suggested correlations between high nitrogen dioxide concentrations and cancer in urban populations, but since neither nitrogen dioxide nor nitric oxide has yet been shown to be carcinogenic it is possible that the culprit may be *N*-nitrosamines rather than the oxides of nitrogen. Certainly, oxides of nitrogen can readily be converted to nitrous acid under quite general atmospheric conditions, as discussed in Chapter 5.

Up till recently analysis of *N*-nitrosamines at the $\mu g\ m^{-3}$ concentration required for atmospheric monitoring was not possible. Success has been achieved by Fine and Rounbehler [18] who have detected 1 $\mu g\ m^{-3}$ of dimethylnitrosamine in the urban Baltimore atmosphere. The procedure adopted [19] represents a modified version of the NO_x chemiluminescence method described in the previous section.

A schematic diagram of the equipment is shown in Fig. 4.6. The *N*-nitrosamine compound dissolved in a suitable solvent is injected into a catalytic pyrolysing chamber at 300°C containing a mixture of WO_3 and $W_{20}O_{58}$. This tungsten oxide catalyst selectively ruptures N—NO bonds and releases nitrosyl radicals (NO). Organic compounds, solvent and fragmentation products either pass directly into the reaction chamber or are by-passed through the cold trap at −150°C. At this temperature the vapour pressure of the nitrosyl radical is greater than 1 atm, whereas the vapour pressure of almost all organic compounds except methane, ethane, acetylene etc. are substantially less than 1 atm and are condensed out.

Hydrocarbons, alcohols, ketones, amines, nitro- and chloro-compounds do not interfere. Organic and inorganic nitrites, particularly 2,2′, 4,4′,

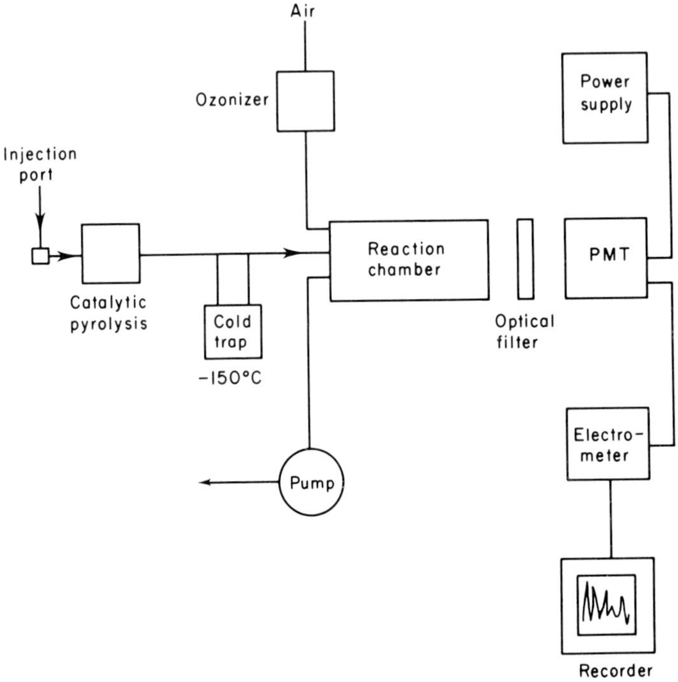

FIG. 4.6 Schematic layout for the analysis of *N*-nitrosamines

6,6′-hexanitrodiphenylamine, sodium nitrate and nitric acid also give a positive response.

In order to determine *N*-nitrosamines in the atmosphere it is necessary to draw about 200 l of air through a cold-trap at a flowrate of roughly 2 l min^{-1}. The contents of the trap are extracted into a suitable solvent, such as dichloromethane; the sample concentrated to a volume of 0·5 ml and 10–20 μl of this solution are then injected into the pyrolysis chamber of the analyser. By including a gas chromatograph with the equipment specific nitroso-compounds may be identified [20].

4.1.7.3 Measurement of sulphur dioxide by chemiluminescence

The method depends upon the chemiluminescence produced when the dichlorosulphitomercurate complex formed in the West–Gaeke estimation of sulphur dioxide is oxidized in solution with potassium permanganate, cerium (IV) sulphate or hydrogen peroxide. The chemiluminescence is detected by a PMT, amplified and recorded [21]. The detection limit found from a plot of PMT response plotted against SO_2 concentration in ng ml^{-1} is about 2 ng ml^{-1}.

4.1.7.4 Measurement of vinyl chloride monomer by gas chromatography and chemiluminescence

Recent disclosures suggesting a causal relationship between angiosarcoma of the liver and exposure to vinyl chloride monomer has centred interest on improved methods of detection and analysis of the compound. The usual analysis of vinyl chloride monomer depends upon adsorption of the gas on activated charcoal or some other adsorbent followed by removal or desorption from the adsorbent and chromatographic analysis with detection by fid (see section 3.3.2). Instead of using fid it is possible to react vinyl chloride with ozone and measure the light emission from the products of the reaction with a photomultiplier tube [22]. Under optimum conditions of gas flow, reaction chamber design and pump speed, concentrations of 0·05 parts 10^{-6} of vinyl chloride monomer are measurable.

4.1.7.5 Determination of ethylene and/or ozone

Measurement of ethylene by chemiluminescent reaction with ozone, first reported by Nederbragt *et al.* [23], has been developed [24] into a standard method useful for either the estimation of ethylene or ozone. The chemiluminescence of the reaction has an emission peak at 440 nm. The signal from the photomultiplier tube gives a linear response between 10^{-7} and 10^{-10} A for concentrations of ethylene from 1 to 10^3 parts 10^{-8}. NO_2, SO_2 and Cl_2 do not interfere.

Kummer *et al.* [25] have shown that besides the examples mentioned there are other useful chemiluminescent reactions with ozone that could be used for pollutant measurements including: hydrogen sulphide, dimethyl sulphide, methanethiol, trimethylethylene, tetramethylethylene, *cis*- and *trans*-butene-2; 2,3-dimethylbutadiene and 2,5-dimethylfuran.

4.1.8 Derivative spectroscopy

The application of derivative spectroscopy to trace gas analysis is a concept that enables gases to be analysed from their absorption spectra in the visible or ultraviolet spectral regions [26]. The principle of the method is outlined in Fig. 4.7. The absorption by a compound at a precise

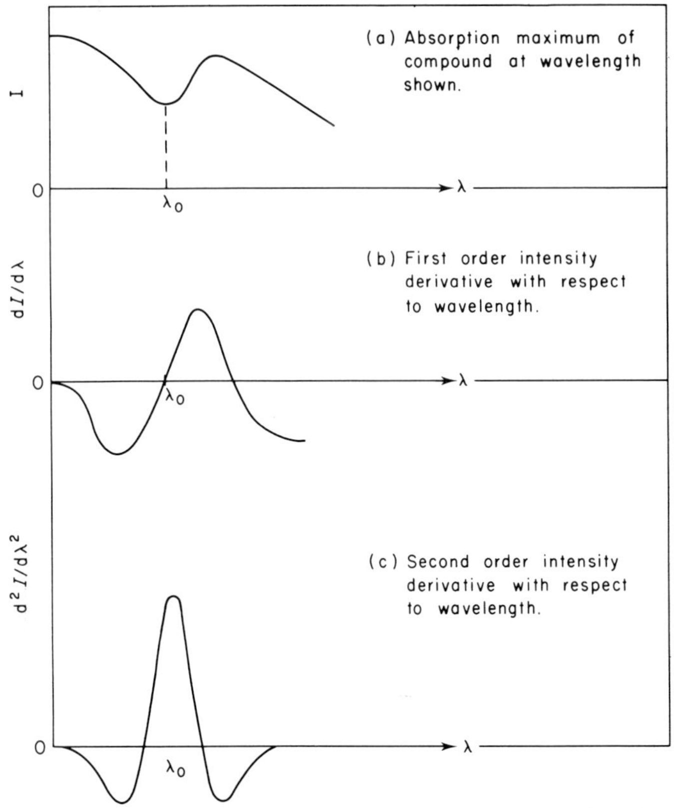

FIG. 4.7 The relationship between absorption maxima at a wavelength with plots of first- and second-order derivatives $dI/d\lambda$ and $dI^2/d\lambda$ in the λ_0 wavelength region. (a) Absorption maximum of compound at wavelength. (b) First-order intensity derivative with respect to wavelength. (c) Second-order intensity derivative with respect to wavelength

wavelength, λ_0, is shown in Fig. 4.7(a). Figure 4.7(b) has been obtained by drawing tangents to this I *vs* λ plot to obtain the first-order derivative $dI/d\lambda$ at wavelength values immediately either side of λ_0. This exercise is repeated on the first-order derivative plot $dI/d\lambda$ *vs* λ shown in Fig. 4.7(b) to give the second-order derivative plot $d^2I/d\lambda^2$ *vs* λ indicated in Fig. 4.7(c). A derivative spectrometer is designed to deliver and display the second-order derivative signal at the precise wavelength associated with maximum absorption. It can be shown that the peak height of the second-order derivative is a function of the concentration of the absorbing compound and simultaneously the compound can be identified by the location of the wavelength at which maximum absorption occurs.

Figure 4.8 shows the arrangement of the components required in the construction of a derivative spectrometer. Radiation from either the u.v. or the visible source is dispersed by a grating monochromator. Movement of this grating allows the spectrum to be scanned over the desired range. The monochromatic radiation from the grating is modulated through movement of the entrance slit, by an electro-mechanical sinusoidal displacement at a frequency of 45 hz. An amplitude oscillation of $\pm 0{\cdot}5$ mm will produce

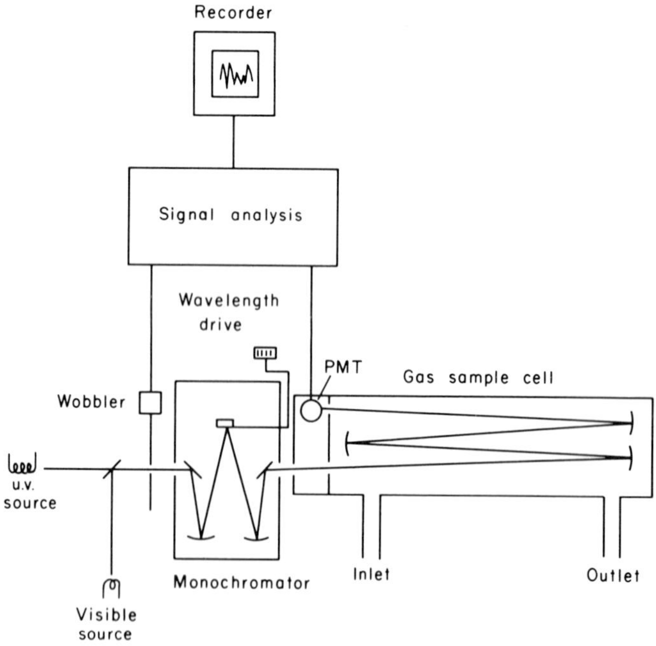

FIG. 4.8 Second-order derivative spectrometer for gas analysis. Source: Lear Siegler Inc., SM 400, Environmental Technology Division, 1, Inverness Drive East, Englewood, Colorado 80110

a wavelength modulaton of ±1·5 nm. Light from the monochromator enters the sample cell and is reflected between the mirrors to give a convenient path length: e.g. 12 reflections within a 1 m sample cell will be equivalent to a path length of 12 m. Eventually, light from the sample cell is focused onto a photomultiplier tube.

The output signal from the photomultiplier tube at twice the input frequency of 45 hz is required for the signal to be related to the second-derivative function. This can be accomplished electronically. A reference signal is exactly matched in phase and frequency with the wavelength-modulated input by coupling the wobbler to a coil mounted in a magnetic field. This device allows the required dc output signal from the PMT to be identified. The output signal when expressed as a fraction of the total output from the PMT is proportional to the gas concentration, i.e. $d^2I/d\lambda^2/I$ concentration of compound identified by the λ_{max} value.

A typical second derivative spectrum of a gas mixture is shown in Fig. 4.9 and some examples of the minimum detectable concentrations are presented in Table 4.1.

4.1.9 Gas chromatography

Gas chromatography is the most widely used analytical technique in chemistry today. Any substance which can be made to exert the necessary

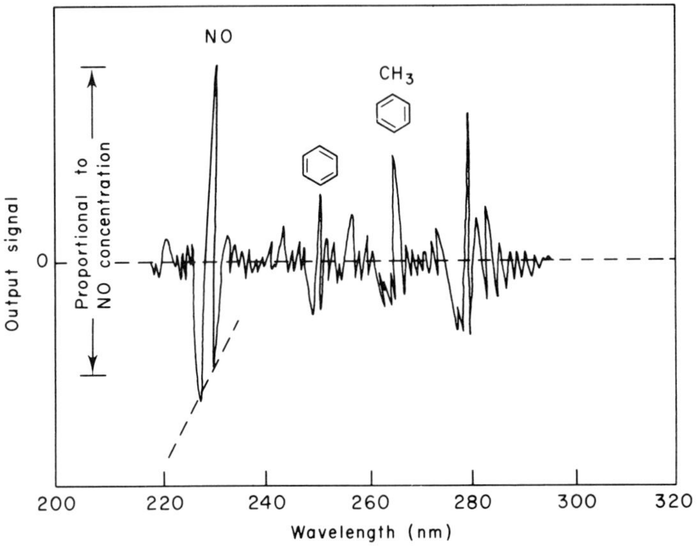

FIG. 4.9 Second-derivative absorption spectrum of a gas mixture containing nitric oxide, benzene and toluene. Source: Lear Siegler Inc., Environmental Technology Division, 1, Inverness Drive East, Englewood, Colorado, 80110

vapour pressure can be passed through a gas-chromatographic column in an inert carrier gas stream. Provided decomposition of the compound does not occur, separations achieved on the column can be detected and quantitative analyses performed on as little as nanogram quantities.

TABLE 4.1 *Minimum concentrations detected by second-derivative spectroscopy of some well known pollutant gases and vapours* [26]

Compound	Concentration (parts 10^{-9})
NO	5
NO_2	40
SO_2	1
O_3	40
NH_3	1
Hg vapour	0·5
Benzene	25
Toluene	50
Xylene	100
Styrene	100
Formaldehyde	200
Benzaldehyde	100
Acetaldehyde	400

The column itself comprises a thermostatted tube, usually made of glass or stainless steel packed with a porous solid which has been impregnated with a few percent by weight of a stationary phase. This stationary phase frequently consists of a high molecular weight polymeric material such as silicone oil or silcone gum. Different stationary phases have different separating capabilities and, therefore, a phase is selected which will accomplish the task in hand most effectively.

Literally thousands of reports and papers have been written and published giving details of gas-chromatographic analyses, and only brief mention will be made here of the applications of the method to atmospheric pollutants. Table 4.2 contains some examples of gas-chromatographic analyses of pesticides, freons, PAN, bis(chloromethyl)ether, dimethyl selenide, etc. Dimethyl selenide has been included to illustrate the use of atomic absorption in gas chromatography (see section 4.2.4). After separation on the g.c. column the gas stream passes into a carbon furnace of an atomic absorption spectrophotometer. Analysis of an element is then carried out by the usual AA procedure. In this instance the 196 nm Se line is employed and quantities of Se as low as 10 ng are determinable.

TABLE 4.2 *Gas-chromatographic analysis of pesticides, freons, phosgene and PAN*

Compounds	Collection method	g.c. Conditions	Ref.
Aldrin; BHC; chlordane; 2,4-D; *p,p'*-DDT; *o,p'*-DDT; DDD (TDE); *p,p'*-(DDE); *o,p'*-(DDE); dieldrin; endrin; heptachlor; (*a*) parathion (*o,o*-diethyl-*o*-*p*-nitrophenyl-phosphorothioate); methylparathion (*o,o*-dimethyl-*o*-*p*-nitrophenyl-phosphorothioate) malathion(diethyl-mercaptosuccinate, *S*-ester with *o,o*-dimethylphosphoro-dithioate).	Draw air through GFA filter backed by impinger containing 2-methyl-2,4-pentanediol as absorbing liquid and an alumina column. Pesticides extracted from all three absorbing media and the chlorinated compounds separated from the organo-phosphates by column chromatography on Florisil	6 m × 6 mm glass column packed with 5% SE-30; 100–120 mesh on gas chrom Q at 185°C; 120 ml min^{-1} N_2 or 1 m × 6 mm glass column packed with 2·5% SE-30 3·5% OV-17 on 100–120 mesh gas chrom Q at 185°C; 100 ml min^{-1} N_2. Ecd at 200°C. (*a*) For organophosphates: 1 m × 6 mm glass column packed with 5% QF-1 on 100–120 mesh gas chrom Q at 140°C; 100 ml min^{-1} N_2 or 1 m × 6 mm glass column packed with 5% OV-1 on 100–120 mesh gas chrom Q at 140°C; 100 ml min^{-1} N_2. Flame photometric detector at 160°C.	27
CCl_3F; CCl_4; CH_3CCl_3; CH_3I; $CHCl{=}CCl_2$; $CCl_2{=}CCl_2$.	Up to 10 ml ambient air.	1·3 m × 6 mm stainless steel column packed with 10% DC-200 on 30–60 mesh chromosorb W at 23°C. Ecd coulometry used.	12
Phosgene.	Direct injection using gas syringe conc. range 10–167 parts 10^{-9}.	25 cm × 6 mm Al column packed with 30% didecylphthalate on 100–120 mesh chromosorb P at ambient temp. Ecd coulometry.	28

TABLE 4.2—(*continued*)

Compounds	Collection method	g.c. Conditions	Ref.
Bis(chloromethyl)ether $ClCH_2OCH_2Cl$ (carcinogenic).	Draw air through 5 cm tube packed with Poropak Q.	2·7 m × 4 mm glass column packed with 30% polyethylene glycoladipate on 100–120 mesh celite. He carrier gas 40 ml min^{-1} at 150°C. Retention time ~6 min. Mass spectrometer detection using $m/e = 78{\cdot}9950$ due to $ClC_2H_4O^+$.	29
Dimethylselenide.	Draw air through 25 cm length tube packed with 3% OV-1 on chromosorb W at −80°C with air flow 130–150 ml min^{-1}.	1·8 m × 6 mm glass column packed with 3% OV-1 on 80–100 mesh chromosorb W. N_2 carrier gas 70 ml min^{-1}. Temp. programme 40°C for 2 min then 15°C min^{-1} until 120°C. Atomic absorption detection.	30
N_2O.	Draw air through 11 g of activated 5 Å molecular sieve contained in a stainless steel tube 20 cm × 12 mm. Displace N_2O with He saturated with water vapour and pass gas stream through $CaSO_4$ and Ascarite traps to remove water before collecting N_2O on silica gel at −80°C. Desorb N_2O by heating to 140° and after 10 min transfer to gc.	3·7 m × 6 mm stainless steel column packed with Poropak Q; 80–100 mesh. Ambient temp. Retention time 6·5 min. Average air conc. monitored 285 parts 10^{-9} SD = ±5 parts 10^{-9}. Thermal conductivity detector used.	31
PAN.	Ambient sample gas syringe injection.	2·7 m × 3 mm glass column packed with 10% Carbowax 600 on gas chrom Z. Retention time ~9 min. Sensitivity ~25 parts 10^{-9}, ecd (at 25°–30°). Purged overnight at 150°C.	32

The synthesis of volatile chelate complexes of metals stable at temperatures in excess of 250°C introduces the possibility of analysing aerosol particulates for metals by gas chromatography. The toxic element beryllium, for example, is finding increasing application in industry and in space research where it occurs in rocket fuels. Analytical procedures have been developed [33, 34] so that atmospheric concentrations can be monitored. The important chemical development which enables this to be done involves the formation of a chelate of trifluoroacetylacetone (tfa) with beryllium. An outline of the collection and extraction sequence of operations for the analysis is shown in Fig. 4.10.

Collect atmospheric dust from
2000 m^3 of air onto 18 × 23 cm^2
GFA filter using a high volume pump.

Ash filter at 150°C at 1 torr
pressure in flowing oxygen at
50 ml min^{-1}.

Residue

Extract residue with aqua regia
concentrate and centrifuge.

Supernatant liquid

1 ml extract + EDTA-sodium acetate buffer, pH adjusted to 5·5–6·0
with 3 N NaOH. Heat to 90°C for 10 min. Cool.

Add 1 ml of 0·16 M H(tfa) in benzene and stir for 15 min. Destroy
excess reagent by addition of 2 ml 0·1 N NaOH shake and allow
aqueous and organic layers to settle.

$Be(CF_3{-}\overset{\overset{\displaystyle O}{\|}}{C}{-}CH{=}\overset{\overset{\displaystyle O}{|}}{C}{-}CH_3)_2$ in benzene

Inject 1 μl aliquots of benzene layer onto 1·2 m × 1·52 mm Teflon
column packed with 5% SE 52, 60–80 mesh on gas chrom Z at 80°C.
50 ml min^{-1} N_2 carrier gas. Ecd at 200°C.

Detection limit
4×10^{-13} g

FIG. 4.10 Scheme illustrating the analytical stages for the estimation of beryllium by gas chromatography [33, 34]

The method has been extended [35] to the determination of Cu, Ni, Pd and Pt by preparing the tetradentate chelates of these metals with ligands of bis-acetylacetoneethylenedi-imine and its fluorine derivatives:

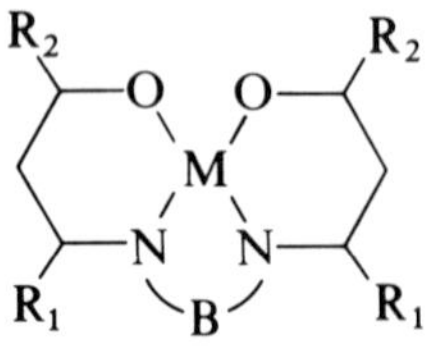

where R_1 and $R_2 = CH_3$, CF_3, etc., and $B = -(CH_2)_2-$, $-(CH_2)_3-$, $-CH_2-CH(CH_3)-$. A linear response at the nanogram level is found on a 1·83 m × 3·2 mm column packed with 3% QF 1 (fluoropropyl silicone oil) on Varaport 30, 100/110 mesh at 215°C with He carrier gas at 50 ml min^{-1} and ecd at 250°C.

4.1.10 Calibration procedures

No matter how sensitive an analytical instrument may be, the final accuracy of the quantity determined will depend on the precision of the calibration. When dealing with atmospheric pollutants the quantities of interest are often in the range of 1 part in 10^6–10^9 but can be even less. Modern instruments have this capability, but the analyst can be frustrated because comparable calibration standards are not available. The problem requires a different approach from that adopted in the conventional procedures of volumetric and gravimetric analysis.

4.1.10.1 Permeation tubes

A gas contained within a Teflon tube gradually permeates the tube wall and diffuses into the atmosphere. The first investigations by O'Keeffe and Ortman [36] showed that the leakage of gas into a dry atmosphere remain constant. This leakage rate is easily measured by periodically determining the weight loss gravimetrically. The method is most useful for gases whose critical temperatures are greater than 300 K, e.g. SO_2, NH_3, NO_2, Cl_2, benzene.

Some typical leak rates are given in Table 4.3. The permeation rate is conveniently expressed in ng cm^{-1} min^{-1} and this enables a calculation to be made of the length of tube x cm necessary to produce the standard required. For a flow of diluent air F_D(ml min^{-1}) passing over a permeation tube at 1 atm pressure liberating gas at a rate of P_R (ng cm^{-1} min^{-1}) and C is the concentration of gas required (parts 10^{-6}) for calibration purposes at a temperature T, then

$$C = \frac{22{\cdot}4 \times P_R T}{273\, F_D M} \tag{4.13}$$

where M is the molecular weight of the gas in the permeation tube.

TABLE 4.3 *Permeation rates of gases through fluorinated ethylene propylene resin (FEP Teflon)* [36]

Compound	Temperature °K	Tube dimensions (cm) Internal diameter	Wall thickness	Permeation rate P_R (ng cm^{-1} min^{-1})
SO_2	287	0·1575	0·0305	111
	293	0·1575	0·0305	203
	302	0·1575	0·0305	396
NO_2	287	0·4762	0·0762	605
	293	0·4762	0·0762	1110
	302	0·4762	0·0762	2290
CH_4[a]	302	0·4762	0·0762	400
C_2H_2[b]	304	0·3651	0·0381	30
$CH_3CH{=}CH_2$	304	0·1575	0·0762	290
trans-$CH_3CH{=}CHCH_3$	304	0·1575	0·0762	90
H_2S	293	0·1575	0·0762	140

[a] Solution in USP mineral oil at 550 psi pressure.
[b] On firebrick at 145 psi pressure.

Example: What flowrate of dry nitrogen at 1 atm pressure will be needed to calibrate a hydrocarbon analyser at the 1·0 part 10^{-6} level at 31°C with methane using a permeation tube 20 cm long having the permeation characteristics given in Table 4.3.? From Table 4.3:

$$P_R \text{ for methane} = 400 \text{ ng cm}^{-1} \text{ min}^{-1}$$

Since

$$C = 1{\cdot}0 \text{ part } 10^{-6}$$
$$x = 20 \text{ cm}$$
$$T = 304 \text{ K}$$

Then using equation (4.13) gives

$$F_D = \frac{22{\cdot}4 \times 20 \times 400 \times 304}{273 \times 1 \times 16} \text{ ml min}^{-1}$$

$$= 12\,472 \text{ ml min}^{-1}$$

$$F_D = 12{\cdot}47 \text{ l min}^{-1}$$

Permeation tubes are conveniently prepared by first selecting Teflon tubing of the desired length, bore, and wall thickness. The ends of the tube are sealed by inserting a steel ball at each end. The diameter of the steel ball should be 1·5 times the diameter of the Teflon tubing. The Teflon tube around one of the seals is distorted by applying pressure with rubber covered pliers whilst the tube is filled with gas or liquid required. The ball is then repositioned in another part of the tube so that once more a perfect seal is obtained. Tubes should be stored in dry atmospheres and dry gas must be used for calibration as moisture affects the permeation rates.

4.1.10.2 Exponential dilution

When a gas A is continuously diluted by gas B in a well stirred sample flask (Fig. 4.11), the concentration of gas A decreases exponentially with time:

$$C_A = C_{A_0} \exp(-F_D t/V) \qquad (4.14)$$

where

C_{A_0} = initial concentration,
F_D = volume flowrate of diluent gas B,
V = volume of the flask,
t = time from the start of dilution.

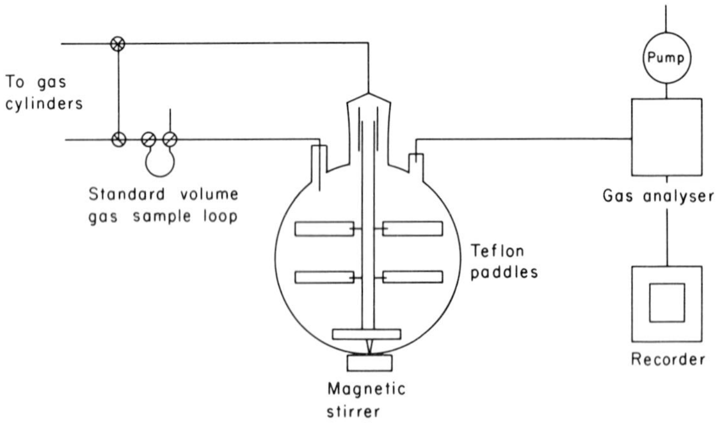

FIG. 4.11 Schematic diagram showing typical layout for exponential dilution calibration

The gas mixture leaving the flask is fed to the instrument requiring calibration. Instrument performance at lower and lower concentrations can readily be followed by plotting the logarithm of the instrument response against the dilution time. This should be a straight line (Fig. 4.12) [37]. The initial concentration can be obtained by introducing a known volume of pure gas into the dilution vessel from a sample loop of standard volume or by injection of a sample with a standard syringe through a rubber septum.

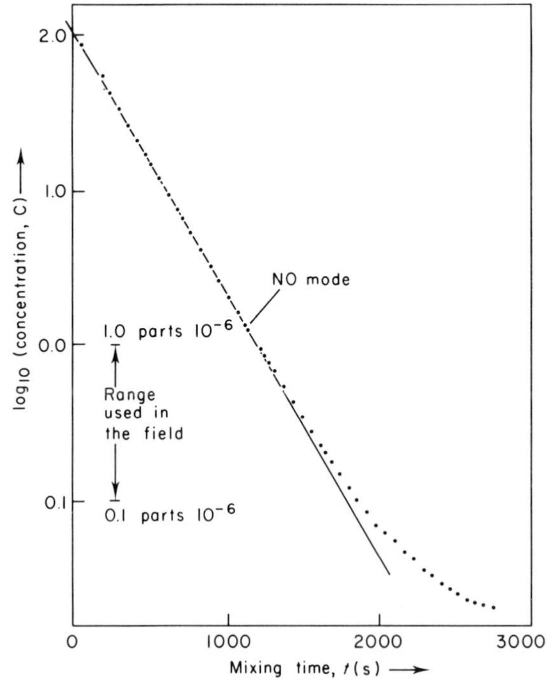

FIG. 4.12 Linearity check of a NO chemiluminescent gas analyser by exponential dilution

4.1.10.3 Gas blending

Precise dilution by a factor of around 10^5 requires good pressure regulation and flow control. A two-stage dilution process is usually required for dilutions of this order of magnitude. The first stage is illustrated in Fig. 4.13: diluent gas nitrogen is blended with gas under examination using calibrated rotameters. The gas mixture from this first stage then enters the second stage dilution system shown in Fig. 4.14. The diluent gas flow is regulated by a rotameter as before, but the gas flow from the first stage is controlled by the needle valve and monitored by the manometer and soap bubble meter. The extent of dilution at any stage can be written:

$$C_g = \frac{F_g}{F_D \times F_g} \times 10^6 \qquad (4.15)$$

where C_g is the v/v concentration (parts 10^{-6}) of gas needed for calibration, F_g and F_D are the flowrates of the gas being diluted and the diluent gas, respectively.

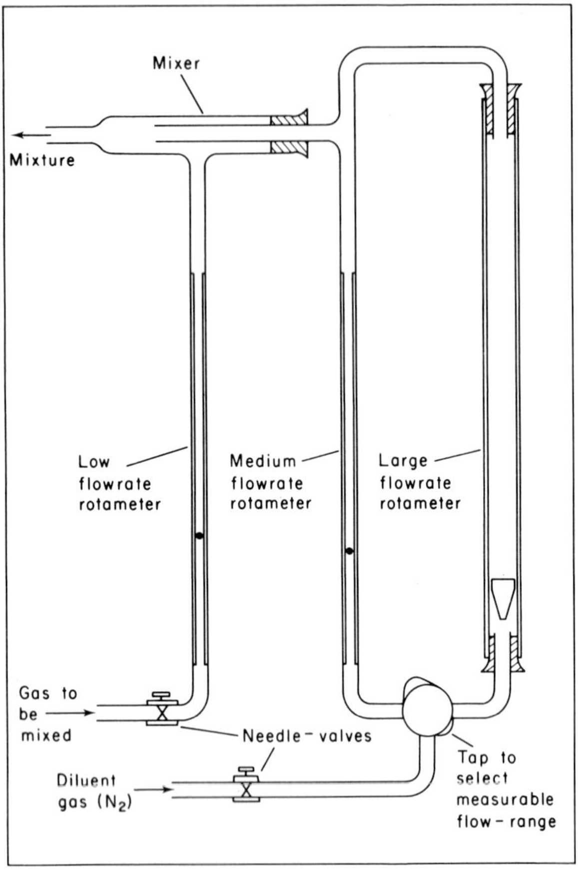

FIG. 4.13 First-stage dilution apparatus [37]

The technique is more difficult to operate than either the permeation tube or exponential dilution methods. The latter two methods are recommended especially when equipment is being calibrated "on site".

On-site calibrations are often performed, in fact, by withdrawing gas from a standard bag sample with a syringe and then injecting into a second bag containing a metered volume of dry nitrogen. In the case of PAN determinations (Table 4.2) standard bag samples can be calibrated by infrared absorptivity [38]. Tedlar (PVF) bags should be used for all these operations.

4.2 Trace metal analysis by instrumental methods

Physico-chemical analytical procedures have been adopted very successfully in trace metal determinations of particulate samples because general chemical separation of elements in the collected sample is usually not

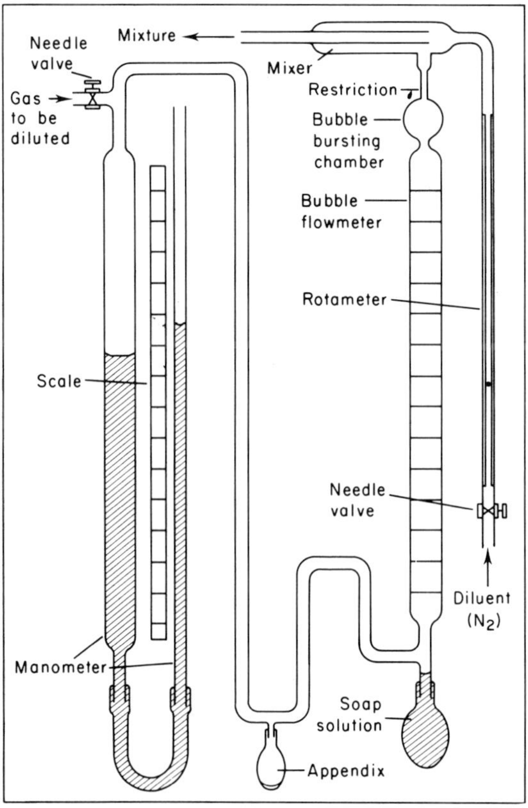

FIG. 4.14 Second-stage dilution apparatus [37]

required. These methods tend to depend upon fundamental properties of elements such as the unique behaviour of extranuclear electrons of atoms when excited in some way. Under ideal circumstances these properties can be exploited through appropriate resolution techniques to prevent interference effects and hence make possible quantitative analysis of small quantities of substances in the presence of larger amounts of other material. Among the most widely used methods embodying these principles are:

(1) neutron activation;

(2) X-ray fluorescence;
(3) atomic emission spectroscopy/atomic emission flame photometry;
(4) atomic absorption flame photometry;
(5) mass spectrometry.

As is always the case, each method has advantages and disadvantages over its rivals. Consideration of detection limits, sensitivity, time required for analysis, state of the sample, must be weighed against availability of equipment, cost, and whether destructive or non-destructive analysis of the sample is desirable.

4.2.1 Neutron activation analysis

When an element is bombarded by neutrons, neutrons can be captured by the element which is thereby transformed into an unstable or radioactive element called a radionuclide. Radionuclides are characterized by

(1) their rate of radioactive decay or half-life;
(2) the type of radiation emitted;
(3) the energy of the emitted radiation.

These properties are specific and may be used for identification purposes. Examples of some useful nuclear reactions together with the characteristic half-life and γ-ray energy of nuclides are given in Table 4.4.

TABLE 4.4 *Nuclear reactions, half-lives and γ-ray energies of nuclides*

		Product Nuclide	
Element	Reaction	Half-life	γ-Ray energy (keV)
V	$^{51}V(n, \gamma)\,^{52}V$	3·7 min	1435·3
Mn	$^{55}Mn(n, \gamma)\,^{56}Mn$	2·6 h	846·7
Cu	$^{63}Cu(n, \gamma)\,^{64}Cu$	12·8 h	511·0
As	$^{75}As(n, \gamma)\,^{76}As$	26·4 h	559·0
Cr	$^{50}Cr(n, \gamma)\,^{51}Cr$	27·8 days	320·1
Fe	$^{58}Fe(n, \gamma)\,^{59}Fe$	45·6 days	1292·0
Hg	$^{202}Hg(n, \gamma)^{203}Hg$	46·9 days	279·1
Cd	$^{114}Cd(n, \gamma)^{115}Cd$	53·5 h	528·0
Sb	$^{123}Sb(n, \gamma)^{124}Sb$	60·3 days	1691·0
Ni	$^{58}Ni(n, p)\,^{58}Co$	71·3 days	811·0
Zn	$^{64}Zn(n, \gamma)\,^{65}Zn$	243 days	1116·0
Co	$^{59}Co(n, \gamma)\,^{60}Co$	5·26 years	1333·0

Analysis consists of a five-stage process, irradiaton of the sample, detection of γ-emission, amplification, pulse height analyser, digital printout/analog recorder. These stages are shown schematically in Fig. 4.15.

The bombarding neutrons required for the irradiation stage can be obtained from nuclear reactors, particle accelerators or from isotopic sources. Sample treatment, therefore, utilizes specialized equipment

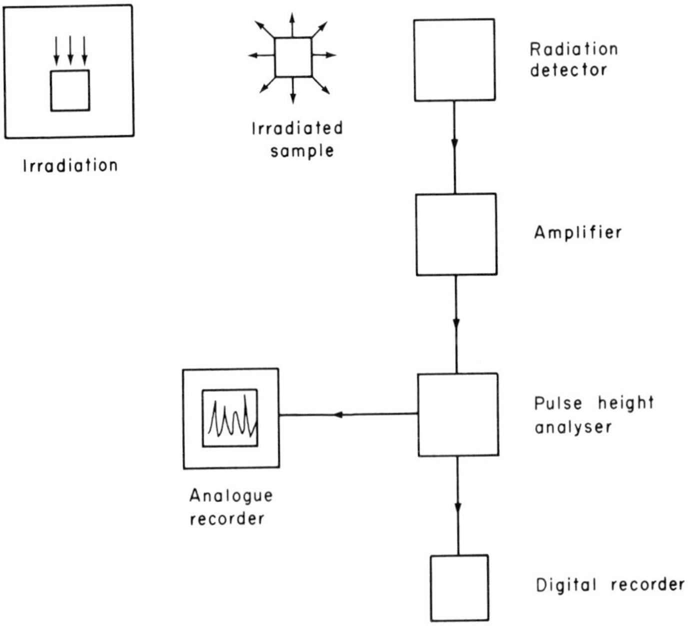

FIG. 4.15 Schematic diagram showing stages required for analysis by neutron activation [39]

restricted by government licence to a limited number of centres. The handling of the samples both at the irradiation and detection stages of the analysis requires specially trained personnel. These are obvious disadvantages which restrict the selection of the method, although nowadays registered laboratories offer neutron activation analysis on a service basis. Among the advantages may be counted, little or no sample preparation, high sensitivity for many elements, relatively short time for analysis and non-destruction of sample. The latter feature means that the sample can be analyzed in another manner and an independent assessment between two analytical procedures can be obtained. Table 4.5 shows some results of analyses carried out by neutron activation of Co, Cr, Fe, Sb and Zn in urban atmospheres in various parts of the world [39].

4.2.2 X-ray fluorescence

Extranuclear electrons of atoms are accommodated in discrete energy levels, K ($n = 1$); L ($n = 2$); M ($n = 3$); N ($n = 4$) etc. according to the Bohr concept of the atom, where the principal quantum number $n = 1, 2, 3, 4, \ldots$ is denoted by the K, L, M, N, ... shells. In excited atoms, electrons

TABLE 4.5 *Concentrations of some metallic contaminants in urban air samples* ($\mu g\ m^{-3}$) *estimated by neutron activation analysis* [39]

Element	UK Bradford	Japan Sakai	Japan Osaki	USA Cambridge (Mass)
Co	5×10^{-3}	2×10^{-3}	$5{\cdot}8 \times 10^{-3}$	2×10^{-4}
Cr	6×10^{-2}	$2{\cdot}2 \times 10^{-2}$	$9{\cdot}6 \times 10^{-2}$	—
Fe	5·5	3·9	12·9	1·0
Sb	$1{\cdot}3 \times 10^{-2}$	$1{\cdot}3 \times 10^{-2}$	$6{\cdot}3 \times 10^{-2}$	5×10^{-4}
Zn	0·5	1·1	5·8	0·1

have been promoted from lower energy levels nearest the atomic nucleus to higher energy levels more remote from the nucleus by the absorption of resonance radiation. If sufficient energy is provided an electron can be ejected from the atom altogether to produce an ion. An excited atom is unstable because it has energy above the ground state. In order to return to the stable condition promoted electrons revert to the lower energy levels left vacant in the initial promotion process. Such transitions from higher to lower energy levels are accompanied by emission of radiation which falls in the X-ray spectral region. This fluorescence X-radiation is characteristic of the excited atom; its detection and measurement form the basis of X-ray energy spectroscopy.

In practice there are two possible ways of monitoring the fluorescence radiation, either by measuring the wavelength or the energy of the dispersive X-rays. The former method employs a conventional optical spectrometer in which X-rays are spatially dispersed by diffraction with a crystal goniometer. This technique is limited because only one element at a time can be analysed. X-ray energy dispersive analysis is more versatile and will be described in more detail.

Initial excitation of the sample is accomplished by using X-rays, a radioactive source, or an electron beam. The important component of an X-ray energy spectrometer is the semiconductor detector which converts the energy of the X-radiation emitted from the sample into electrical pulses [40]. The semiconductor used is silicon doped with lithium across which a bias voltage is applied. X-rays absorbed by the detector result in ionization and the production of electrons and positive holes within the semiconductor. The number of electrons and positive holes created within the device is proportional to the energy of the incident X-ray beam. The statistical precision with which this conversion is accomplished is a measure of the resolution of the detector. The detector operates at liquid nitrogen temperatures to reduce the influence of thermal electrons in the semiconductor and also to restrict the mobility of lithium ions in the silicon matrix. Because of this cooling requirement and the need for absolute cleanliness, the detector is housed in its own vacuum system with a thin window to permit entrance of X-rays. Windows made from a thin sheet (7–8 μm thickness of beryllium) are employed in most commercial instruments. X-rays having energies less than about 1 keV cannot penetrate this window and this, in practice, prevents the detection of elements below sodium, i.e. atomic number 11 of the periodic table. Table 4.6 lists the prominent X-ray lines (keV) which are useful for element identification purposes [41].

Figure 4.16 shows a typical component layout of an X-ray energy spectrometer [40]. Pulses from the Si(Li) detector are pre-amplified and then amplified before being fed to an analog-to-digital converter. This converter translates the energy signal of incident X-rays detected by the semiconductor into a number of pulses. These pulses are fed to the multi-channel analyser for storage. The multi-channel analyser will usually contain about 10^3 channels of memory with a capacity of something like 10^6 counts per

TABLE 4.6 *Prominent X-ray line energies (keV) useful for element identification purposes* [41]

Z		Kα	Kβ
11	Na	1·041	1·071
12	Mg	1·254	1·302
13	Al	1·487	1·557
14	Si	1·740	1·836
15	P	2·013	2·139
16	S	2·307	2·464
17	Cl	2·622	2·816
19	K	3·313	3·590
20	Ca	3·690	4·013
21	Sc	4·089	4·460
22	Ti	4·509	4·932
23	V	4·950	5·427
24	Cr	5·412	5·947
25	Mn	5·895	6·490
26	Fe	6·400	7·058
27	Co	6·925	7·649
28	Ni	7·472	8·265
29	Cu	8·041	8·905

TABLE 4.6—(*continued*)

		Lα, β	Kα	Kβ
30	Zn	1·019	8·631	9·572
31	Ga	1·107	9·243	10·262
32	Ge	1·198	9·876	10·980
33	As	1·294	10·532	11·723
34	Se	1·392	11·209	12·493
35	Br	1·496	11·909	13·288
37	Rb	1·713	13·375	14·956
38	Sr	1·828	14·143	15·830
39	Y	1·947	14·933	16·732
40	Zr	2·070	15·747	17·661
41	Nb	2·196	16·584	18·614
42	Mo	2·327	17·444	19·599
43	Tc	2·462	18·328	20·609
44	Ru	2·600	19·236	21·646
45	Rh	2·743	20·169	22·711
46	Pd	2·889	21·125	23·805
47	Ag	3·040	22·105	24·927
48	Cd	3·195	23·110	26·078
49	In	3·354	24·140	27·257

		L1	Lα₁	Lβ₁	Lβ₂	Lγ₁	Lγ₂	Kα₂	Kα₁,₂	Kα₁	Kβ₁,₂
50	Sn	3·045	3·444	3·663	3·903	4·131	4·377		25·196		28·787
51	Sb	3·189	3·604	3·844	4·101	4·348	4·600		26·276		30·046
52	Te	3·336	3·769	4·031	4·302	4·571	4·829		27·382		31·335
53	I	3·485	3·938	4·221	4·509	4·800	5·067		28·514		32·655
55	Cs	3·795	4·286	4·620	4·936	5·280	5·542	30·625		30·973	35·388
56	Ba	3·854	4·466	4·828	5·156	5·531	5·797	31·817		32·194	36·799

		Mα, β	L1	Lα₁	Lβ₁	Lβ₂	Lβ₃	Lγ₁	Lγ₂
73	Ta	1·738	7·174	8·146	9·343	9·652		10·895	11·217
74	W	1·805	7·388	8·398	9·671	9·960		11·286	11·608
75	Re	1·874	7·604	8·652	10·000	10·275		11·685	12·010
76	Os	1·944	7·822	8·912	10·355	10·598		12·096	12·422
77	Ir	2·016	8·046	9·175	10·709	10·920		12·513	12·842
78	Pt	2·088	8·268	9·442	11·071	11·250		12·942	13·270
79	Au	2·163	8·494	9·713	11·443	11·585	11·610	13·382	13·709
80	Hg	2·238	8·721	9·989	11·823	11·924	11·995	13·830	14·162
81	Tl	2·315	8·953	10·270	12·213	12·271	12·390	14·292	14·625
82	Pb	2·393	9·184	10·551	12·614	12·623	12·792	14·764	15·100

Reprinted by permission of the American Society for Testing and Materials, Copyright.

channel. The channel number is assigned to a particular X-ray energy and hence the spectrum is gradually built up for display.

An interesting development in applying X-ray fluorescence analysis to atmospheric dust samples collected on filters or impactor films has been described [42]. The electron beam of a scanning electron microscope (SEM) is used to generate fluorescence in the sample. This technique enables the magnified image portrayed by the SEM to be identified chem-

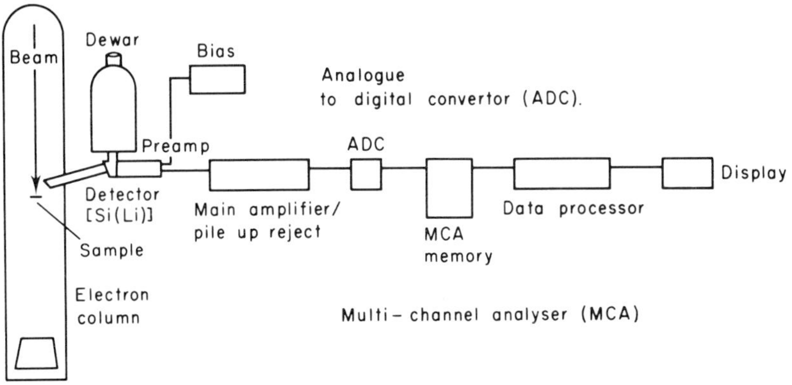

FIG. 4.16 Schematic diagram showing the component layout of an X-ray energy spectrometer coupled to a scanning electron microscope

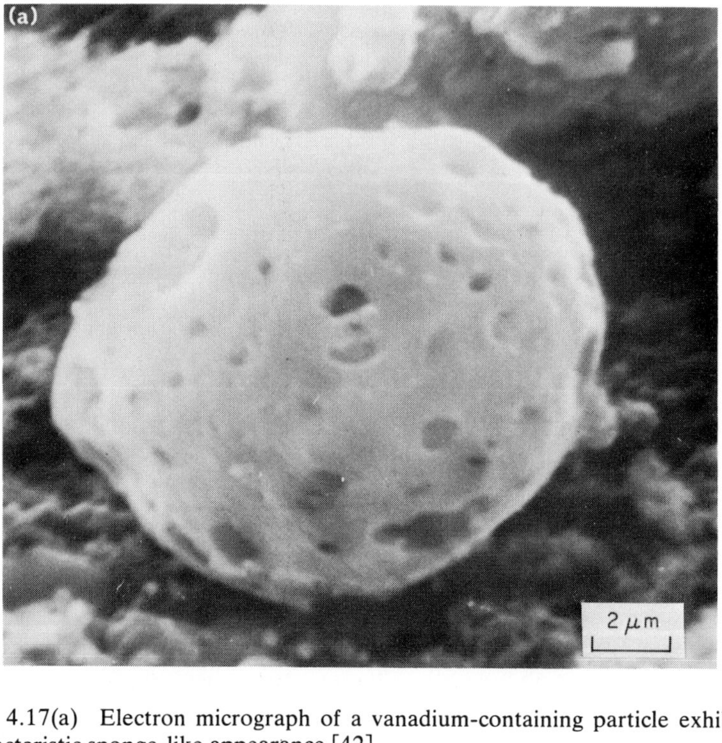

FIG. 4.17(a) Electron micrograph of a vanadium-containing particle exhibiting characteristic sponge-like appearance [42]

(b)

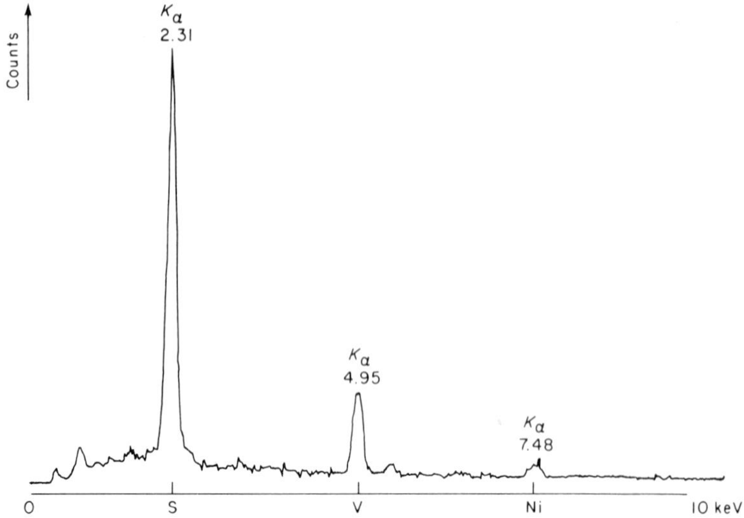

FIG. 4.17(b) X-ray fluorescence spectrum of the particle shown above in (a) [42].

ically through the X-ray energy spectrum emitted by the image. Figure 4.17(a) shows an electron micrograph of a rather distinctive, easily recognizable particle having a porous sponge-like appearance. Figure 4.17(b) gives the X-ray energy fluorescence spectrum of this particle. The spectrum shows energies $K\alpha$ at about 2·3, 4·9 and 7·5 keV which are attributable to sulphur, vanadium and nickel, respectively. As pointed out by the authors, filters recovered in the vicinity of heavy vehicular traffic were frequently associated with this type of particle. Undoubtedly, we are witnessing in this instance the conversion of vanadium and nickel compounds present in petroleum products (section 2.3.1.2) into atmospheric pollutants. In the light of the evidence already cited in section 1.4.1.2 regarding the statistical possibility of a relationship between vanadium and diseases of the heart, it would be prudent to investigate the implications of these observations further.

4.2.3 Atomic emission spectroscopy and atomic emission flame photometry

Atoms excited thermally or electrically emit light at definite wavelengths to produce characteristic spectra. The emitted radiation is resolved optically and recorded either photographically or photoelectrically. The location of the spectral lines can be used to identify elements present in the sample and the intensity of these lines is a quantitative measure of the amount of material present. Three stages are involved in the analysis, excitation, optical resolution and detection. Sample excitation can be accomplished by d.c. or a.c. arc or high voltage a.c. spark discharge. In either case the sample vaporizes, producing predominantly excited neutral atoms with the d.c. arc and excited atomic ions with the a.c. spark. Radiation from these excited species is resolved by a diffraction grating or a prism and finally recorded on a photographic plate acting as the detector (Fig. 4.18a). Alternatively, the radiation can be filtered through slits and detected by photomultiplier tubes (Fig. 4.18b). Immediate printout of the result is now possible so that the time-consuming exposure and development stages needed for the photographic system are eliminated.

The principal disadvantages of the method are the strong matrix effects and a low sensitivity for easily vaporized elements. Among the advantages are the ability to analyse simultaneously about 70 elements. For quantitative analysis careful calibration is essential. Once this has been carried out, however, continuous routine sample analysis is possible.

Atomic emission flame photometry differs from atomic emission spectroscopy or photometry merely in the manner of sample excitation. As the name implies, emitted radiations are derived from the sample by burning in a flame (Fig. 4.18c). The technique is very sensitive, rapidly accomplished, but spectral interferences are possible and flame stability can influence accuracy. The sample must be in solution.

4.2.4 Atomic absorption flame photometry

Currently, atomic absorption spectrophotometry is one of the most popular methods employed for elemental analysis. For specific details on the analysis of many elements Slavin [43] should be consulted. A schematic diagram showing the main components of an atomic absorption spectrophotometer is given in Fig. 4.19.

The sample in solution is atomized in a burner; this ruptures chemical bonds and breaks down molecules into atoms. These atoms are not excited: they exist mainly in their ground state. Atoms in this condition will absorb characteristic resonance radiation of precisely the same wavelength that they would emit if they were excited. This property is utilized by allowing resonant radiation of a metallic element generated in a hollow cathode lamp source to pass through the flame vaporized sample. The extent of absorption of radiation, at a particular wavelength characteristic of the metal, is proportional to the concentration of the element being aspirated in the flame. The combined effects of flame instability and solution aspiration tend to produce a slightly unsteady signal, so that other methods of sample vaporization have been sought. Some success in this direction has

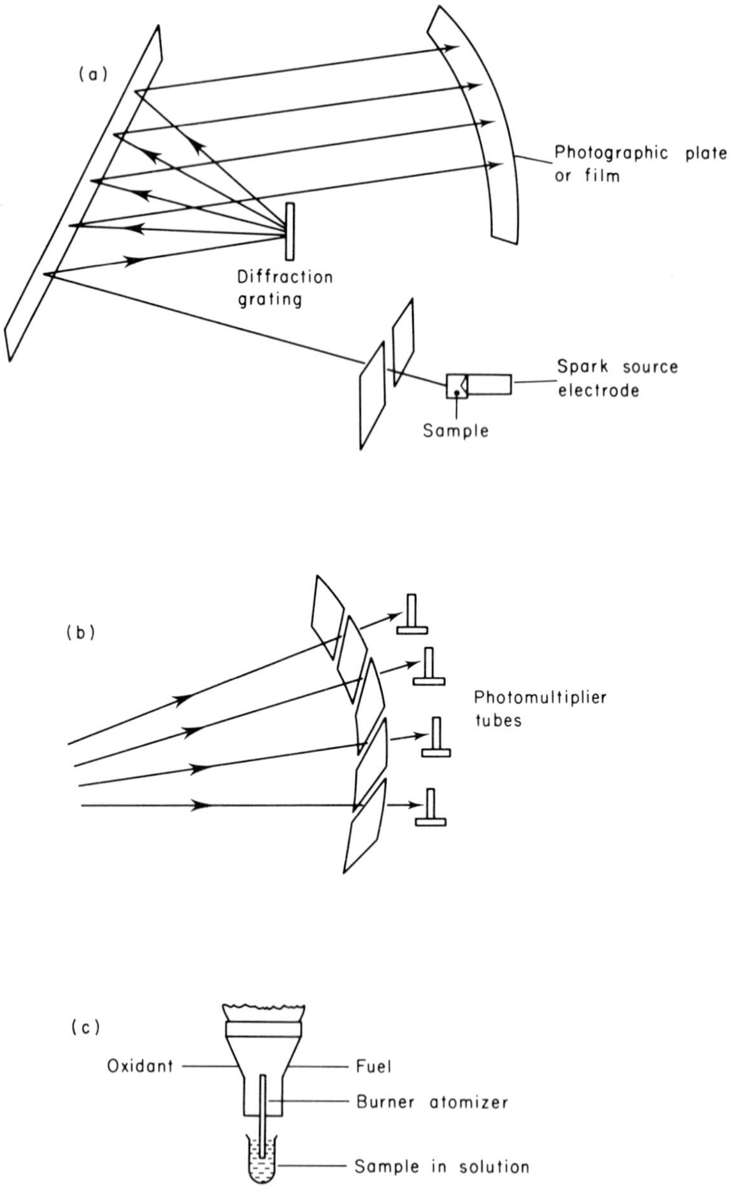

FIG. 4.18 Schematic diagram showing the excitation, optical resolution and detection stages of atomic emission spectroscopy. (a) Atomic emission spectrograph; (b) Detector system for atomic emission photometry; (c) Excitation for atomic emission flame photometry.

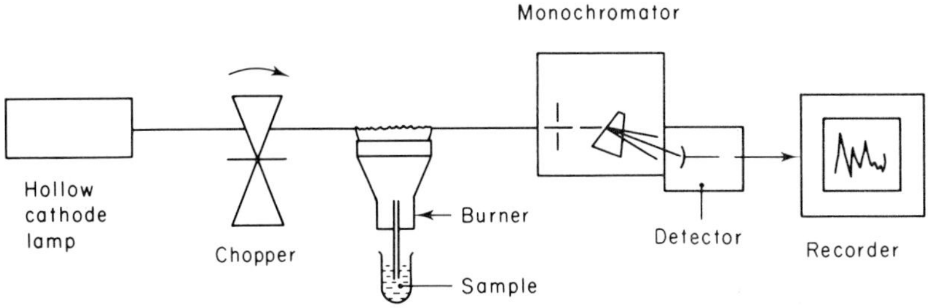

FIG. 4.19 Atomic absorption spectrophotometer

been achieved by Massmann [44] who introduced the sample into a carbon cell that was maintained in an inert argon atmosphere. Flameless atomic absorption spectrophotometers designed along these lines are generally preferred nowadays, since the sample vaporization stage can be more easily and reproducibly controlled.

Non-flame atomic absorption has been used to determine lead in airborne particulate matter using the 217 nm lead line with an absolute sensitivity of 0·1 μg Pb m^{-3} for a 200 ml air sample [45]. Ranweiler and Moyes [46] have described an analytical procedure for the determination of metals in atmospheric particulate matter by atomic absorption and a flow-chart of their method is given in Fig. 4.20. Table 4.7 lists the practical lower limits of detection for elements determined in this manner from particulate collected on polystyrene filters.

TABLE 4.7 *Practical detection limits for the atomic absorption analysis of elements present in airborne particulate matter collected on polystyrene filters* [46]

Element	Concentration (μg m^{-3}) in air sample	Element	Concentration (μg m^{-3}) in air sample
Si	0·8	Sr	0·003
Al	0·4	Ni	0·002
Ca	0·03	V	0·006
Fe	0·07	Mn	0·0003
K	0·04	Cr	0·0003
Na	0·03	Rb	0·0005
Mg	0·01	Li	0·0003
Pb	0·007	Bi	0·002
Cu	0·003	Co	0·0008
Ti	0·07	Cs	0·00004
Zn	0·03	Be	0·00004
Cd	0·0003		

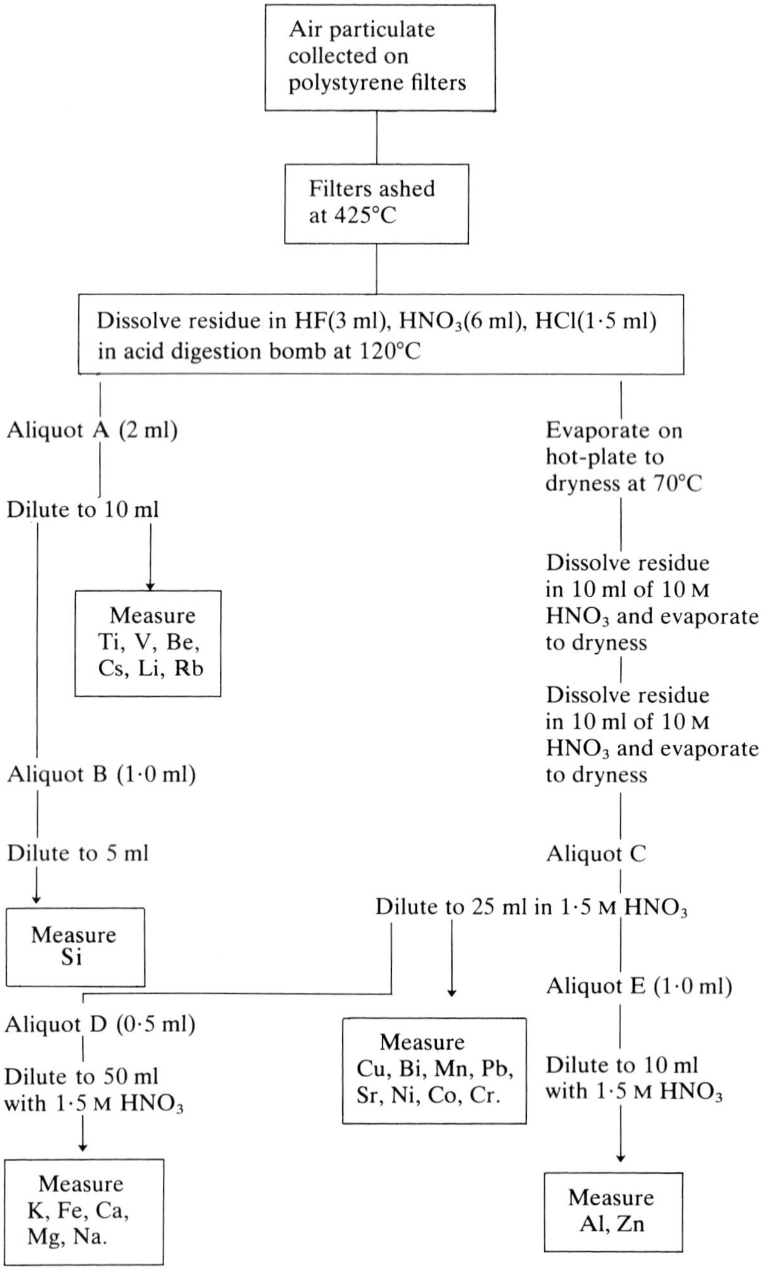
Air particulate collected on polystyrene filters
Filters ashed at 425°C
Dissolve residue in HF(3 ml), HNO_3(6 ml), HCl(1·5 ml) in acid digestion bomb at 120°C
Aliquot A (2 ml)
Dilute to 10 ml
Measure Ti, V, Be, Cs, Li, Rb
Aliquot B (1·0 ml)
Dilute to 5 ml
Measure Si
Evaporate on hot-plate to dryness at 70°C
Dissolve residue in 10 ml of 10 M HNO_3 and evaporate to dryness
Dissolve residue in 10 ml of 10 M HNO_3 and evaporate to dryness
Aliquot C
Dilute to 25 ml in 1·5 M HNO_3
Aliquot D (0·5 ml)
Dilute to 50 ml with 1·5 M HNO_3
Measure K, Fe, Ca, Mg, Na.
Measure Cu, Bi, Mn, Pb, Sr, Ni, Co, Cr.
Aliquot E (1·0 ml)
Dilute to 10 ml with 1·5 M HNO_3
Measure Al, Zn

FIG. 4.20 Flow-chart of the processing stages for the analysis of metals in the atmosphere by atomic absorption [46]

4.2.5 Spark source mass spectrometry

The mass spectrometer resolves molecular and atomic ions on the basis of their mass-to-charge ratio. Mass spectrometers are generally built in five stages (Fig. 4.21): (1) a sample compartment into which the material for analysis is injected; (2) an ion source and ionization chamber; (3) an accelerating system comprising a series of slits maintained at different potentials; (4) a mass analyser which discriminates between the masses of the ions and focuses them onto (5) a detector. The detector signal is amplified and used to drive some suitable display or recording system.

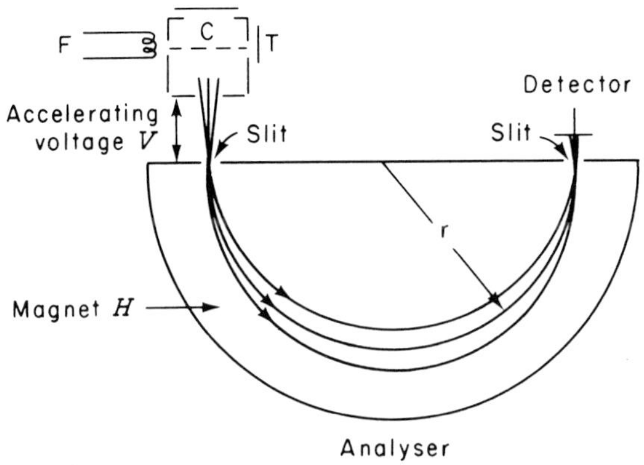

FIG. 4.21 Schematic diagram showing the principle of a 180° deflection mass spectrometer. Electrons from the filament source F enter the ionization chamber C and generate positively charged ions by bombardment of sample molecules present in the chamber. The residual electron beam is collected by the trap plate T. A potential V accelerates the positively charged species out of the ionization chamber into the magnetic field H. The beam is bent through 180° in the magnetic field and is focused onto the detector.

The molecules in the ionization chamber are bombarded by electrons of 70 eV energy and these cause ionization of species under investigation m by the reaction:

$$\mathrm{m} + e \rightarrow \mathrm{m}^{+} + 2e \qquad (4.16)$$

The positive ions generated in the chamber are accelerated out of the chamber by an accelerating potential *V*. Since the positive ions carry unit charge the kinetic energy they acquire is given by

$$eV = \tfrac{1}{2}\mathrm{m}v^{2} \qquad (4.17)$$

where v is their velocity. In the simplest instruments the positive ion beam passes into a magnetic field of strength *H* where it is deflected through a radius of curvature R onto the collecting plate. During this deflection the centripetal force must equal the centrifugal force exerted on the particles in the beam, so that

$$Hev = \frac{mv^2}{R} \tag{4.18}$$

Eliminating v from equations (4.17) and (4.18) and rearranging gives the fundamental equation of the mass spectrometer:

$$\frac{m}{e} = \frac{H^2R^2}{2V} \tag{4.19}$$

Since $e = 1$ (almost invariably) and R is fixed by the geometry of the instrument (otherwise, ions would not be in focus on the collecting plate), different masses are focused either by varying the potential V and keeping H fixed or by varying the magnetic field strength H at constant accelerating potential V.

When used for elemental analysis the instrument requires modification. The inlet and ion source compartments are replaced by electrodes so that the sample can be vaporized by a high-voltage radio-frequency spark. By making the electrodes hollow, liquid nitrogen can be circulated through them as a coolant. This allows films of vaporized sample to adhere to the electrodes. With the exception of hydrogen and helium all elements in the periodic table can be analysed by scanning mass numbers from lithium-7 upwards with uniform sensitivity.

4.3 Chemical analysis of polynuclear hydrocarbons (PNHs)

With the possible exception of naturally high atmospheric concentrations of terpenes in the vicinity of forested regions of the earth's surface, the majority of complex organic molecules containing ring systems are derived from the partial combustion of organic matter. In Chapter 1 the carcinogenic nature of PNHs was discussed and in Chapter 2 some sources were identified. The acknowledgement of this widespread distribution of a potentially harmful class of compounds has lead to strenuous efforts to improve chemical analytical methods for their accurate estimation in air. Most of the procedures employ various forms of chromatography at some stage of the analysis. This section contains a summary of these methods. Table 4.8 lists the names, gives the structures and indicates the systematic nomenclature that has been adopted for the compounds that are considered.

TABLE 4.8 *Structures and systematic nomenclature of PNHs found in air*

Compound	Structure	Carcinogenicity
Anthracene $C_{14}H_{10}$ m.p., 216°C; b.p., 340°C	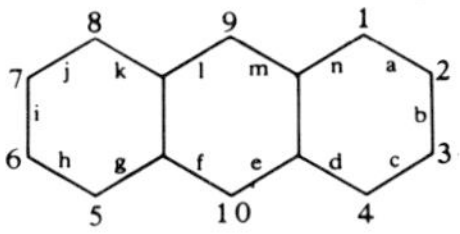	
Phenanthrene $C_{14}H_{10}$ m.p., 101°C; b.p., 340°C	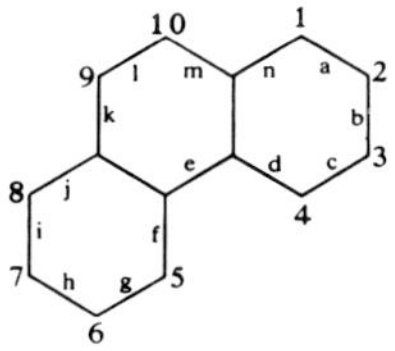	
Fluorene $C_{13}H_{10}$ m.p., 116°C; b.p., 293°C	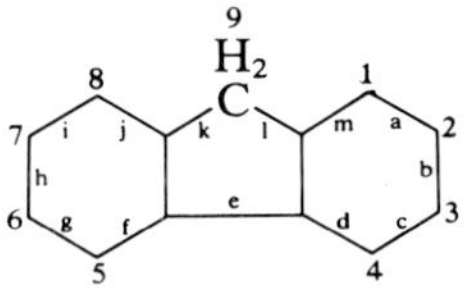	
Fluoranthene $C_{16}H_{10}$ m.p., 110°C; b.p., 393°C	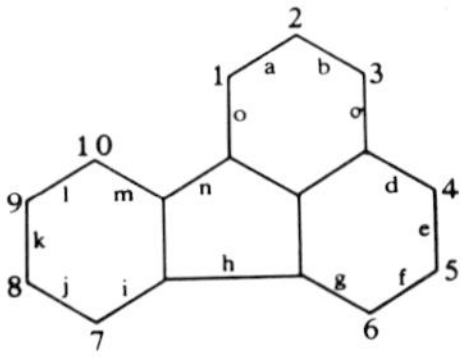	
Aceanthrylene $C_{16}H_{12}$ m.p. 113°C	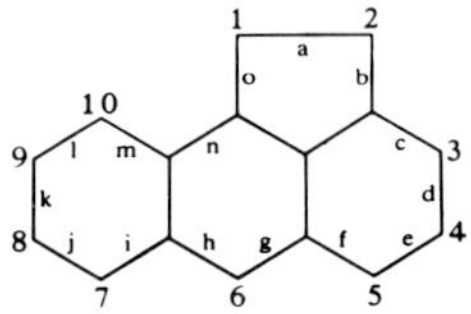	
Naphthacene = benz(b)anthracene $C_{18}H_{12}$ m.p., 341°C; sublimes	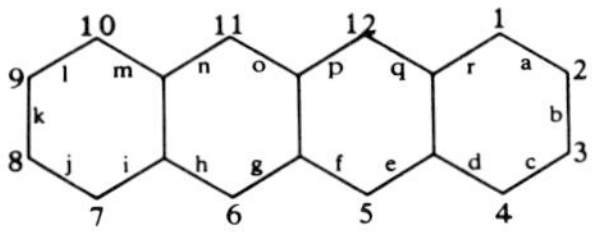	
Chrysene = 1,2-benzophenanthrene $C_{18}H_{12}$ m.p., 254°C; b.p., 448°C	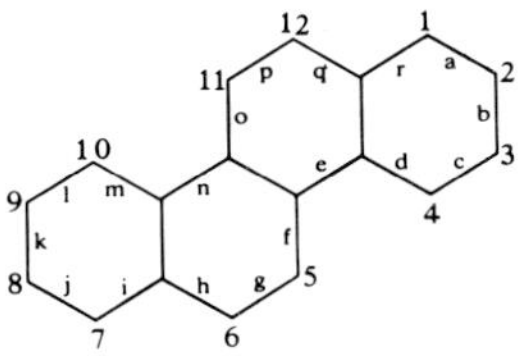	

TABLE 4.8—(*continued*)

Compound	Structure	Carcinogenicity
Pyrene $C_{16}H_{10}$ m.p., 150°C; b.p., >360°C	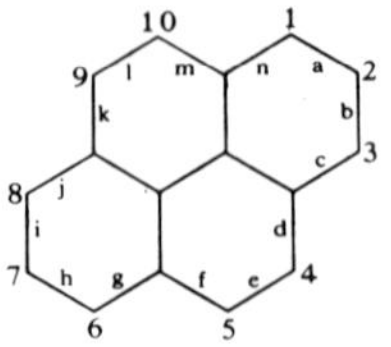	
Perylene $C_{20}H_{12}$ m.p., 273°C; b.p., *c* 500°C	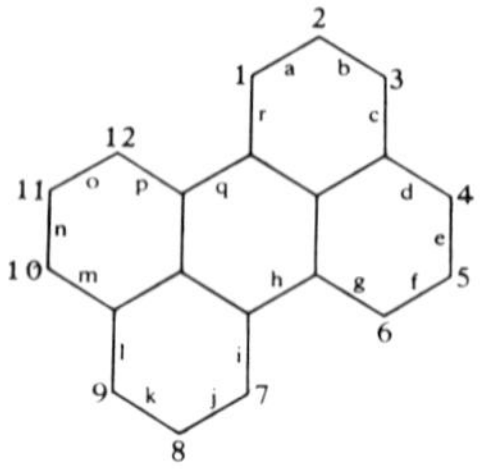	
Carbazole $C_{12}H_9N$ m.p., 246°C; b.p., 355°C	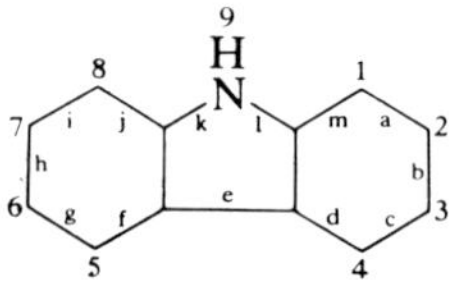	
Acridine $C_{13}H_9N$ m.p., 111°C; b.p., >360°C	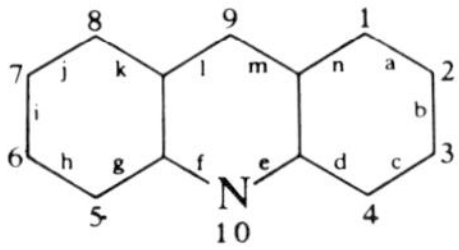	
Benz(a)anthracene $C_{18}H_{12}$ m.p., 158°C; sublimes (1,2-benzanthracene)	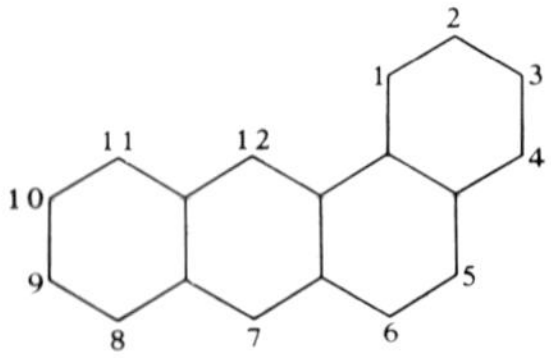	+
7,12-Dimethyl-benz(a)anthracene (9,10-dimethyl-1,2-benzanthracene)	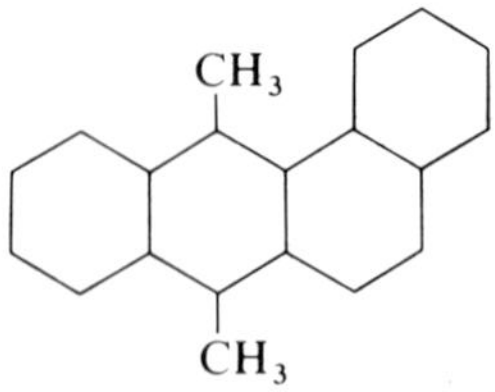	++++
Dibenz(a,j)anthracene (1,2–7,8-dibenzanthracene)	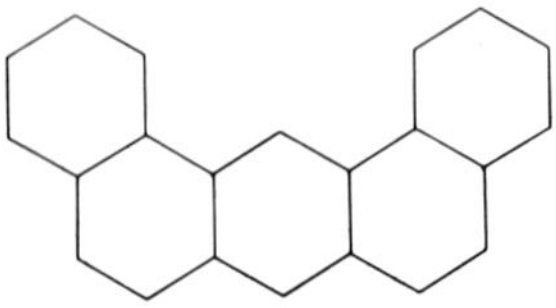	+

TABLE 4.8—(*continued*)

Compound	Structure	Carcinogenicity
Dibenz(a,h)anthracene (1,2–5,6-dibenzanthracene)	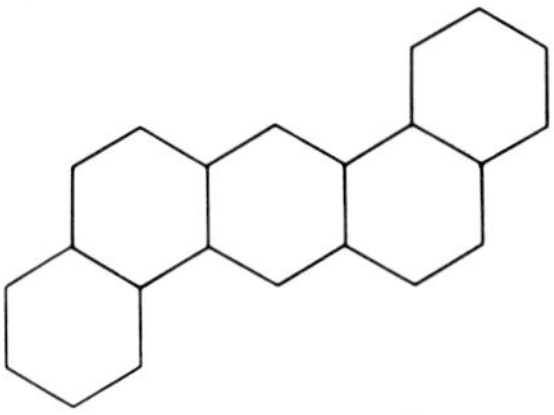	+++
Dibenz(a,c)anthracene (1,2–3,4-dibenzanthracene)	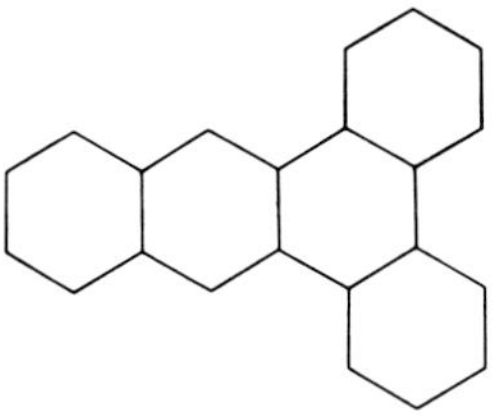	+
Benzo(c)phenanthrene (3,4-benzphenanthrene)	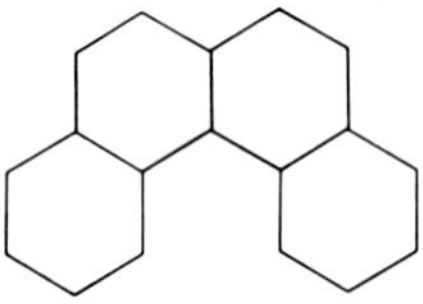	+++
Benzo(a)pyrene (1,2-benzpyrene) (3,4-benzypyrene*)	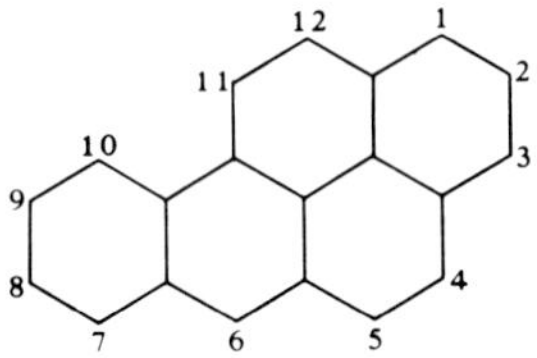	+++
Benzo(e)pyrene (4,5-benzpyrene) (1,2-benzpyrene*)	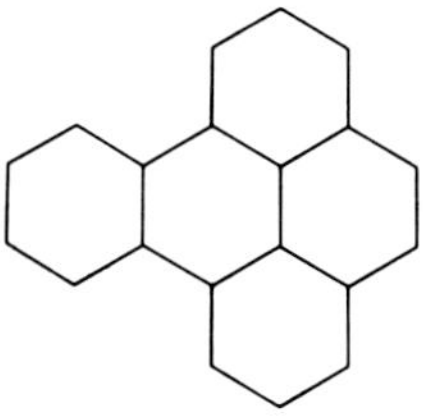	–
Dibenzo(a,1)pyrene (2,3–4,5-dibenzpyrene) (1,2–3,4-dibenzypyrene*)	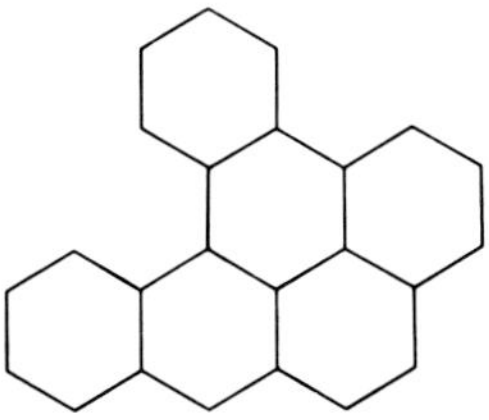	±

* Richter nomenclature (obsolete)

TABLE 4.8—(*continued*)

Compound	Structure	Carcinogenicity
Dibenzo(a,h)pyrene (1,2-6,7-dibenzpyrene) (3,4–8,9-dibenzpyrene*)	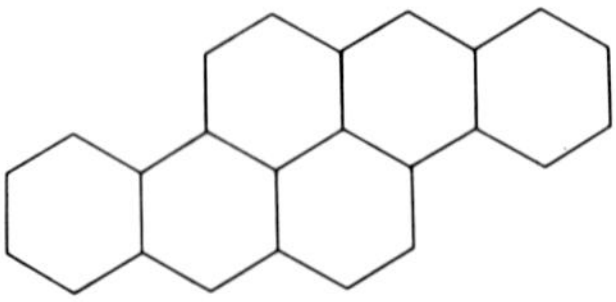	+++
Dibenzo(a,i)pyrene (2,3–6,7-dibenzpyrene) (4,5–8,9-dibenzypyrene*)	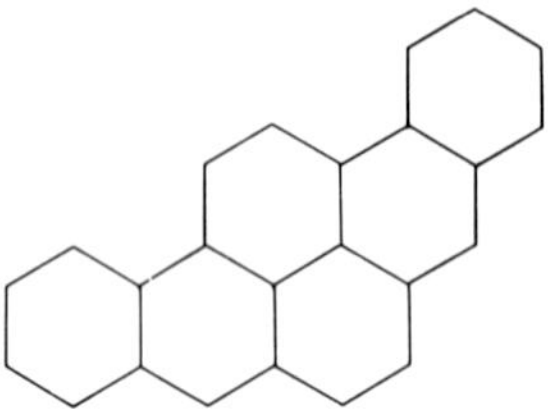	+++
Dibenzo(cdjk)pyrene = anthanthrene	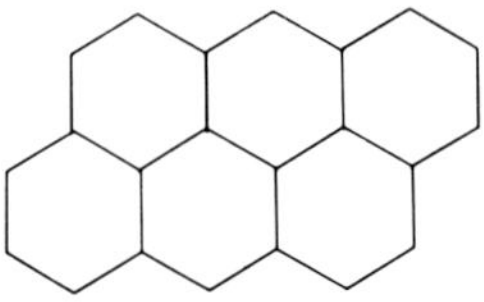	–
Indeno(1,2,3-cd)pyrene (*O*-phenylenepyrene)	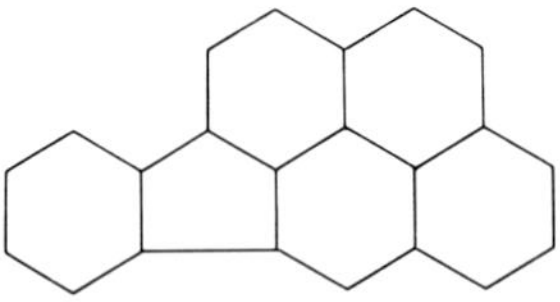	+
Benzo(a)fluorene (1,2-benzfluorene)	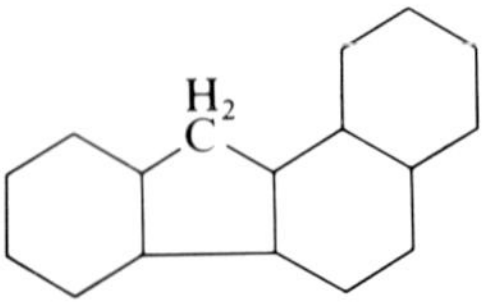	–
Benzo(b)fluorene (2,3-benzfluorine)	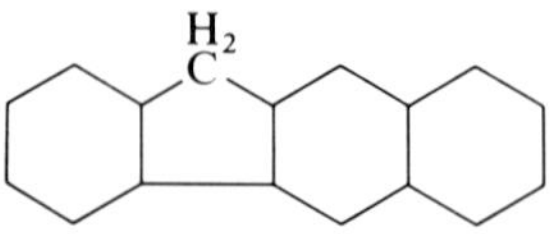	–
Dibenzo(a,h)fluorene (1,2–6,7-dibenzfluorene)	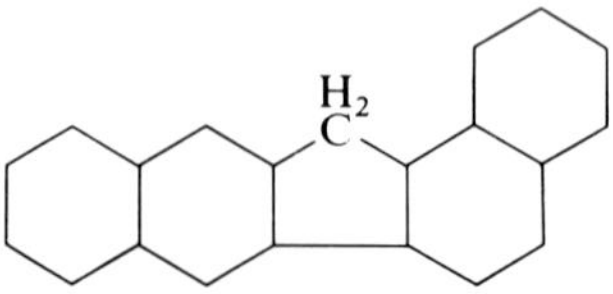	±

TABLE 4.8—(*continued*)

Compound	Structure	Carcinogenicity
Dibenzo(a,g)fluorene (1,2–5,6-dibenzfluorene)	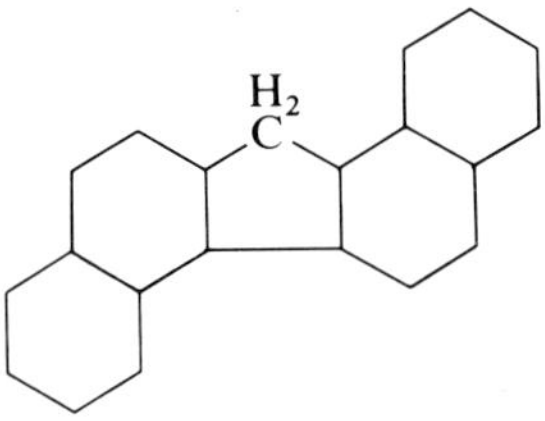	+
Benzo(c)fluorene (3,4-benzfluorene)	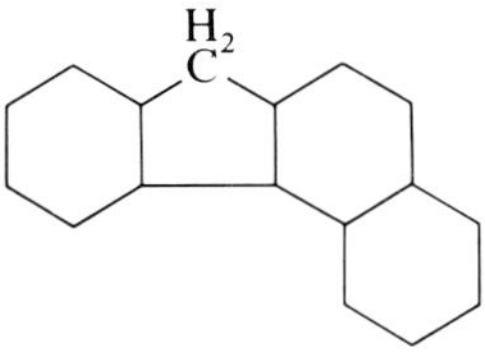	−
Dibenzo(a,c)fluorene (1,2–3,4-dibenzfluorene)	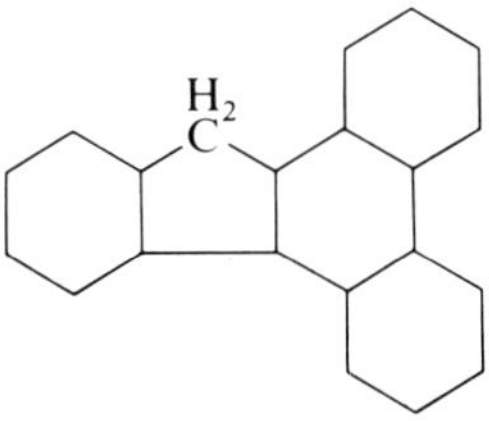	±
Benzo(b)fluoranthene (2,3-benzofluoranthene)	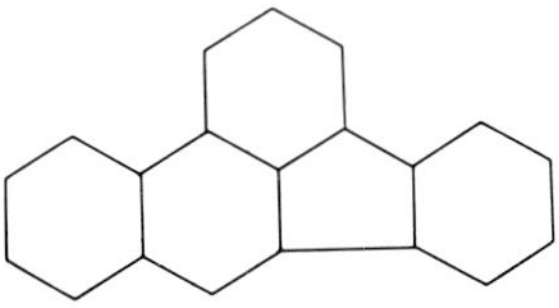	++
Benzo(j)fluoranthene (7,8-benzofluoranthene)	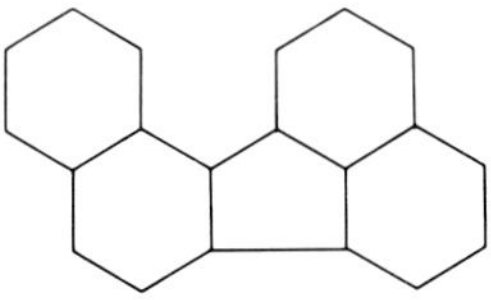	++
Benzo(k)fluoranthene (8,9-benzfluoranthene)	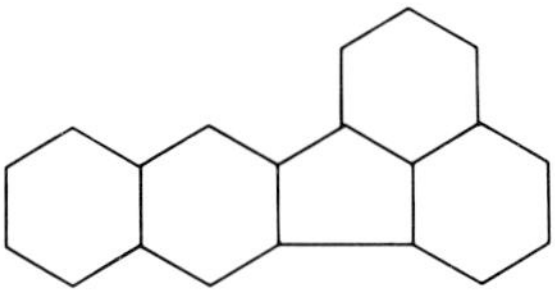	−

TABLE 4.8—(*continued*)

Compound	Structure	Carcinogenicity
Benzo(ghi)fluoranthene	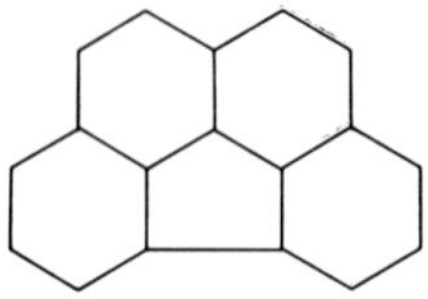	–
Dibenzo(def,mno)chrysene = anthanthrene = dibenzo(cdjk)pyrene	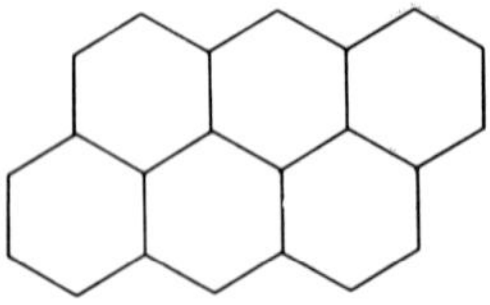	–
Dibenzo(b,def)chrysene = dibenzo(a,h)pyrene (3,4–8,9-dibenzpyrene*)	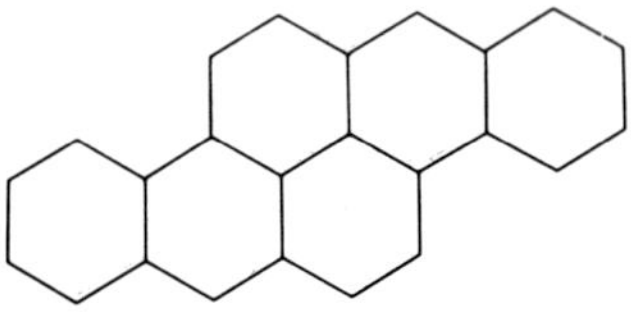	++
Dibenzo(def,p)chrysene = dibenzo(a,l)pyrene (1,2-3,4-dibenzpyrene*)	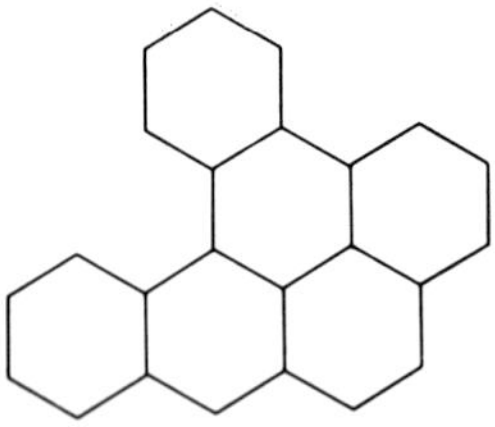	+
Naphtho(2,1,8-qra)naphthacene = naphtho(2,3-a)pyrene (2′,3′-naphtho-1,2-pyrene)	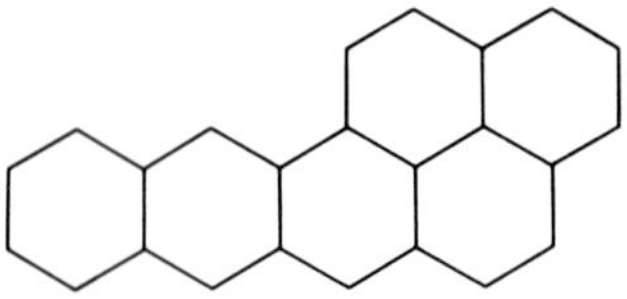	–
Benzo(ghi)perylene	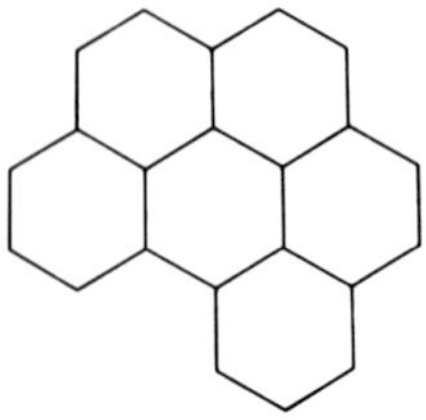	–

TABLE 4.8—(*continued*)

Compound	Structure	Carcinogenicity
Coronene $C_{24}H_{12}$ m.p., 438°C; b.p., 525°C	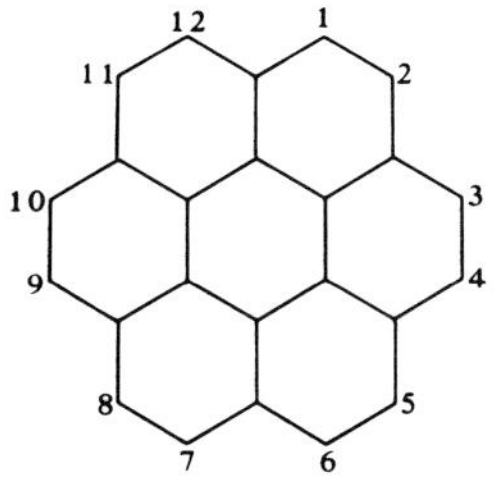	–
3-Methylcholanthrene	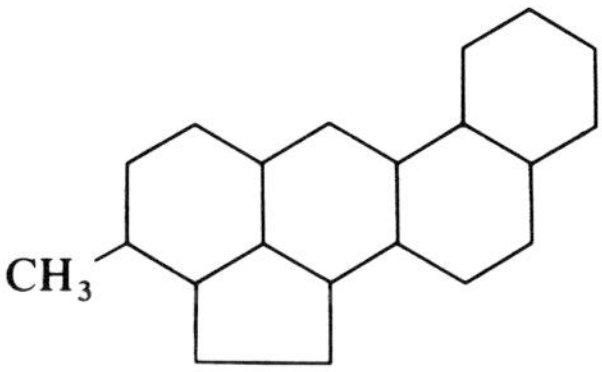	++++
Benz(j)aceanthrylene = cholanthrene	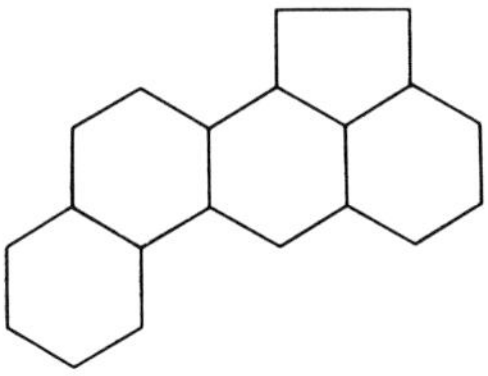	++
Benzo(a)carbazole (1,2-benzcarbazole)	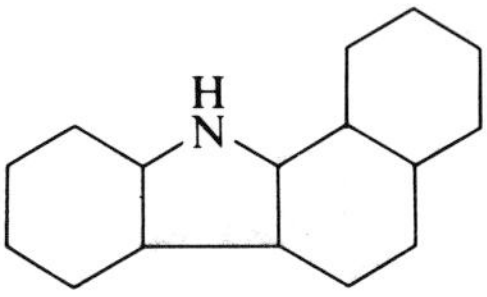	±
Dibenzo(a,g)carbazole (1,2–5,6-dibenzcarbazole)	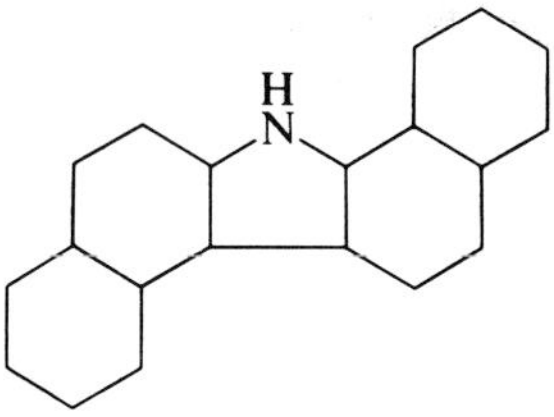	±
Dibenzo(c,g)carbazole (3,4–5,6-dibenzcarbazole)	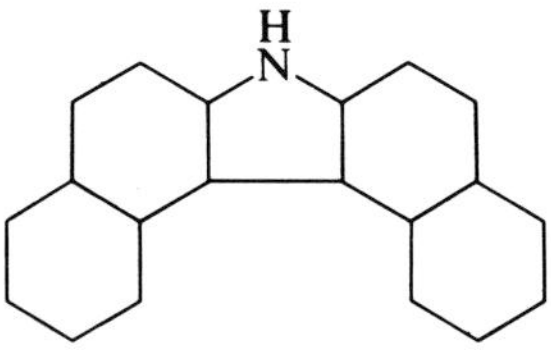	+++

TABLE 4.8—(*continued*)

Compound	Structure	Carcinogenicity
Dibenzo(a,i)carbazole (1,2–7,8-dibenzcarbazole)	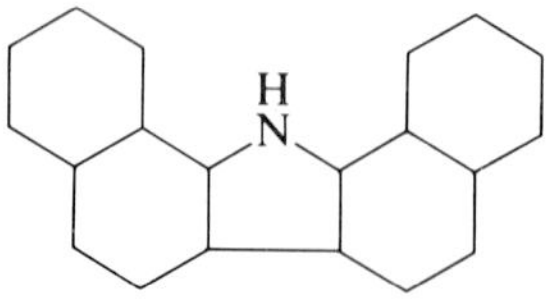	±
Dibenz(a,j)acridine (1,2–7,8-dibenzacridine)	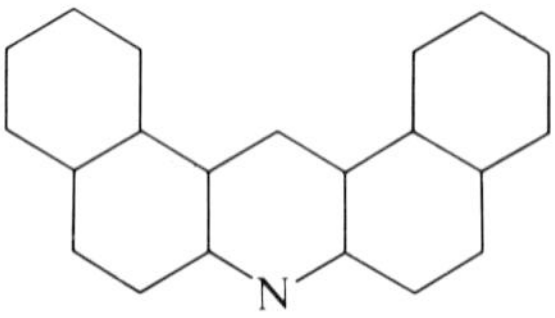	++
Dibenz(a,h)acridine (1,2–5,6-dibenzacridine)	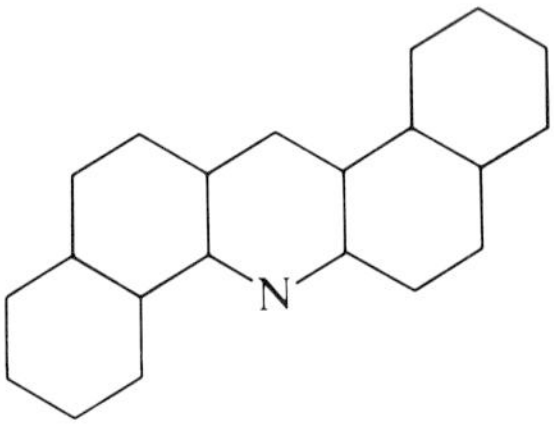	++
Dibenz(c,h)acridine (3,4–5,6-dibenzacridine)	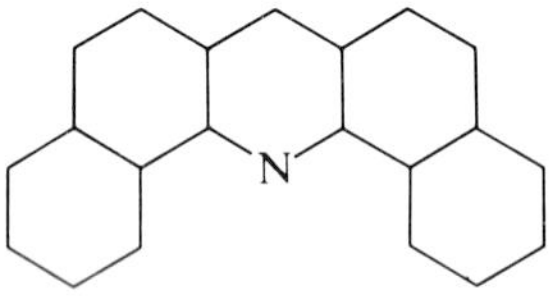	±

4.3.1 Collection of PNHs present in air—thermodynamic considerations

PNHs occur in air in ng m^{-3} quantities. They are solids at ambient temperatures and most probably exist in air in an adsorbed condition attached to dust particles and soot or coexisting in droplets with tar and other organic residues. Filtration techniques, in which air is drawn through a filter for known time periods and flowrates have generally been used for collection purposes. In order to collect sufficient samples for chemical analysis to be possible, between 2000 and 3000 m^3 of air need to be filtered. This is done by sampling at a high flowrate (about 10^3 l min^{-1}) for 24 h or at a lower flowrate for periods of days or weeks.

There is evidence suggesting that neither of these methods is satisfactory, at least for the more volatile PNHs. Work by Commins and

Lawther [47] and Commins [48] has noted that loss of benzo(a)pyrene can occur at the sample collection stage, particularly if warm air is filtered. These observations may now be interpreted in quantitative terms since Murray *et al.* [49] have succeeded in measuring the enthalpies and entropies of sublimation of some of these compounds (Table 4.9). Murray *et al.* [50] have used this information to calculate the equilibrium vapour concentrations (EVCs) from an equation having the form

TABLE 4.9 *Enthalpies and entropies of sublimation at the average temperatures of the vapour pressure measurements of some typical PNH atmospheric pollutants* [49]

Compound	m.pt.°C	Temp. range	$\Delta H^0_{sublimation}$ (kJ mol^{-1})	$\Delta S^0_{sublimation}$ (J mol^{-1} K^{-1})
Benz(a)anthracene	158–159	57–117	113·4±2·2	192·3±6·3
Benzo(a)pyrene	176·5–177·5	85–158	118·2±2·1	183·5±5·4
Benzo(e)pyrene	178–179	86–150	119·0±2·3	186·5±5·4
Benzo(k)fluoranthene	217	90–157	130·0	208·1
Benzo(ghi)perylene	272–273	116–195	127·6±2·3	181·8±5·4
Coronene	~440	154–237	135·8±3·1	174·3±6·7

$$\log_{10}(\text{EVC}) = -\frac{\text{A}}{\text{T}} - \log_{10} T + C \qquad (4.20)$$

where EVC is the equilibrium vapour concentration expressed in μg 10^{-3} m^{-3}, T is the absolute temperature and the constants A and C for the various PNHs are given in Table 4.10 ($\mu g 10^{-3}$ m^{-3} ≡ ng m^{-3}).

As an example we can calculate the EVC of pyrene at 25°C. From Table 4.10:

$$A = 4840 \qquad C = 23{\cdot}586$$

TABLE 4.10 *Constants* A *and* C *in the equation* $\log_{10} EVC = -A/T - \log_{10} T + C$, *where EVC is the equilibrium vapour concentration in* $\mu g\ 10^{-3}\ m^{-3}$ *and* T *is the absolute temperature* [50]

Compound	*A*	*C*
Pyrene	4840	23·586
Benz(a)anthracene	5926	25·489
Benzo(a)pyrene, Benzo(e)pyrene	6182	25·089
Benzo(k)fluoranthene	6792	26·373
Benzo(ghi)perylene	6674	25·046
Coronene	7100	24·674

Using $T = 298$ K in equation (4.20) gives

$$\log \text{EVC} = -\frac{4840}{298} - \log 298 + 23{\cdot}586$$

$$= -16{\cdot}241 - 2{\cdot}474 + 23{\cdot}586$$

$$= 4{\cdot}871$$

$$\text{EVC} = 7{\cdot}43 \times 10^4 \ \mu\text{g}\ 10^{-3}\ \text{m}^{-3}$$

$$= 74{\cdot}3\ \mu\text{g m}^{-3}$$

Clearly, for this type of collection an amount (in μg 10^{-3} of throughput) equal to the vapour concentration will not be collected on the filter. In addition clean air passing through a filter containing these substances will carry away amounts equal to or less than the EVC because of sublimation from the filter. Thus, it must be recognized that EVC values represent upper limits for losses during collection on particulate filters. In practice, the actual loss can be less than the EVC because,

(1) The evaporative loss may be governed by kinetic factors which prevent the attainment of the equilibrium value, and
(2) the calculation of EVCs assumes that the substances are pure, but in fact they are adsorbed on a substrate. Bonding between the PNH and the substrate may prevent or restrict sublimation.

The kinetic factors will have an influence mainly on the rate of loss of material already collected on the filter. For a slow flow of air through a filter, the air will become saturated with the pollutant and the rate of loss per unit throughput will approach the EVC value. At faster flows, the rate of loss will be limited by the rate of sublimation. The rate of sublimation is proportional to the EVC and to the exposed area, so that at higher flowrates the rate of loss per unit throughput will be less than the EVC. At a fixed temperature, flowrate and surface area, the loss per unit throughput is constant. The percentage loss, on the other hand, willl depend upon the pollutant concentration in the air. Losses from the filter will increase, therefore, as pollutant concentrations decrease.

Another factor which adds to the difficulty of obtaining authentic samples of PNHs from the air concerns the stability of the material in the adsorbed state towards photochemical reaction. Thomas *et al.* [51] have studied the effect of light on benzo(a)pyrene adsorbed on soot particles. Benzo(a)pyrene adsorbed on soot particles was passed through a tubular reactor approximately 3·3 m long and exposed to light of intensity about one-quarter of that usually experienced at midday during the month of July. The amount of benzo(a)pyrene per unit weight of soot at the entrance and exit points of the reactor under both light and dark conditions are compared in Table 4.11. The results show that there is a 6-fold increase in acidity and a 58% decrease in the benzo(a)pyrene content of the soot after exposure to light. It was also shown that the extent of deterioration of benzo(a)pyrene was dependent upon the surface area of the substrate. The rapid deterioration of 58% reported was attributed to the high surface areas associated with soot particles.

TABLE 4.11 *Effect of light on production of acid products and reduction of benzo(a)pyrene content of soot*

Light source	Flowrate ($l\ min^{-1}$)	Time in reactor (min)	Benzo(a)pyrene ($\mu g\ mg^{-1}$)	Acid ($\mu g\ mg^{-1}$)
Blank			3·77	4·34
Off	1·5	40	3·44	4·14
On	1·5	40	1·62[a]	28·44[b]

[a] Benzo(a)pyrene decrease 58%.
[b] Acid fraction increase 6-fold.

PNHs show intense absorption bands in the infrared between 700 cm^{-1} and 900 cm^{-1} characteristic of the out-of-plane C—H bending vibrations. Studies in this spectral region reveal differences between freshly recrystallized benzo(a)pyrene and benzo(a)pyrene that has been exposed to sunlight for several hours (Fig. 4.22a). By analogy with the well known photostationary equilibrium condition of anthracene in the mono- and dimer forms, it has been suggested that benzo(a)pyrene may behave in a similar manner [51].

Undissociated benzo(a)pyrene which gives the original spectrum shown in Fig. 4.22(a) (**C**) is very reactive and readily enters into equilibria involving complex formation. The C—H out-of-plane vibrations occur randomly on both sides of the benzo(a)pyrene planar configuration, but when activated by light or heat a complex is formed, possibly a dimer, which restricts the C—H out-of-plane vibrations to one side of the plane, as envisaged in Fig. 4.22(b).

A number of similar hydrocarbons tested did not give spectral changes in the infrared after exposure to light—e.g. anthracene, pyrene, benz(a)anthracene, benzo(e)pyrene, perylene, dibenz(a,h)anthracene, dibenzo(a,h)pyrene and dibenzo(a,i)pyrene. This result and the foregoing discussion emphasize the inherent differences in the chemistry of these types of compounds. Differences which have not always been taken into consideration by many of the atmospheric monitoring programmes conducted in the past. It seems on this evidence that the estimation of the

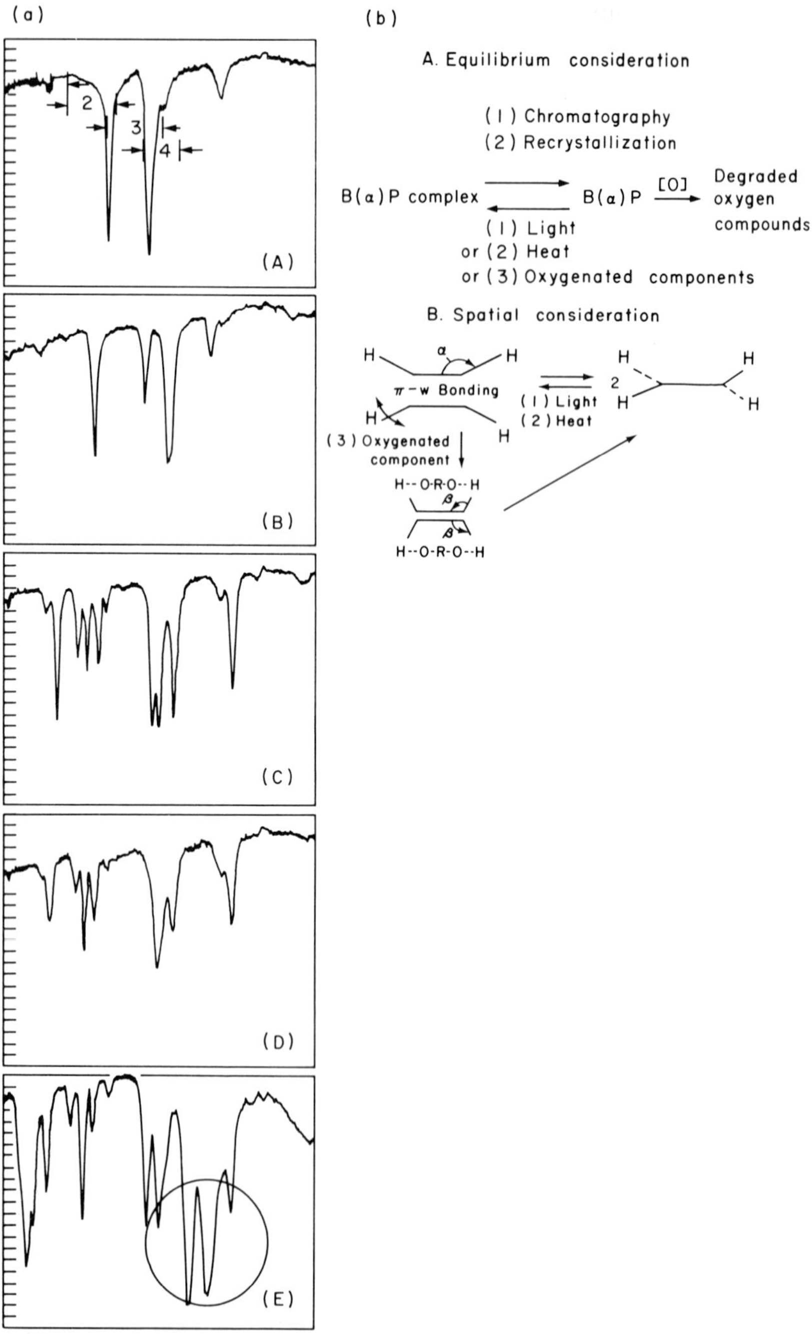

FIG. 4.22 [51] (a) Infrared curves of B(a)P isomers. **A**. Perylene: 1. Solo; 2. Duo; 3. Trio; 4. Quartet. **B**. B(e)P; **C**. B(a)P; **D**. B(a)P + light; **E**. B(a)P + complexing agent. (b) Hypothetical presentation of B(a)P complex formation as explanation for changes in infrared spectra. **A**. Equilibrium consideration; **B**. Spatial consideration.

benzo(a)pyrene content of the air and certainly the pyrene content may be in error, even at the first stage of the analysis (i.e. the sample collection stage), because of:

(1) loss of sample due to sublimation, and

(2) loss of sample due to photochemical reaction and oxidation.

It is interesting to speculate that for the photochemical degradation of PNHs singlet molecular oxygen may be involved (see section 5.3.3.3).

The first of these problems, especially for the estimation of PNHs that have four- or fewer-membered ring systems, has been recognized by Brockhaus [52]. To prevent sublimation of the sample on the filter Brockhaus advocates treating the filter with glyceryltrioctanoate. Glass fibre filters sprayed with a chloroform solution of this ester (9 : 1) to give a coverage of 3 mg cm^{-2} show an improvement in filter efficiency for pyrene and fluoranthene from 32–72% to 95–97%.

4.3.2 Preparation of PNHs for analysis

Once collected, samples are returned to the laboratory for chemical analysis. In order to separate, identify and estimate the numerous compounds present, a variety of procedures and precautions have been devised. These routines are familiar to chemists: the analytical methods employed are adaptions of well tried established techniques. For this reason greater reliability can be placed on this aspect of the analysis compared with the collection stage already discussed.

Before proceeding with the actual chemical analysis, however, the compounds retained on the filter must be extracted into solution. The choice of solvent used for the extraction of organic matter from a filter is of prime importance. The results of a systematic investigation by Gordon [53] on the effectiveness of different solvents for dissolving airborne particulate matter are given in Table 4.12. They show good correlation with the solvent strength, e^0, and on this basis Gordon recommends that for adequate recovery of PNHs solvents with e^0 values of 0·3 or greater should be used. The use of cyclohexane which has been widely used for solvent extraction purposes in air pollution fails this test and should not be employed. The US National Air Surveillance Network's choice of benzene with an $e^0 = 0{\cdot}32$ satisfies this criterion. If it is required to extract oxygenated and nitrogenous organic matter besides hydrocarbons, e.g. carboxylic acids, ketones, aldehydes, phenols and nitrates, then a more polar solvent should be used. Grosjean [54] maintains that it is not possible to select any one solvent capable of recovering the total organic matter from a filter; to do this a binary system of a nonpolar–polar combination of solvents is necessary.

When benzene is used as the extracting solvent, PNHs, tars and epoxides are extracted [55, 56, 57]. This benzene-soluble portion is often reported on a weight basis of the total particulate and is intended to represent the

TABLE 4.12 *Influence of solvent on the extraction of atmospheric particulate matter from a filter* [53]

Solvent	Extract yield ($\mu g\,cm^{-2}$)	Solvent strength[a] e^0
n-Hexane	187	0·01
Cyclohexane	163	0·04
Carbon tetrachloride	228	0·18
Toluene	264	0·29
Benzene	263	0·32
Diethyl ether	204	0·38
Dichloromethane	282	0·42
1,2-Dichloroethane	350	0·49
Acetone	630	0·56
Ethyl acetate	328	0·58
Nitromethane	520	0·64
Acetonitrile	447	0·65
Pyridine	682	0·71
2-Propanol	585	0·82
Ethanol	616	0·88
Benzene-methanol (60:40 v/v)	673	0·89
Methanol	717	0·95

[a] For alumina with 4% water referred to *n*-pentane, $e° = 0$. Snyder, L. R. (1968). "Principles of Adsorption Chromatography". Edward Arnold, London, and Marcel Dekker, New York, p. 194.

organic content of airborne particulate. Instead of benzene, Hoffmann and Wynder [58] have described an ether extraction procedure which separates organic components into neutral, acidic and basic fractions (Fig. 4.23). According to Cautreels and van Cauwenberghe [59], however, such procedures cause losses in that 10% of the PNHs can be found in the acidic fraction and up to 50% of the bases can be missing from the basic fraction.

Several methods have been used for the separation of the aliphatic and aromatic compounds present in the neutral fraction. Rosen and Middleton [60] have used a 10 cm-long column packed with silica gel with 85 ml of isooctane to elute the aliphatics followed by 85 ml of benzene to obtain aromatics and finally with a (1 : 1) methanol–chloroform mixture to elute the neutral oxygenated components, ethers, esters, aldehydes and ketones.

If spectroscopic methods are to be employed after these separations for identification and estimation then it is important that spectrographic grade reagents are employed. Hoffman and Wynder [61] and Candeli *et al.* [62], on the other hand, favour the use of nitromethane. The nitromethane

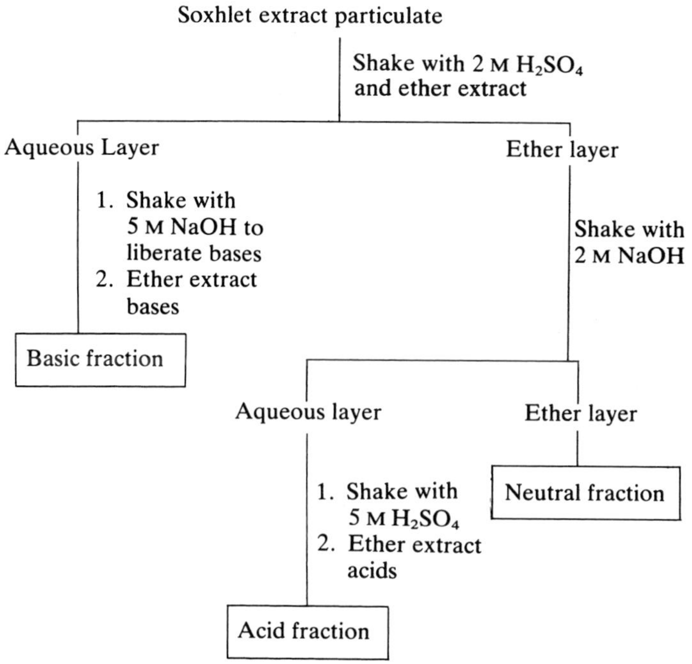
Soxhlet extract particulate
Shake with 2 M H_2SO_4 and ether extract
Aqueous Layer
Ether layer
1. Shake with 5 M NaOH to liberate bases
2. Ether extract bases
Shake with 2 M NaOH
Basic fraction
Aqueous layer
Ether layer
1. Shake with 5 M H_2SO_4
2. Ether extract acids
Neutral fraction
Acid fraction

FIG. 4.23 Separation scheme for the isolation of neutral, acidic and basic portions of organic matter collected from the atmosphere by filtration onto glass fibre media

method exploits the fact that PNHs are more soluble in nitromethane than aliphatic compounds (Table 4.13). The first stage in the analysis of PNHs by this method depends upon partition between methanol–water (4 : 1) and cyclohexane. This treatment separates hydrophilic compounds into the aqueous methanol layer and concentrates aliphatic and aromatic compounds into the cyclohexane layer. As shown in Table 4.13, it is possible to retain more than 99% of the PNHs in the cyclohexane layer after three extractions. Separation of aliphatics from PNHs can be achieved by partition between cyclohexane and nitromethane. A disadvantage of the nitromethane technique is that this solvent suppresses fluorescence of PNHs and hence prohibits the use of fluorescence methods in the final stages of the analysis.

From our experience in this laboratory [63], the use of dimethylsulphoxide is preferred to nitromethane. Furthermore, since dimethylsulphoxide is miscible with water, PNHs can readily be recovered by

TABLE 4.13 *Partition coefficients of PNHs at 20°C* [58]

Hydrocarbon	C_c/C_m	C_n/C_c
Pyrene	150·0	4·40
Benzo(a)pyrene	14·9	1·53
Benzo(e)pyrene	14·5	1·68
Fluoranthene	140·0	1·80
Benzo(b)fluoranthene	14·7	1·94
Benzo(j)fluoranthene	14·8	1·75
Benzo(k)fluoranthene	14·9	1·76
Benz(a)anthracene	18·6	1·75
Dibenz(a,h)anthracene	14·9	1·80
Perylene	14·9	1·69
Benzo(ghi)perylene	14·6	1·77
Chrysene	14·6	1·65

C_c = concentration in cyclohexane; C_m = concentration in methanol water (4:1); C_n = concentration in nitromethane; coefficients are estimated by spectroscopic determination of concentrations.

partitioning between an aqueous dimethylsulphoxide layer and a hydrocarbon layer, say *n*-pentane, so that solubility of PNH in the hydrocarbon layer is favoured. An outline of such a scheme is shown in Fig. 4.24.

Ideally, all manipulative operations should be carried out in light of wavelength greater than 450 nm to avoid photochemical decomposition, water baths should not exceed 45°C, and evaporations should be conducted under nitrogen atmospheres; but even these precautions will not prevent some losses. According to Murray *et al.* [50] the rate of sublimation per unit area dn/dt, into a vacuum ($<10^{-4}$ atm) is given by

$$\frac{dn}{dt} = \mathrm{EVC}\left(\frac{RT}{2\pi M}\right)^{1/2} \tag{4.21}$$

which, using $R = 8{\cdot}314 \times 10^7$ dyn cm K^{-1}, gives for the EVC in $\mu g\ 10^{-3}\ m^{-3}$; dn/dt in $\mu g\ s^{-1}$ or

$$\frac{dn}{dt} = 3{\cdot}637 \times 10^{-6}\mathrm{EVC}\left(\frac{T}{M}\right)^{1/2} \tag{4.22}$$

where T is temperature and M the molecular weight, respectively.

At higher pressures the rate of diffusion of the vapour into the air controls the rate of sublimation. The form of the resulting rate equation depends upon the shape of the container into which the vapour diffuses. In the case of a beaker the sublimation rate is given by Jost [64]:

$$\frac{dn}{dt}=\frac{EVC}{h}D \tag{4.23}$$

where h is the height of the beaker and D the diffusion coefficient.

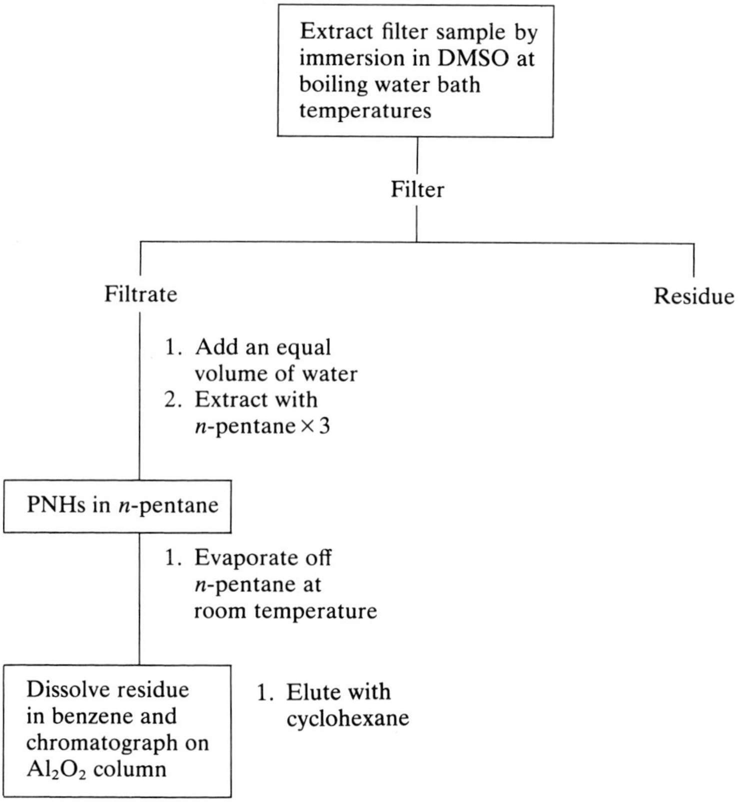
Extract filter sample by immersion in DMSO at boiling water bath temperatures
Filter
Filtrate
Residue
1. Add an equal volume of water
2. Extract with n-pentane × 3
PNHs in n-pentane
1. Evaporate off n-pentane at room temperature
Dissolve residue in benzene and chromatograph on Al₂O₂ column
1. Elute with cyclohexane

FIG. 4.24 Separation scheme for the recovery of PNHs in atmospheric particulate matter using dimethylsulphoxide (DMSO)

Estimating $D_{298\,K}^{1\,atm} \sim 0{\cdot}03\ cm^2\ s^{-1}$ the rates given by equation (4.23) become comparable at about 10^{-4} atm. Sublimation rates calculated at 20 torr for a beaker of 5 cm^2 cross-section and 10 cm height are compared with those measured by Grimmer and Hildebrandt in Table 4.14. Clearly, evaporative processes during analysis of PNHs should be avoided. This is especially true when analysis of pyrene is contemplated.

So far in this section we have considered the recovery of PNHs from particulate matter by solvent extraction, but this is not the only method

TABLE 4.14 *Calculated and measured (Grimmer and Hildebrandt[a]) losses during vacuum sublimation of PNHs. Loss rates in* $\mu g\ h^{-1}$ [50]

Compound	Calculated loss rate at 20 torr, 25°C $D = 1{\cdot}14\ cm^2\ s^{-1}$	Measured loss rate, 30°C
Pyrene	116	10·24
Benz(a)anthracene	2·9	1·62
Benzo(a)pyrene	0·16	0·12
Benzo(e)pyrene	0·16	0·10
Benzo(ghi)perylene	0·0035	0·14

[a] Grimmer, G. and Hildebrandt, A. (1965). Kohlenwasserstoffe in der Umgebung des Menschen. 1. mitt. eine Methode zur simultanen Bestimmung von dreizehn polycyclischen Kohlenwasserstoffen. *J. Chromatog.* **20**, 89–99.

that has been used. It has been shown by Sakabe [65] that benzo(a)pyrene adsorbed on furnace oil carbon black can be recovered by vacuum sublimation. This method exploits the sublimation properties of PNHs discussed in the previous paragraph as a means of separation. After collection on glass fibre filter media the sample is placed in an electrically heated furnace tube attached to a vacuum pump and manometer. The tube is evacuated to 10^{-2} torr and heated to 300°C for 3 h. PNHs subliming from the sample are cooled and collected in a receiver. Benzene or other suitable solvent is injected into the receiver from a micro-syringe to dissolve the PNHs preparatory to analysis. Recoveries of 98% of PNHs are claimed.

4.3.3 Chromatographic methods

4.3.3.1 Column chromatography

The application of column chromatography for the separation of the neutral extract from air sampling into aliphatic, aromatic and oxygenated compounds using silica gel has already been mentioned [60]. After the isolation of the aromatic fraction, separation and identification of the individual PNHs has been shown by a number of workers to be possible using alumina (Table 4.15). It is necessary before embarking upon a separation employing "activated" alumina to standardize the material. The hygroscopic nature of alumina gives a variable water content, so before use this must be adjusted to a known value. To do this chromatographic alumina is placed in an oven overnight at 140°C. About 100 g of this alumina are removed from the oven, cooled in a desiccator, and placed in a ground-glass stoppered flask and weighed. Distilled cyclohexane is then added to form a slurry, followed by about 6 ml of water added slowly from a microburette with shaking. This method ensures that water is evenly distributed over the alumina which thus becomes uniformly deactivated to the correct extent. Before use this must be checked by placing 2 g of deactivated alumina in contact with 5 ml of a standard coronene solution (10 $\mu g\ ml^{-1}$) for 15 min with occasional shaking. After allowing the alumina to settle decant the supernatant liquid into a 1·0 cm silica cell and measure the absorbance at 339 nm spectrophotometrically. Compare the absorbance at 339 nm of the original standard coronene solution and express the result as a percentage. Values in the range 70–80% are acceptable.

TABLE 4.15 *Separations of polynuclear hydrocarbons by column chromatography*

Column	Eluting solvent	Reference
Al_2O_3	Cyclohexane followed by cyclohexane + 2% ether and cyclohexane + 4% ether	66
SiO_2-gel 1·2 m	Cyclohexane + 8% benzene	67
Al_2O_3 + 4% water 7·6 m × 6·3 mm	Cyclohexane	68
Al_2O_3 + 13% water 0·58 m × 13 mm	Cyclohexane + ether (0–30%)	69
Al_2O_3 + 2% water 1 m × 4 mm	Cyclohexane	70
Al_2O_3	Pentane–ether (0–64% ether)	71

A 30–40 cm-long column of alumina prepared as above will give at least a partial separation of PNHs (Table 4.16). When the fractions from the column are analysed spectrophotometrically individual compounds can be identified and estimated quantitatively (Fig. 4.25). A selection of useful u.v. absorptions used by Cooper [72], Commins [73], Moore *et al.* [66], and Sawicki *et al.* [74, 75] for identification purposes are shown in Table 4.17. Details of procedures needed to estimate benzo(ghi)perylene in the presence of benzo(a)pyrene and for the quantitative interpretation of spectra using a baseline technique have been described [72, 73]. Benzo(a)pyrene can be accurately measured in the presence of benzo(k)fluoranthene by fluorescence [76].

When column chromatography is employed a correction factor must be used to take into account PNHs that are retained on the column. Recovery factors can be calculated by using standard solutions of PNHs and determining the concentrations eluted in comparison with the original concentrations.

TABLE 4.16 *A typical separation of polynuclear hydrocarbons on an activated alumina column. Since the separation will depend on column activity, length, volume of fraction collected and solvent employed, the fraction numbers shown are used for illustrative purposes only*

Fraction number	Solvent	PNHs Present
0–6	Cyclohexane	None
7–12	Cyclohexane	Pyrene, fluoranthene
13–15	Cyclohexane	Pyrene, fluoranthene, 4-methylpyrene
16–34	Cyclohexane	Benz(a)anthracene, chrysene
35–60	Cyclohexane + 2% ether	Benzo(a)pyrene, benzo(e)pyrene, benzo(k)fluoranthene, benzo(ghi)perylene
60–65	Cyclohexane + 4% ether	Benzo(ghi)perylene, anthranthrene, coronene

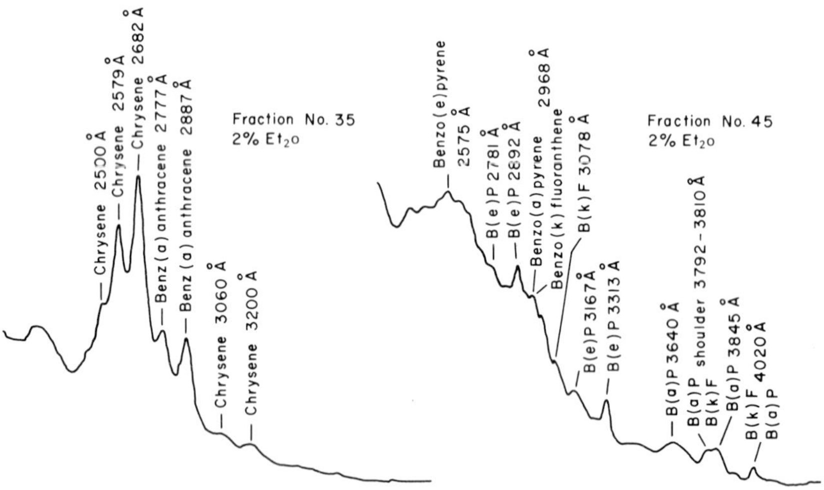

Benz (a)anthracene and chrysene, fraction 35.

Benzo (a) pyrene, benzo (e) pyrene and benzo (k) fluoranthene, fraction 45.

FIG. 4.25 PNHs present in chromatographed fractions identified spectrophotometrically [66]

Column chromatography has been a popular analytical technique in air pollution studies, but the method is exacting and uses fairly large volumes of solvents which need special purification; consequently the analysis is time-consuming from the man-hour operation viewpoint. It seems inevitable that the method will decline in importance and be superseded by other chromatographic developments.

TABLE 4.17 *Ultraviolet absorptions useful in the determination of polynuclear hydrocarbons and aza-heterocycles separated by activated alumina chromatography*

Compound	Absorptions (nm)
Anthracene	366, 376, 386
Pyrene	327, 335, 344
Fluoranthene	284, 287
Benzo(a)pyrene	381, 385, 403
Benzo(e)pyrene	289, 331
Benz(a)anthracene	278, 288,
Chrysene	268, 320
Naphthacene	275, 441, 471
Perylene	409, 436
Benzo(ghi)perylene	299, 384
Anthanthrene	423, 430
Coronene	302, 339
Benzo(h)quinoline	342, 347, 350
Benzo(f)quinoline	341, 345, 350
Acridine	351, 356, 360
Phenanthridine	340, 343, 350
Benz(a)acridine	377, 380, 384
Benz(c)acridine	378, 382, 390
Indeno(1,2,3-i,j)isoquinoline	360, 363, 370
11 H-Indeno(1,2-b)quinoline	337, 340, 344
Benzo(lmn)phenanthridine	366, 370, 374
Dibenz(a,h)acridine	388, 390, 394
Dibenz(a,j)acridine	387, 389, 393

4.3.3.2 High pressure liquid chromatography

The emergence of high pressure liquid chromatography represents such a development. Speeds and efficiencies approaching packed gas chromatographic columns are now possible, since the introduction of special porous layer beads as packing materials, e.g. Corasil [77] and Zipax [78]. These materials comprise an impenetrable spherical core with a thin porous outer surface layer. Generally, the packing is contained in stainless steel tubes of internal diameter 1–3 mm and length 1–3 m. Some examples of their application for the separation of PNHs and aza-heterocycles are given in Table 4.18. Figure 4.26 shows the effect of changing the composition of the mobile phase between methanol and water from (70 : 30) to (75 : 25) with a Corasil/C_{18} packing [80] indicated in Table 4.18.

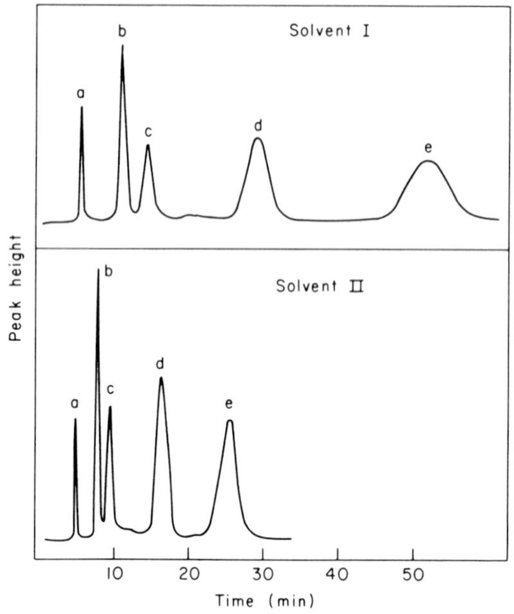

FIG. 4.26 High pressure liquid chromatography separation of PNHs [8]. Effect of solvent composition on the liquid chromatographic separation of polynuclear hydrocarbons. a, Naphthalene; b, anthracene; c, fluoranthene; d, 1,2-benzanthracene; e, 3,4-benzofluoranthene. Column, Corasil/C_{18}. Solvents: I, methanol–water (70 : 30); II, methanol–water (75 : 25); flow-rates, both 0·5 ml/min.

4.3.3.3 Thin-layer and paper chromatography

Thin-layer chromatography (t.l.c.) has been extensively used by Sawicki and coworkers [86, 87, 88] for the analysis of PNHs in air. The method can be very valuable for the analysis of complex multi-component mixtures and can often be selective when specific analysis of one or only a few compounds are required, e.g. the determination of benzo(a)pyrene in a benzene-soluble fraction extracted from a filter (Fig. 4.27). Although the t.l.c. procedure itself is relatively simple and inexpensive, quantitative interpretation employs spectrophotofluorometry (Fig. 4.28). Preferably the spectrophotofluorometer should be coupled to a plate scanning device so that eluted spots on the plate can be analysed *in situ*. Some wavelengths of fluorescence excitation and emission spectra useful for the identification of

TABLE 4.18 *Separation of polynuclear hydrocarbons and azaheterocycles by high pressure liquid chromatography*

Column packing and dimensions	Mobile phase	Application	Ref.
2,4,7-trinitro-fluorenone on Corasil 3 m × 1·6 mm	*n*-heptane and water saturated with *n*-heptane	3–6 PNH fused ring systems	79
Porapak T	iso-octane + 30% methanol	3–5 PNH fused ring systems	80
Porasil A	iso-octane		
Corasil/C_{18} 0·9 m × 2·2 mm	methanol : water (70 : 30)		
p-nitrophenyl-isocyanate bonded to Corasil 1 m × 2·2 mm	1% acetonitrile in n-hexane	3–5 aza-heterocyclic fused ring systems	81
Bentone-34 (dimethyldiocta-decylammonium montmorillonite) on Zipax 50 cm × 2·1 mm	methanol : water (2 : 1)	3–5 PNH fused ring systems	82
Porasil T 2·4 m × 3·2 mm	*n*-heptane	2–5 PNH fused ring systems	83
Zorbax 25 cm × 2.1 mm	methanol : water (65 : 35)	3–6 PNH fused ring systems	84
Partisil 10–ODS 25 cm × 4·6 mm	methanol–water	2–6 PNH fused ring systems	85
Bondapak C_{18} 30 cm × 4·0 mm	methanol–water	2–6 PNH fused ring systems	85

PNHs are given in Table 4.19 [89]. An example showing the tlc separation and identification of benzo(a)pyrene and benzo(k)fluoranthene by spectrophotofluorometry is given in Fig. 4.29. The fluorometric excitation spectra and the emission spectra are indicated by A and B, respectively, in Fig. 4.29(b).

As an alternative the location of the spots on the plate can be made visible by viewing the plate in u.v. light, spraying with concentrated sulphuric acid or exposure to the fumes of trifluoroacetic acid. The latter reagents cause fluorescence, the colour of which can be used for identification purposes.

The fluorescence of PNHs can be suppressed by nitromethane and nitrobenzene. Sawicki *et al.* [86] have used this quenching property to

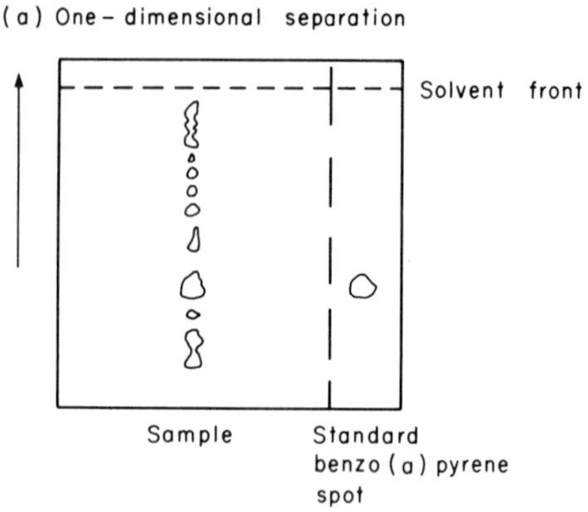

The components of the benzene soluble fraction are shown on the left hand side of the plate. The benzo(a)pyrene spot in this mixture can be located by comparison with the standard benzo(a)pyrene spot on the right-hand side of the plate which has been run at the same time.

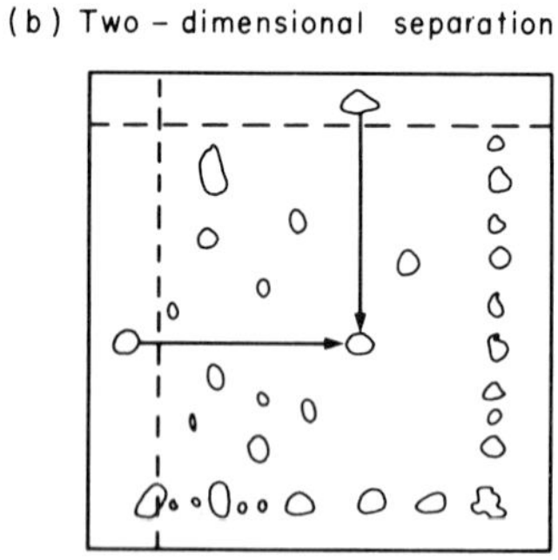

After running the sample as shown in (a) the plate is turned through 90°, respotted with standard benzo(a)pyrene and chromatographed in the same or another solvent.

FIG. 4.27 Examples of the identification of benzo(a)pyrene in the benzene soluble fraction of an air pollution sample.

assist in the identification of PNHs and their derivatives. Table 4.20 lists compounds whose fluorescence is quenched in nitromethane solution. Since fluoranthenic hydrocarbons are not quenched in nitromethane solution the method has been employed to separate and identify benzo(k)fluoranthene in the "benzpyrene fraction" of atmospheric samples. Sawicki *et al.* [87] have extended the method for the separation

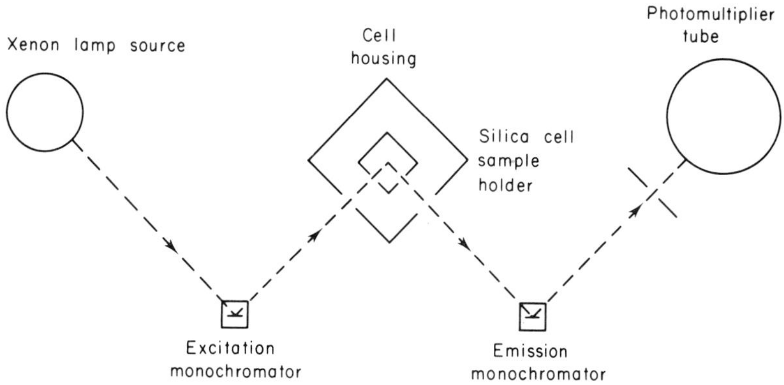

FIG. 4.28 Component layout of the Aminco–Bowman spectrophotofluorometer

TABLE 4.19 *Fluorescence excitation and emission spectra of PNHs* [89]

Compound	Wavelength (nm) fluorescence excitation spectra	Wavelength (nm) fluorescence emission spectra
Benzo(e)pyrene	274, 284, 300, 324, *329*, 349, 358, 366	*389*, 397, 410, 421
Benzo(a)pyrene	267, 286, 297, 333, 347, 375, *381*	*402*, 407, 414, 424, 452
Benzo(b)fluoranthene	280, 293, *301*, 340, 349, 368	404, *424*, 446
Benzo(k)fluoranthene	267, 294, *305*, 323, 338, 350, 357, 377, 390	*402*, 409, 418, 425, 435, 454
Perylene	365, 384, *407*	*438*, 444, 465, 500
Anthanthrene	263, 293, *302* 360, 377, 396, 402	*429*, 436, 462, 466 488
Benzo(ghi)perylene	288, 298, 329, 346, 362, *382*	406, 415, *419*, 429, 438, 444
Naphtho(1,2,3,4-def)chrysene	274, 292, *304*, 330, 340 356, 374	*396*, 402, 407, 419, 429, 445
Benzo(rst)pentaphene	285, 297, 316, 332, 354 373, *393*	*432*, 436, 447, 458, 492
Dibenzo(b,def)chrysene	299, *312*, 378, 398, 421	*451*, 480, 514
Naphtho(2,1,8-qra)naphthacene	294, 317, *332*, 383, 404, 427	*457*, 483, 523
Dibenzo(def,p)chrysene	275, 292, 300, 316, 330, 380, *401*	466, 473, *502*

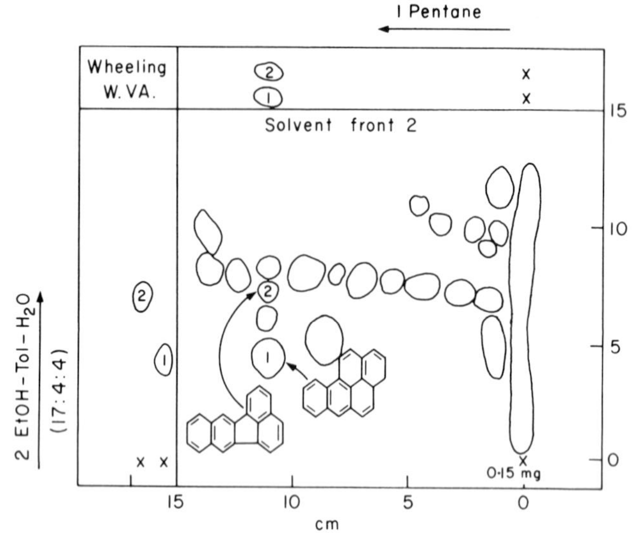

—Two-dimensional thin-layer chromatogram on alumina-cellulose acetate (2:1) of 0·15 mg of benzene-soluble fraction of the airborne particulates from the downtown area in Wheeling, West Virginia. One-dimensional runs of standards.

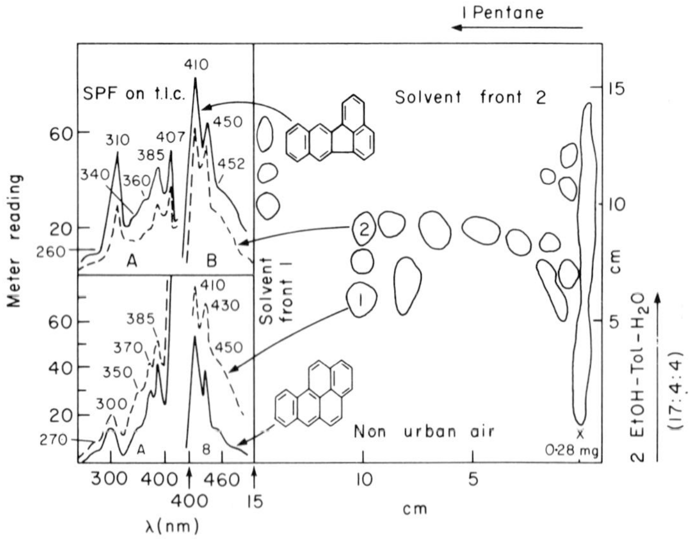

—Two-dimensional thin-layer chromatogram on alumina-cellulose acetate (2:1) of 0·28 mg of a benzene-soluble fraction of the airborne particulates from a non-urban atmosphere. Fluorometric excitation (A) and emission spectra of two separated spots (---) and of benzo(a)pyrene and benzo(k)fluoranthene on the plate (——).

FIG. 4.29 Separation and identification of benzo(a)pyrene and benzo(k)fluoranthene [97]. (a) Compounds identified by running standards as spots along the sides of the plate. (b) Compounds identified using fluorometric excitation–emission spectra technique. [97]

and spectral analysis of polynuclear aromatic amines and heterocyclic imines.

For the qualitative analysis of benzo(a)pyrene in particulate collected by filtration of about 20 m^3 of air Sawicki *et al.* [88] recommend the following procedure:

Extract particulate six times with methylene chloride allowing about 3 min for each extraction. Evaporate off the methylene chloride solution and dissolve residue in 0·2 ml of cyclohexane. Chromatograph the sample on an alumina–cellulose acetate (2:1) t.l.-plate with ethanol:toluene:water (17:4:4) as developer. At the same time run a standard benzo(a)pyrene spot (0·02 μg). Carry out these stages in semi-darkness. Remove from the developing tank and spray plate with preserving medium. Scan the plate at F300/430, i.e. a fluorescence excitation wavelength of 300 nm and an emission wavelength of 430 nm. To determine the benzo(a)pyrene present either in ng g^{-1} of original sample weight or ng m^{-3} in air the equations

$$\text{ng B(a)P g}^{-1} = \frac{W_s \times A_x \times R_x}{1000 \times W_t \times A_s \times R_s} \tag{4.24}$$

and

$$\text{ng B(a)P m}^{-3} = \frac{W_s \times A_x \times R_x}{V \times A_s \times R_s} \tag{4.25}$$

may be used, where W_s denotes weight in ng of benzo(a)pyrene used as

TABLE 4.20. *Compounds whose fluorescence is quenched in nitromethane solution* [86]

Acenaphthylene	4-H-Cyclopenta(def)phenanthrene
Anthranthrene	Dibenz(ac)anthracene
Anthracene	Dibenz(ah)anthracene
Benz(a)anthracene	Dibenz(aj)anthracene
Benzo(b)chrysene	Dibenzo(gp)chrysene
11-H-Benzo(a)fluorene	Dibenzo(brst)pentaphene
11-H-Benzo(b)fluorene	Dibenzo(ae)pyrene
7-H-Benzo(c)fluorene	1-Methylpyrene
Benzo(c)phenanthrene	Naphthacene
Benzo(ghi)perylene	Naphtho(2,1,8-qra)naphthacene
Benzo(a)pyrene	Perylene
Benzo(e)pyrene	Phenanthrene
9-Bromoanthracene	Picene
1-Bromopyrene	Pyrene
Chrysene	Triphenylene

Particulate sample from 20 m^3 of urban air

Extract (x6) with CH_2Cl_2

Methylene chloride extract

Evaporate

Residue

Dissolve residue in cyclohexane

Cyclohexane extract

(i) Spot solution onto alumina-cellulose acetate (2:1) t.l.c plate.

(ii) Develop in EtOH-ϕCH_3-H_2O (17:4:4) solution.

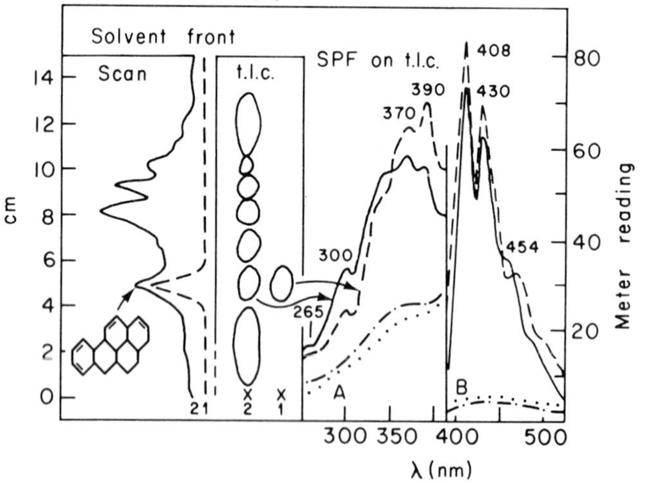

Thin-layer chromatographic separation of 50 μg of the benzene-soluble fraction on alumina cellulose acetate with ethanol–toluene–water (17:4:4) as developer. The spectra obtained at F 300/470 and meter multiplier reading (MM) of 0·1, the scan at F 300/430 and MM of 0·01. The spots from the benzene-soluble fraction (——); the standard benzo(a)-pyrene spot (– – –).

On the right the fluorescence excitation (A) and emission (B) spectra obtained directly from the plate. Spot from t.l.c., run two, with same R_f as BaP (——); appropriate blank (– · – ·); pure BaP spot from t.l.c., run one (– – –); appropriate blank (. . . .).

FIG. 4.30 Separation and identification procedure for the determination of benzo(a)pyrene in an air sample according to Sawicki *et al.* [88]

standard, W_t the weight in mg of original particulate sample, A_s and A_x are the areas of the standard spot and the unknown spot on the plates, R_x and R_s are the product of the meter multiplier and transmittance readings at the maximum of the unknown and standard, respectively, and V is the volume of air analysed in (m^{-3}). Figure 4.30 summarizes the complete operation including a portrayal of the fluorescence excitation and emission spectra obtained from the actual air sample and the standard sample.

The technique has been applied to a whole range of PNHs and some of their aza-homologues, nitro-compounds, aldehyde, ketone, quinones and amine derivatives, that have been discovered in atmospheric samples. These are summarized in Table 4.21.

Paper chromatography has been employed to separate PNHs and their derivatives and these methods are also included in Table 4.21. For example, with the aqueous formamide–Whatman No. 1 paper system used by Sawicki and Pfaff [99] the aza-heterocyclic compounds are separated in the following order of decreasing adsorbability: dicyclic, tricyclic, tetracyclic peri condensed, tetracyclic linear or angular, pentacyclic peri condensed, pentacyclic linear or angular and hexacyclic compounds. The last two named types of compounds are best separated by aqueous dimethylformamide. The parent compounds may readily be separated from their monoalkyl and dialkyl derivatives e.g. the R_F values of benz(c)acridine and the 7-methyl and 7,10-dimethylbenz(c)acridine have values of 0·44, 0·34 and 0·22, respectively. Polynuclear aromatics such as benzo(a)pyrene and perylene remain at the origin in the aqueous formamide system and hence do not interfere.

4.3.3.4 Gas-liquid chromatography

Separation of PNHs by gas–liquid chromatography has been developed largely from experience obtained in the analysis of motor vehicle exhaust gases. A summary of column packings that have been used is given in Table 4.22. None of these methods has proved entirely satisfactory for the separation of mixtures of anthracene and phenanthrene, benzo(a)pyrene and benzo(e)pyrene or perylene and benzo(k)fluoranthene.

A recent development which overcomes these difficulties by using nematic liquid crystal as a stationary phase has been described by Janini *et al.* [105]. Nematic crystals exhibit an ordered molecular arrangement within a defined temperature range. This property can be used to differentiate between molecules on the basis of their shape and linearity. Generally, the more rod-like a solute molecule is, the more readily it can spatially approach and hence more easily interact with a nematic crystal.

TABLE 4.21 *Identification and determination of polycyclic hydrocarbons and derivatives present in the atmosphere by t.l.c. methods*

Compound	Plate composition	Developer solvent	Reference
Separation of azaheterocycles from PNHs and paraffins	Silanized silica gel	Benzene–hexane–pyridine	90
5- and 6-membered ring PNHs	Alumina and acetylated cellulose	Hexane : ether (19 : 1) for alumina. *n*-propanol : acteone : water (2 : 1 : 1) for cellulose	89
Benzo(a)pyrene in "Benzene-soluble fraction"	Alumina–cellulose acetate (2 : 1)	Ethanol : toluene : water (17 : 4 : 4)	88
Benzo(a)pyrene	Silica gel	Hexane : benzene (1 : 1)	62
Carbazole and polynuclear carbazoles	Alumina	Pentane : ether (19 : 1) followed by pentane : chloroform (3 : 2)	91
Aza-heterocycles, PNHs	Alumina	Pentane : nitrobenzene (9 : 1). Pentane : 2-nitropropane (19 : 1) and (9 : 1). Pentane : ether (19 : 1)	92, 93
Polynuclear ring carbonyl compounds	Alumina	Pentane : ether (19 : 1) or toluene	94
	Cellulose	Dimethylformamide : water (65 : 35)	94
Phenalen-1-one (perinaphthalenone) and 7 H-benz(de)anthrac-7-one (benzanthrone)	Alumina	Pentane : ether (19 : 1)	95
	Cellulose	Dimethylformamide : water (65 : 35)	95
Polycyclic quinones	Polyamide 11	Acetic acid : water (19 : 1)	96
	$Mg(OH)_2$	Chlorinated solvents	96
	Alumina	Pentane : chloroform (3 : 2)	96
PNHs carbazole aza-heterocycles	Alumina : cellulose acetate (2 : 1)	For PNHs Pentane and ethanol : Toluene : water (7 : 4 : 4). For aza-heterocycles *n*-Pentane and dimethylformamide: water (35 : 65)	97
PAPER CHROMATOGRAPHY			
Aza-heterocyclic aromatic hydrocarbons	Whatman No. 1 or 3MM	Aqueous solvents, e.g. 1% acetic acid, 6 N HCl, conc. NH_3, 10% pyridine	98

TABLE 4.21—(*continued*)

Compound	Plate composition	Developer solvent	Reference
Aromatics	Whatman No. 1	Formamide : water (35 : 65) Dimethylformamide : water (35 : 65)	99
Aza-heterocyclic hydrocarbons	Whatman No. 3MM	Aqueous acetic acid, 6 N HCl, conc. NH_3 10% pyridine	100
PNHs	Whatman No. 1 or 31 ET impregnated with liquid paraffin (10% in petroleum ether)	Methanol saturated with medicinal paraffin for 5 h at room temperature	101

TABLE 4.22 *Column packing used in the separation of PNHs by g.c. and g.l.c.*

Dimensions	Column packing	Temperature (°C)	Ref.
	(*a*) *Packed Columns*		
5·4 m × 6 mm	1% OV-7 on 80/100 mesh chromosorb W	170–280	102
0·7 m × 2·3 mm	Graphitized carbon black	350	103
0·95 m × 2 mm	15% Graphitized carbon black on chromosorb W	460	104
6 m × 3·5 mm	0·05 w/w Apiezon L on soda-lime glass beads	254	62
1·22 m × 3·2 mm	2·5% BMBT on chromosorb W	185–265	105
	(*b*) *Capillary Columns*		
35 m × 0·35 mm	SE 30	200	106
30 m	SE 30 (methyl silicone)	200	107
50 m	SE 52(methylphenyl silicone)	100–300	107
35 m	XE 60(cyanoethylmethyl silicone)	max. 200	107
60 m × 0·5 m	SE 30	300	108

To illustrate the principle Fig. 4.31 shows the separation of 1-methylnaphthalene from 2-methylnaphthalene using *N,N'*-bis(*p*-methoxybenzylidene)-α,α'-bi-*p*-toluidine, abbreviated to BMBT*, as the liquid crystal phase. Based upon boiling point differences alone the 1-methyl isomer should elute last but the reverse is observed. This is due to the greater length-to-breadth ratio for the 2-methyl isomer which prolongs its retention by the liquid crystal phase.

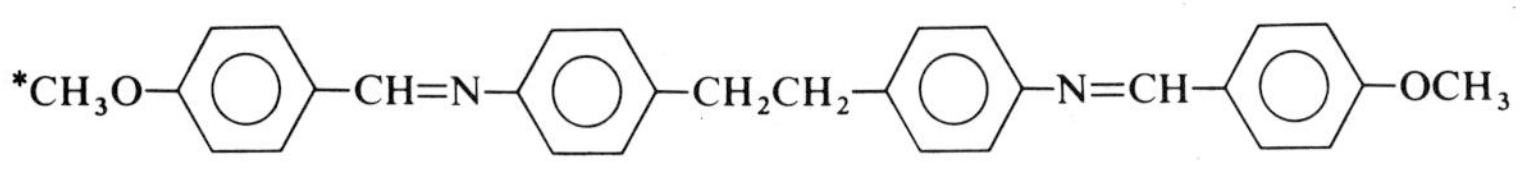

Figure 4.32 shows a chromatogram of the separation of 16 PNHs using BMBT liquid crystal phase (2·5% w/w) on HP Chromosorb W in a 1·22 m column programmed between 185° and 265°C. Comparison of Fig. 4·32 with Fig. 4·33 which illustrates a chromatogram of 20 PNHs on a 5·4 m column of Chromosorb W impregnated with OV-7 (1% w/w) [102] reveals

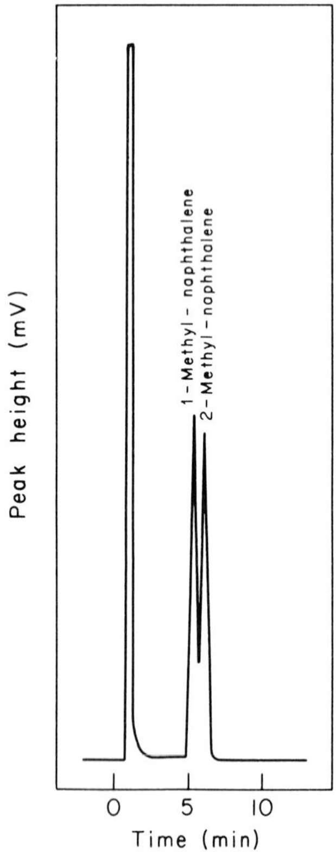

FIG. 4.31 Separation of 1-methyl naphthalene by g.l.c. using a nematic crystal as stationary phase [105]

that there has been a change in the elution order. The isotropic liquid phase gives the elution order benzo(a)pyrene before benzo(e)pyrene before perylene, whereas the nematic crystal phase gives benzo(e)pyrene before perylene before benzo(a)pyrene. In a similar manner, the isotropic liquid phase yields dibenz(ah)anthracene before benzo(ghi)perylene but this is reversed when the nematic crystal is employed.

Since demonstrating the value of nematic crystals in g.l.c. separations of PNHs Janini *et al.* [109] have synthesized a range of related compounds in

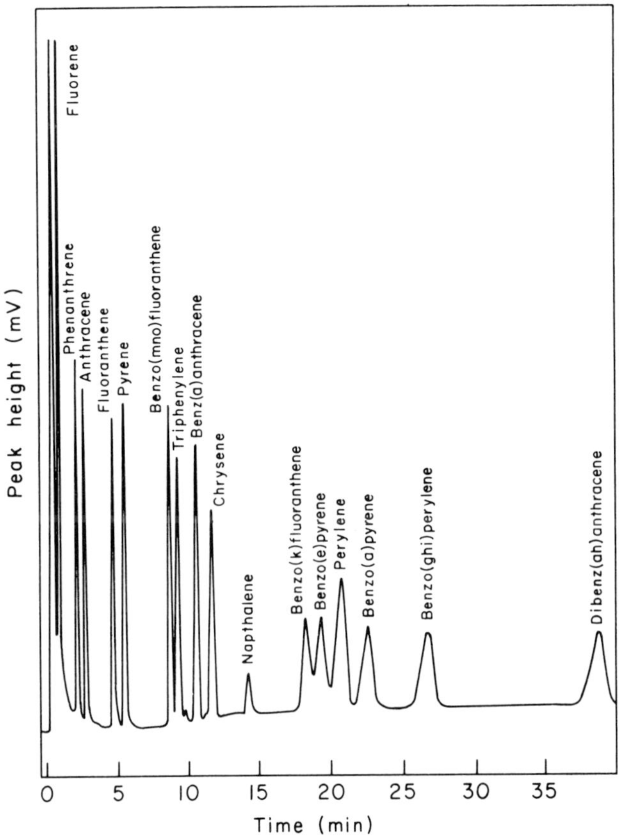

FIG. 4.32 Separation of PNHs by a temperature programmed g.l.c. procedure using a nematic crystal BMBT stationary phase [105]

an effort to find a compound which has a superior thermal stability compared with BMBT. Their results show that *N,N'*-bis(*p*-butoxybenzylidene)-α,α'-bi-*p*-toluidine (i.e. BBBT) fulfils this requirement.

The g.l.c. methods reviewed have used flame ionization detector systems, but the electron-capturing properties of PNHs suggest that the use of ecd's can be advantageous. Since aliphatic hydrocarbons do not respond to ecd, interference effects from these compounds can be eliminated when this detector is employed for PNH analysis.

4.3.3.5 Coupled gas chromatography and mass spectrometry

Advances in instrumental techniques have led to the coupling together of two powerful analytical procedures, namely, gas chromatography or gas–liquid chromatography and mass spectrometry [110, 111, 112, 113]. After chromatographic separation, compounds eluted from the column are split between a detector, which records the chromatogram by input to a pen recorder in the normal way, and a mass spectrometer. Operational pressures within a mass spectrometer are around 10^{-5}–10^{-6} torr so an interface has to be provided to allow leakage of sample into the mass spectrometer without causing a detrimental increase in pressure. One way of doing this, described by Markey [114], is to position a glass-frit between the chromatograph and the spectrometer.

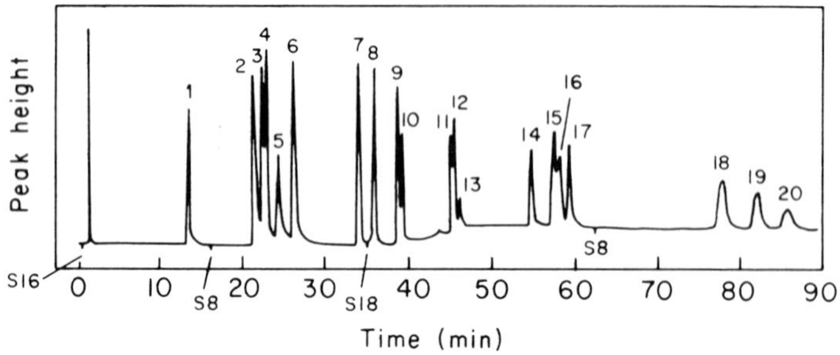

Chromatogram of a known mixture of polycyclics

Peak No.	Compound	Peak No.	Compound
1	Fluorene	11	Benz(a)anthracene
2	9-Fluorenone	(11)	Triphenylene
3	Phenanthrene	12	Chrysene
4	Anthracene	13	Naphthacene
5	Acridine	14	Benzo(k)fluoranthene
6	Carbazole	15	Benzo(a)pyrene
7	Fluoranthene	16	Benzo(e)pyrene
8	Pyrene	17	Perylene
9	11*H*-Benzo(a)fluorene	18	Dibenz(a,h)anthracene
10	11*H*-Benzo(b)fluorene	19	Benzo(ghi)perylene
(1)	Benz(a)anthracene	20	7*H*-dibenzo(c,g)carbazole

FIG. 4.33 Separation of PNHs by a temperature-programmed g.l.c. using an isotropic liquid stationary phase (1% OV-7) on Chromosorb W [102]

Once the sample enters the spectrometer ionization takes place, as described in section 4.2.5 according to equation (4.16). In the case of an organic compound these m^+ molecular ions, or parent ions as they are called, are unstable and breakdown into smaller fragments. The molecular structure and chemical composition of the parent ion influences the way in which the smaller fragments are produced. Every molecule gives its own characteristic fragmentation pattern in a mass spectrometer. Thus, this mass spectrum can be interpreted [115, 116] and the original parent compound identified. Mass spectra tables have been constructed for many molecules and should be used to verify an analysis whenever possible [117]. An example illustrating the mass spectrum of benzo(a)pyrene is shown in Table 4.23.

TABLE 4.23 *Mass spectrum of benzo(a)pyrene,* $C_{20}H_{12}$, *molecular weight 252·30. Electron accelerating voltage 70 V* [117]

m/e	Relative intensity[a]
112	6·92
113	9·43
124	6·52
125	15·30
126	23·00
250	16·20
252	100·00
253	21·40

[a] Only those relative intensities >6·0 are shown in the table.

Thus, the coupled g.c.–m.s. system enables the peaks on the chromatogram due to compounds eluting from the column to be simultaneously analysed in the mass spectrometer. Incredibly complex mixtures of organic compounds can now be analysed. As an example of such an analysis the identity of the 49 compounds separated by gas chromatography of an airborne coal tar sample (Fig. 4.34) are shown in Table 4.24.

4.3.3.6 *Specific chemical reagents for analysis of PNHs*

These procedures, which were developed in the 1950s and 1960s have now been almost completely superseded by the instrumental methods discussed in the previous sections. They depend on the use of specific reagents which react with PNHs and their derivatives to give coloured compounds, the absorbance maxima of which can be measured spectrophotometrically. Sawicki and Barry [118] have shown, for example, that aromatic aldehydes can be made to react, particularly with peri condensed systems, i.e. systems which have a carbon atom associated with three aromatic rings, to give diarylmethane dyes:

$$ArCHO + PCl_5 \rightarrow ArCHCl_2 \xrightarrow[POCl_3 \text{ or } CF_3COOH]{\text{pyrene}} Ar\overset{+}{C}H-$$

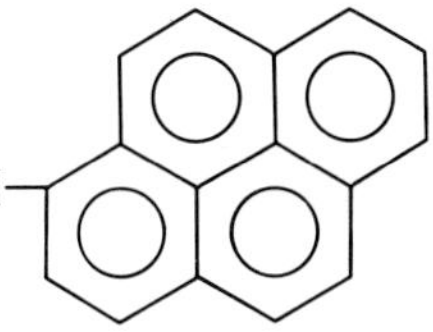

(4.26)

The phosphorus oxychloride indicated in equation (4.26) serves as a catalyst in the reaction. Various aldehydes have been investigated as reagents, e.g. furfural, 2-thenaldehyde, 3-nitro-4-dimethylaminobenzaldehyde, 9-anthraldehyde and indole-3-aldehyde.

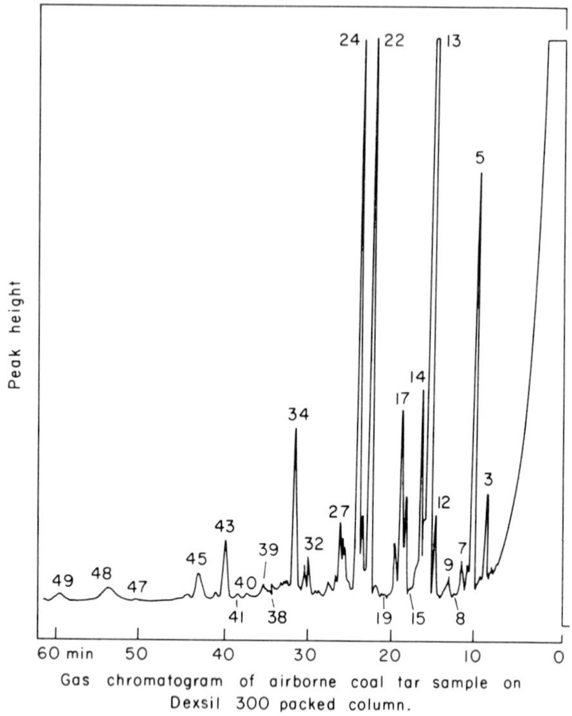

FIG. 4.34 Gas chromatogram of airborne coal tar sample on Dexsil 300 packed column [110]. The identification of the 49 compounds shown is given in Table 4.24.

1 ml of test solution containing PNH with 1 ml of aldehyde-phosphorus pentachloride reagent solution plus 0·5 ml of phosphorus oxychloride followed by dilution to a standard volume of 10 ml with trifluoroacetic acid results in the development of the colour. The coloured solution is then transferred to a 1 cm silica cell and the absorption maxima recorded in the normal manner against the reagent blank contained in another silica cell.

Table 4.25 lists the wavelengths at which maximum absorption occurs for different reagents with some typical PNHs [119]. In the case of the reagents anthraldehyde and 3-nitro-4-dimethylaminobenzaldehyde the

TABLE 4.24 *Separation of airborne coal tar sample on Dexsil 300: Compounds identified mass spectrometrically* [110]

Peak number	Compound	Concentration mg g^{-1} of sample	Concentration $\mu g\ m^{-3}$ of air sampled
3, 4	Benzindene	1·44	51
5	Fluorene	2·41	85
6, 7	Dihydrophenanthrene and dihydroanthracene	0·68	24
8, 9, 10, 11	Methylfluorenes	0·74	26
12, 13	Phenanthrene and anthracene	142·9	5020
14,	Acridine	8·44	297
15, 16	Methylphenanthrene and methylanthracene	2·18	77
17, 18	Ethylphenanthrene and ethylanthracene	3·88	136
19	Octahydrofluoranthene	0·15	5
20	Octahydropyrene	0·15	5
21	Dihydrofluoranthene and dihydropyrene	0·15	5
22	Fluoranthene	144·85	5090
23	Dihydrobenzofluorenes	8·85	311
24	Pyrene	105·47	3705
25, 26, 27	Benzofluorenes	24·53	862
28	Methylfluoranthenes	3·91	137
29	Methylpyrenes	3·88	136
30, 31	Trimethylfluoranthene and trimethylpyrene	1·29	45
32	Benzo(ghi)fluoranthene	3·29	115
33	Dihydrobenz(a)anthracene, dihydrochrysene and dihydrotriphenylene	2·65	93
34	Benz(a)anthracene, chrysene and triphenylene	45·74	1607
35, 36, 37	Dihydromethylbenz(a)anthracene, chrysene and triphenylene	0·50	18
38, 39	Methylbenz(a)anthracene, methylchrysene and methyltriphenylene	4·26	150
40	Dimethylbenz(a)anthracene, chrysene and triphenylene	0·41	15
41	Benzo(f)fluoranthene	0·12	4
43	Benzo(k)fluoranthene	9·96	350
44	Methylbenzo(k)fluoranthene and methylbenzo(j)fluoranthene	2·47	88
45	Benzo(a)pyrene and benzo(e)pyrene	13·83	486
46	Methylbenzo(a)pyrene and methylbenzo(e)pyrene	2·15	76
47	Dibenzanthracene	0·41	14
48	Benzo(b)chrysene and *o*-phenylenepyrene	3·88	137
49	Benzo(ghi)perylene and anthanthrene	2·62	92

TABLE 4.25 *Wavelengths at which maximum absorptions occur for coloured compounds formed between PNHs and various aldehydes. Compiled from data of reference 118*

PNH compound	Reagent aldehyde: Furfural (nm)	2-Thenaldehyde (nm)	Indole-3-aldehyde (nm)	9-Anthraldehyde (nm)	3-Nitro-4 dimethylamino-benzaldehyde (nm)
Fluoranthene		585			748
Acenaphthene		623		790	618
Pyrene	535, 630, 670	630s,[a] 670	690	850	702
Benzo(a)pyrene	685, 743	705, 750	750	910	770
Benzo(e)pyrene	573, 668	578, 684	685	860	710
Benzo(ghi)perylene	820	760		910	770
Anthranthrene	750s, 787	560, 775	790	955	845
Perylene	820	825	830	965	860
3-Methylcholanthrene					743

principal absorption lies outside the visible spectral region. The colours associated with these reagents are due to weak secondary bands which fall in the visible region. A modified version of this test employs piperonal and phosphorus pentachloride as the reagent (Table 4.26) [120].

Isatin tests for aromatic hydrocarbons and phenols have also been used [121]. The reaction with pyrene may be cited as an example:

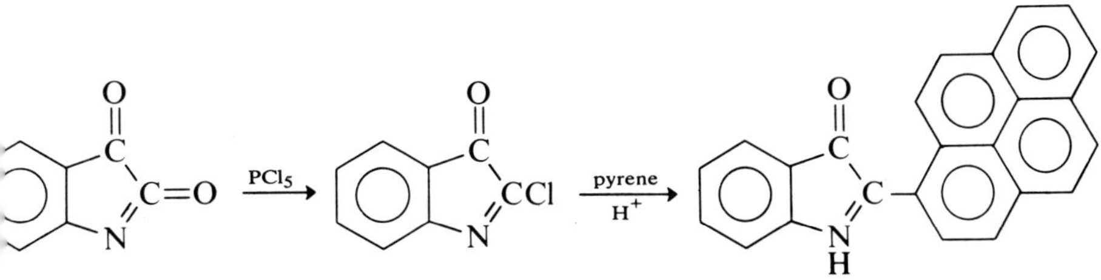

(4.27)

TABLE 4.26 *Colour and wavelength maxima of piperonal chloride reaction products with PNHs. Table compiled from data of reference 119*

Compound	Colour	Wavelength λ_{max} (nm)
Indole	red	502
Phenol	red	524
Phenyl ether	red	537
Dibenzofuran	violet	546
Fluorene	purple	580
Naphthalene	blue	590
Fluoranthene	blue	625
Pyrene	green	680
Benzo(e)pyrene	green	705
Chrysene	green	730
9-Methylanthracene	red	737
Benzo(a)pyrene	yellowish-brown	755
Benzo(ghi)perylene	red	760
3-Methylcholanthrene	yellow-brown	768
Anthanthrene	yellowish-brown	792

Once more the wavelength maximum typical of the derivative can be measured spectrophotometrically.

Somewhat similar procedures have been described by Sawicki *et al.* [122] for the detection of aniline, α- and β-naphthylamines, α- and β-anthramines as well as their *N*-alkyl and *N,N*-dialkyl derivatives using the coupling reaction with 4-azobenzenediazonium fluoborate. The process may be illustrated using *N,N*-dimethylaniline by the following scheme:

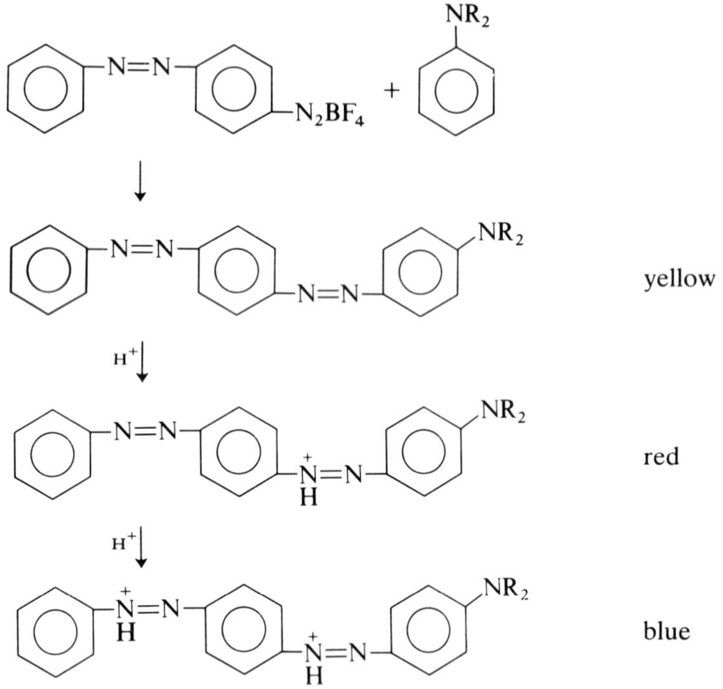

The compound is analysed spectrophotometrically at the blue stage since the blue colour of the dication is more intense than the red monocation.

Polynuclear hydrocarbons containing the fluorenic methylene group develop red, blue, violet or green colours when the fluorene moiety is treated with 0·5 ml of dimethylformamide and one drop of 10% aqueous tetraethylammonium hydroxide solution followed by 20 mg of powdered potassium borohydride with boiling. Upon cooling the solution reverts to a yellow colour [123]. Polycyclic *p*-quinones and fluorenones also give a positive test but these carbonyl compounds can be distinguished because they yield coloured derivatives without the addition of tetraethylammonium hydroxide [124].

Finally, a spot test procedure has been devised for pyrene and its derivatives [125]:

The sample is treated with 0·1 ml of 1:1 mixture of acetic acid-fuming nitric acid and heated on a steam bath for 5 min. After evaporating to dryness 1 drop of dimethylformamide is added to dissolve the residue. One drop of 25% aqueous tetraethylammonium hydroxide solution is placed on a spot-plate and then into the centre of this spot place 1 μg of the dimethylformamide solution of the residue. A blue spot indicates the presence of pyrene. 1-Chloro, 1-bromo, 1-acetamido and 1-trifluoracetamide derivatives give negative results.

4.3.4 Reflectance methods

Filters used for the collection of atmospheric particulates become blackened after exposure by the accumulation of dust on their surfaces. The extent of this blackening is obviously related to the concentration of airborne particulate matter that has been filtered from the air sampled. This has been expressed by Sanderson *et al.* [126] as

$$W = aD^b \tag{4.28}$$

where W is the weight deposited, D is the absorbance of the stain $\times$ 100 and a and b are constants. The constants a and b vary from day to day because they are dependent upon the characteristics of emission sources in the vicinity monitored and upon dispersion influences controlled by meteorological parameters.

The blackness index of a filter is measured by reflectance using a reflectometer and is defined as 100 less the reflectometer reading. A standard curve may be prepared relating blackness index with smoke concentration so that from a particular blackness reading the corresponding smoke concentration can be estimated. Since PNHs are generally adsorbed on soot and particulate matter, attempts have been made to correlate the blackness index with the concentration of a particular polynuclear hydrocarbon, e.g. benzo(a)pyrene present in the air [127]. If reliable, the method would be useful for the routine estimation of PNHs because the blackness of the filter may be measured with a reflectometer in a few minutes, whereas the conventional analysis of a PNH requires much more time and effort.

Figure 4.35 shows a typical plot of the blackness index against the amount of benzo(a)pyrene analysed on filters collected from a monitoring site. Although such a curve is reasonably self-consistent for a location within a limited time duration, another location will show a different pattern and, hence, unless extensive calibrations are performed at each location little trust can be put in the method.

An exception to this conclusion concerns the possible application of the method for the analysis of seasonal trends in atmospheric PNH content. Figure 4.36 shows the results from a monitoring station in which the benzo(k)fluoranthene content of the filtered particulate has been plotted against the blackness index. The figure shows that results are dependent on the month in which the sample is taken: the benzo(k)fluoranthene in March in this example is 17 times higher than the August value.

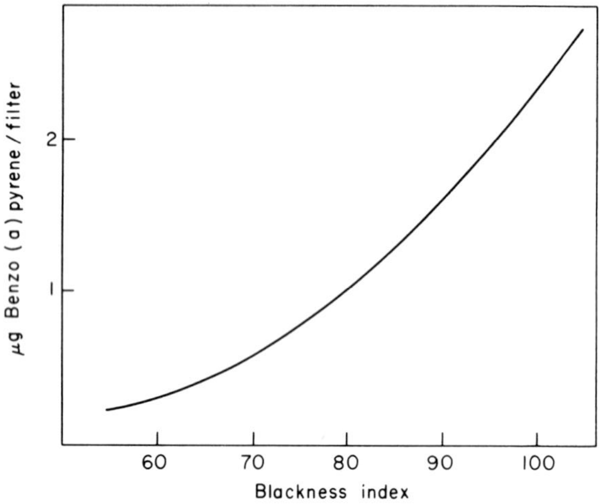

FIG. 4.35 Typical plot of the blackness index measured with a reflectometer and the amount of benzo(a)pyrene in μg analysed on a filter paper.

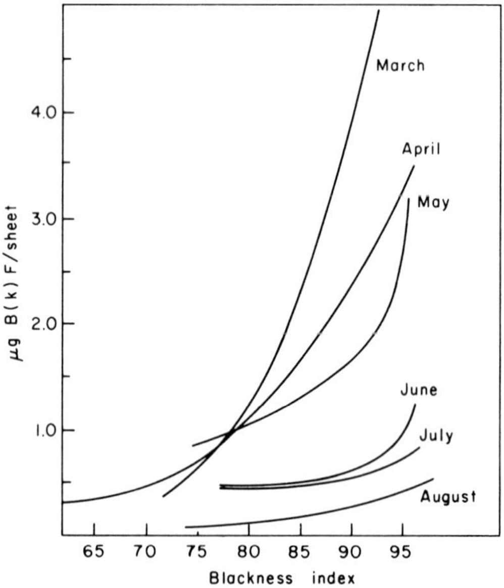

FIG. 4.36 Seasonal variation of the blackness index with the amount of benzo(k)fluoranthene in μg analysed on a filter paper [127].

4.4 Detection and determination of asbestos

Airborne asbestos determinations differ from all other methods discussed in this chapter, in that ultimately a microscope is needed to view, recognize and count asbestos fibres in the sample collected. The method is normally not sensitive enough for ambient air except in the vicinity of a known asbestos source. These are places where occupational exposure may be a problem, as in a factory or a workshop, or possibly on a building construction site where asbestos-containing materials are being used. The results are expressed in terms of the number of fibres present in 1 cm^3 of air (fibres cm^{-3}) where a fibre is defined as being longer than 5 μm and having a length-to-breadth ratio of at least 3:1. There is no upper limit to the length of a fibre but the diameter must not be greater than 3 μm.

4.4.1 Analysis of airborne asbestos: microscope procedure

The sample is collected by air filtration at a flowrate of 200 cm^3 min^{-1} onto a white gridded membrane filter of 25 mm diameter with a pore size of 0·8 μm [128]. The sample volume of air required will depend upon the asbestos dust concentration. Some indications of the volume required are given in Table 4.27. After collection, the dust deposit on the filter is fixed

TABLE 4.27 *A guide showing the sample volumes of air required for the analysis of asbestos dust concentrations (fibres cm^{-3})*

Asbestos dust concentration (fibres cm^{-3})	Sample volume (cm^3)
<2	10 000–20 000
2–4	5 000–10 000
4–12	2 000–5 000

whilst still in the sample holder by spotting several drops of a polymethylmethacrylate solution (0·025% in chloroform) onto the membrane while clean air is being drawn through it by means of a low flowrate pump: a water pump is suitable. Finally, the sample is mounted by placing a few drops of glyceryltriacetate onto a clean microscope slide and allowing the liquid to spread into a circle the same diameter as the filter. The filter is removed from the filter holder with forceps and placed dust-side uppermost on the glyceryltriacetate spot on the plate. After leaving for 3 min a clean 25 mm diameter cover slip is placed over the sample and gentle pressure applied to remove any trapped air bubbles underneath the cover slip. The sample is set aside for about 30 min to allow the filter to clear before starting the count.

When clear, the sample is examined by transmitted light at a magnification of about 500× in a suitable microscope. For an evenly distributed dust deposit, count the number of fibres within a field of view of the microscope. Repeat the count for randomly selected fields. Note the number of fields needed for a total count of about 200 fibres. Sometimes a decision must be made about whether a fibre is within or otherwise a field of view. If more than half the fibre is in the field it is included in the count. If the deposit is sparse or unevenly distributed it is recommended that 100 fields are observed. This can be done by selecting 10 filter grids scattered over the sample area and counting 10 random fields within each grid.

Let

D = the diameter of the dust deposit
d = the diameter of the field of view in the microscope

then the dust concentration is given by:

$$C_f = \frac{D^2}{d^2} \times \frac{N}{n} \times \frac{1}{V} \text{ fibres cm}^{-3} \tag{4.29}$$

where

N = number of fibres counted
n = number of fields examined

and

V = volume of air sample collected

4.4.2 Dye adsorption procedure

In contrast to the method just given, Markham and Wosczyna [129] have described another technique which may be developed further in future. In this method suspended asbestos fibres in water are heated at 50–60°C for 1 h with aluminon dye. Chrysotile and crocidolite forms of asbestos are dyed a purple rose colour by the reagent and the dye concentration of the supernatant liquid becomes depleted. The mixture is centrifuged to separate the dyed asbestos fibres and the absorbance of the supernatant liquid at 525 nm is measured with reference to the original dye solution in a spectrophotometer. Estimation of quantities of chrysotile asbestos down to the 100 μg level are claimed.

4.5 Analysis of airborne radioactivity

The only instruments of a specialist nature required for the estimation of airborne radioactivity are the counters. These are always a feature of an analysis of a radioactive material, airborne or otherwise. Unlike the developments that have taken place for the analysis of other pollutants, the method for monitoring radioactive materials in the air has not changed very much over the last decade or two. The reasons for this, of course, lie in the very nature of the property of the element being determined, namely the radioactivity.

Most of the radioactivity in the atmosphere is associated with rainfall, or dusts and particulates. The usual procedures for the collection of such samples—deposit gauges and filters through which known volumes of air are drawn—may be used. Once collected, the simplest measurement of radioactivity that can be determined is the α- and/or β-content of the sample. If further specific information is needed on individual nuclides then these must be separated and counted independently. The separations are performed chemically, so often many precipitation and filtration stages are required before a sample of sufficient purity can be obtained.

4.5.1 Analysis of total radioactivity

After collection on a filter paper, the activity is counted using a scintillation counter. Scintillation counters tend to be more widely used in air pollution studies than Geiger–Muller detectors. The scintillation detector consists of a phosphor which is activated when radioactive particles hit or pass through the surface. The light flash produced is monitored by a photomultiplier tube. The number of flashes as well as their intensity can be obtained with such a counter.

For such total sample activity, γ-emitters are the most commonly counted. Developments in scintillation counters for specialized applications, have lead to the replacement of traditional phosphors, e.g. ZnS activated by Ag, with semiconductor devices. The introduction of the lithium-doped germanium detector has lead to a 30- to 40-fold improvement in spectral resolution over the sodium iodide scintillator. When coupled to a multichannel analyser, the γ-ray spectrometer (cf. section 4.2.2) will yield a spectrum in which the different nuclides present in the sample can be identified. After allowing for the decay of the shorter half-life material (Ce-141 and Ru-103), it is possible, for example, to discriminate the γ-emission from Cs-137, Sb-125, Ru-106, Ce-144, Zr-95 and Mn-54.

α-Particles have a limited penetrating power and consequently can be absorbed by the material of the filter and by other dust particles that are also on the filter. To help minimize these effects membrane filters are recommended and the filter loading should not exceed about 1 mg cm^{-2}.

After exposure a filter can have appreciable activity due to the decay of radon and thoron escaping from rocks and soil. Immediate counting after sample collection can give high counts due to the presence of short-lived radioactive material. By delaying the count for several days or alternatively doing a series of counts at least 12 h apart a correction can be made. The activity of the long-lived α-emitters in the sample can be obtained from the equation [130]:

$$C = \frac{C_2 - C_1\,e^{-\lambda\Delta t}}{1 - e^{-\lambda\Delta t}} \tag{4.30}$$

Where

C = activity of long-lived material
C_1 = activity on filter at time t_1
C_2 = activity on filter at time t_2
Δt = time between counts $t_2 - t_1$ (hours)
λ = decay constant for $^{212}Pb = 0{\cdot}693/10{\cdot}6 = 0{\cdot}0653\ h^{-1}$

If a single isotope is being counted a plot of log of the activity against time will be linear. For a sample consisting of two different isotopes with different half-lives, the log of the total activity against time will follow a curve. The initial part of the curve is determined by the shorter half-life material. When the air pollutant being monitored comprises the "mixed fission products" from a nuclear explosion a log-log plot of activity *vs* time will be a straight line with slope $-1{\cdot}2$.

When very short-lived radioactive material is being collected the activity of the filter will increase rapidly at first but the count will eventually become constant. This is because an equilibrium is established between new material trapped on the filter and the decay of material already captured by the filter. At equilibrium the activity on the filter A is given by

$$A = \frac{CF\lambda_{1/2}}{0{\cdot}693} \tag{4.31}$$

where

C = air concentratin of radioactive material,
F = flowrate of air through the filter,

and

$\lambda_{1/2}$ = half-life of shorter lived material collected.

During the actual counting of the sample the following procedures must be observed:

(1) The filter paper should be placed in a holder to preserve the same distance between the counter and filter for every count.
(2) Counts should be continued for a sufficiently long time period to ensure a statistically reliable result.
(3) The counter efficiency should be measured using a standard uranium solution for α-emitters and a Cs-137 solution for β-emitters.
(4) Determine the background count for at least as long as the sampling time.

The result is obtained in the following manner [131],

the counter efficiency e is defined by

$$e = \frac{\text{net counting rate of standard (cpm)}}{\text{disintegration rate of standard (dpm)}} \tag{4.32}$$

Net sample disintegration rate (dpm)

$$= \frac{\dfrac{\text{total sample count}}{\text{time taken to count sample (min)}} - \dfrac{\text{background count}}{\text{time taken to count background}}}{\text{counter efficiency}} \tag{4.33}$$

and then the activity (pCi m^{-3})

$$= \frac{\text{sample disintegration rate from equation (4.33) (dpm)}}{\text{volume sampled (m}^3\text{)} \times 2{\cdot}2 \text{ (dpm/pCi)}} \tag{4.34}$$

the standard deviation of the result SD can be quoted from

$$\text{SD (cpm)} = \pm\left[\frac{\text{total counts}}{\text{time taken to count sample}} + \frac{\text{background count}}{\text{time taken to count background}}\right]^{1/2} \tag{4.35}$$

and

$$\text{SD (dpm)} = \pm\frac{\text{SD (cpm)}}{e} \tag{4.36}$$

4.5.2 Separation and activity determinations of specific nuclides

The three most important fission products are strontium-90, caesium-137 and iodine-131. Of these strontium-90 is normally regarded as the most dangerous because of its half-life of 28 years and its ability to replace calcium in bone, as already mentioned in section 1.6.4. Only an outline of the procedures that have been adopted for the determinations of these elements in air pollution samples will be given here. For complete details the reader should consult reference 131 or reference 13 of Chapter 3.

Strontium-90 can be analysed from a solution prepared by either digesting the filter and dust particulate with aqua regia or ashing at 900°C followed by a sodium carbonate fusion. A strontium carrier is added to this solution and the strontium first precipitated as the carbonate and then as the nitrate. Further purification is achieved by precipitation of barium as chromate and treatment with ammonium hydroxide. An yttrium carrier is added to the solution and the yttrium extracted with 2-thenoltrifluoroacetone in monochlorobenzene at pH 5. After strontium-90 and yttrium-90 equilibration, the yttrium-90 is precipitated as the hydroxide converted to the oxalate and the β-particles from yttrium counted. The strontium is precipitated as the carbonate and analysed by flame photometry or by gravimetry to calculate the strontium recovery.

Caesium-137 can be analysed from a rainwater sample by addition of caesium chloride and potassium chloride as carriers, followed by precipitation of the alkali metals as their cobaltinitrites from an acetic acid solution. After separation of the precipitate they are redissolved in 6 N HCl and the caesium reprecipitated as the silicotungstate to isolate caesium from the other alkali metals. Perchloric acid is added and silicon and tungsten oxides are separated by centrifuging the hot solution. Caesium is obtained by addition of alcohol to the supernatant liquor as a precipitate of caesium perchlorate. After collecting the precipitate, washing with alcohol, mounting, drying and weighing, the β-activity is counted.

Controversy concerning plutonium-based economies for fulfilling nations' future energy requirements is currently being voiced. One of the important considerations evolves around the need for adequate and fail-safe containment procedures for plutonium, because it is a potent inhalation carcinogenic agent. Losses to atmosphere and the spread of airborne particulate containing plutonium will cause an increase in lung cancer in human populations. The most dangerous stage from the containment aspect is the escape of plutonium dioxide dust. This can happen at the fuel reprocessing, fuel refabrication and transportation stages between these operations rather than at the nuclear reactor itself. In view of this awareness, a brief outline of the analytical procedures necessary for airborne plutonium analysis are indicated below.

The method depends upon the equilibration of plutonium in particulate collected by filtration of about 10^4 m^3 of air with plutonium–236 added as a tracer. Plutonium is isolated by co-precipitation with cerium and yttrium fluorides which are added as tracers. The plutonium is purified by a two-stage ion exchange technique which is followed by electro-deposition onto a platinum electrode. This electrode is finally counted in an α-spectrometer in which the plutonium isotopes −236, −238 and −239 are resolved.

4.5.3 *Analysis of radioactive vapours and gases*

Iodine-131 analysis is carried out by drawing a known volume of air through a particulate filter backed by a charcoal trap. The charcoal trap can conveniently be prepared from about 3 g of 12–30 mesh activated charcoal packed into a tube about 1 cm diameter and 3 cm long. A flowrate of about $10\,l\,min^{-1}$ for a period of 1 week will be needed to obtain a count equivalent to 0·1 $pCi\,m^{-3}$. Counting is done with an α-scintillation counter coupled to a multi-channel analyser to measure the 0·364 MeV radiation of iodine-131.

Non-particulate airborne radioactivity mainly arises from $^{14}CO_2$, and the rare gas isotopes argon-37 and −39, krypton-81 and −85, radon-220 and −222. $^{14}CO_2$ can be collected away from industrial areas merely by exposing an 8 M KOH solution to the atmosphere for about 1 month. During this time carbon dioxide will be absorbed from the atmosphere and retained as potassium carbonate. The carbon dioxide can be recovered by acidification of the potassium carbonate solution, purified and counted. Alternatively, the regenerated carbon dioxide may be converted to a hydrocarbon, such as methane, ethane or benzene before being counted [132, 133]. Gaseous radioactive materials require a proportional counter for their determination. Such counters are used for the rare gas isotopes argon and krypton after their isolation from air [134].

The hydrogen isotope tritium is present in the atmosphere as water vapour and can be analysed using a liquid scintillation β-counter with a scintillation solution based on 2,5-diphenyloxazol, i.e. *p*-bis-2-(5-phenyl-oxazolyl)benzene and naphthalene in *p*-dioxane [135]. The sample is collected by drawing air at about $150\,ml\,min^{-1}$ through a column of non-indicating silica gel for about 2 weeks. The silica gel column is prepared from 180 g of 6–16 mesh silica gel which has been activated at 150°C overnight before being packed into a tube 30 cm long by 6 cm diameter. After exposure, the tube is sealed with rubber end-caps and returned to the laboratory for analysis. This is carried out by emptying the silica gel into a distillation flask and heating to drive the moisture into a collection vessel. 4 ml of this distillate are added to 20 ml of the scintillation solution contained in a polyethylene phial and the solution counted with a liquid scintillation β-counter. Standards are prepared by addition of tritiated water with distilled tritium free water so that between 10^3 and 10^4 dpm cm^{-3} are obtained.

4.6 Statistical methods for the presentation of results

Submission of results from any monitoring programme should be accompanied by a statistical evaluation of their worth. The analysis of atmospheric particulate lead will serve as an illustration. Such analyses are usually performed by drawing air through a filter for a specific time period, say 12 h, followed by dissolution of the sample and determination of lead by atomic absorption spectrophotometry using either the 217·0 nm or the 283·3 nm lead lines. One month's analyses can be averaged by totalling all the individually analysed samples, Σw and then dividing by the number of samples, N. This procedure does not allow any limits of reliability above or below the quoted value to be given. Some knowledge of these limits can be important because even when the most exacting precautions are taken some variation is bound to occur between one analysis and another.

By way of example, suppose one laboratory is responsible for the simultaneous collection and analysis from two different districts of a city and the results are reported as averages in the manner just described. If the results differ, then is this difference genuinely due to some different environmental factors from the two locations, or are there other reasons? A statistical appraisal is helpful in making a decision.

Let

w = value of each sample in μg Pb m^{-3} of air,
N = number of samples,

and

M = the mean of these N samples,

then the standard deviation SD is given by

$$\mathrm{SD} = \left[\frac{\Sigma w^2}{N} - M^2\right]^{1/2} \tag{4.37}$$

and the standard error of the mean SD (mean) is given by

$$\mathrm{SD\ (mean)} = \frac{\mathrm{SD}}{(N-1)^{1/2}} \tag{4.38}$$

The limits to which the mean value M are subject are given by multiplying SD (mean) by the appropriate t value for the number of samples considered (Table 4.28).

TABLE 4.28 "t" *Values for significance at the 10%, 5%, 1% and 0·2% levels*[a]

Number of samples	$P = 10\%$	$P = 5\%$	$P = 1\%$	$P = 0{\cdot}1\%$
3	t = 2·92	4·30	9·92	22·33
4	2·35	3·18	5·84	10·21
5	2·13	2·78	4·60	7·17
6	2·02	2·57	4·03	5·89
7	1·94	2·45	3·71	5·21
8	1·89	2·36	3·50	4·79
9	1·86	2·31	3·36	4·50
10	1·83	2·26	3·25	4·30
11	1·81	2·23	3·17	4·14
12	1·80	2·20	3·11	4·02
13	1·78	2·18	3·05	3·93
14	1·77	2·16	3·01	3·85
15	1·76	2·14	2·98	3·79
16	1·75	2·13	2·95	3·73
17	1·75	2·12	2·92	3·69
18	1·74	2·11	2·90	3·65
19	1·73	2·10	2·88	3·61
20	1·73	2·09	2·86	3·58
21	1·72	2·09	2·85	3·55
22	1·72	2·08	2·83	3·53
23	1·72	2·07	2·82	3·50
24	1·71	2·07	2·81	3·48
25	1·71	2·06	2·80	3·47
26	1·71	2·06	2·79	3·45
27	1·71	2·06	2·78	3·44
28	1·70	2·05	2·77	3·42
29	1·70	2·05	2·76	3·41
30	1·70	2·05	2·76	3·40
40	1·68	2·02	2·70	3·31
60	1·67	2·00	2·66	3·23
120	1·66	1·98	2·62	3·16
∞	1·64	1·96	2·58	3·09

Adapted from Pearson, E.S. and Hartley, H. O. (1966). "Biometrika Tables for Statisticians", Vol. 1, Table 12. Cambridge University Press.

[a] A value quoted at the 5% significance level, for example, means that the value will fall within the range specified in 95 out of every 100 cases.

To decide whether there is a significant difference between two sets of data, in which the number of samples considered are comparable, then t is calculated from the equation:

$$t = \frac{M_2 - M_1}{\dfrac{N_1(SD_2)^2 + N_2(SD_2)^2}{N_1 + N_2 - 2} \times \dfrac{N_1 + N_2}{N_1 N_2}} \tag{4.39}$$

where the suffixes 1 and 2 denote the different sets of values. If this t value is greater than that given in Table 4.28 for the number of samples considered, then there is a significant difference. When the value is less than the table value the differences are not significant at the 5% level.

Table 4.29 gives the results of atmospheric particulate lead in $\mu g\ m^{-3}$ monitored daily at two different sites in a city. The mean of the 21 determinations is 2·84 $\mu g\ m^{-3}$ at site 1 and 2·69 $\mu g\ m^{-3}$ at site 2. The

TABLE 4.29 *Daily atmospheric lead analyses in* $\mu g\ m^{-3}$ *reported at two different locations in a city on the same days in one month*

Site 1		Site 2	
w_1	w_1^2	w_2	w_2^2
3·12	9·73	5·21	27·14
2·84	8·06	4·73	22·37
3·72	13·84	3·05	9·30
3·98	15·84	2·55	6·50
2·92	8·53	2·68	7·18
2·41	5·81	2·88	8·29
2·09	4·37	3·34	11·15
3·10	9·61	2·20	4·84
3·84	14·74	4·42	19·54
1·74	3·03	3·77	14·21
5·95	35·40	3·38	11·42
0·95	0·90	1·58	2·50
1·79	3·20	0·61	0·37
1·98	3·92	0·79	0·62
2·73	7·45	1·18	1·39
1·02	1·04	1·57	2·46
1·69	2·86	1·24	1·54
2·37	5·62	1·78	3·17
6·36	40·45	7·81	61·00
2·40	5·76	0·88	0·77
2·60	7·29	0·86	0·74

M_1	2·84		M_2	2·69	
M_1^2	8·06		M_2^2	7·24	
Σw_1^2		207·4	Σw_2^2		216·5

$$\text{SD} = \left(\frac{207{\cdot}45}{21} - 8{\cdot}06\right)^{1/2} = 1{\cdot}35 \qquad \text{SD} = \left(\frac{216{\cdot}5}{21} - 7{\cdot}24\right)^{1/2} = 1{\cdot}75$$

$$\text{SD (mean)} = \frac{1{\cdot}35}{20^{1/2}} = 0{\cdot}30 \qquad \text{SD (mean)} = \frac{1{\cdot}75}{20^{1/2}} = 0{\cdot}39$$

Limits $= \pm 2{\cdot}1 \times 0{\cdot}3 = \pm 0{\cdot}63$
Giving $M_1 = 2{\cdot}84 \pm 0{\cdot}63\ \mu g\ m^{-3}$

Limits $= \pm 2{\cdot}1 \times 0{\cdot}39 = \pm 0{\cdot}82$
Giving $M_2 = 2{\cdot}69 \pm 0{\cdot}82\ \mu g\ m^{-3}$

calculations shown in Table 4.29 using equations (4.37 and 4.38) indicate that the value of 2·84 can be $2{\cdot}84 \pm 0{\cdot}63\ \mu g\ m^{-3}$ and that 2·69 can be $2{\cdot}69 \pm 0{\cdot}82\ \mu g\ m^{-3}$. To test whether there is any significant difference between these two sites on this evidence substitution into equation (4.39) gives,

$$t = \frac{2{\cdot}84 - 2{\cdot}69}{\dfrac{21 \times 1{\cdot}35^2 + 21 \times 1{\cdot}75^2}{21 + 21 - 2} \times \dfrac{21 + 21}{21 \times 21}} = 0{\cdot}6$$

As we might have guessed from the results of the first calculation, since $0{\cdot}6 < 2{\cdot}1$ there is no significant difference between the lead pollution concentrations at the two sites.*

For further reading on the application of statistical methods to pollution problems see Pratt [136].

References

1. Hochheiser, S. (1964). "Methods of measuring and monitoring atmospheric sulfur dioxide." US Department of Health, Education and Welfare, August 1964.
2. Hochheiser, S., Santner, J. and Ludmann, W. F. (1966). The effect of analytical method on indicated atmospheric SO_2 concentrations. *J. Air Pollut. Contr. Assoc.* **16**, 266–271.
3. Porter, K. and Volman, D. H. (1962). Flame ionization of carbon monoxide for gas chromatographic analysis *Anal. Chem.* **34**, 748–749.
4. Karmen, A. and Giuffrida, L. (1964). Enhancement of the response of the hydrogen flame ionization detector to compounds containing halogens and phosphorus. *Nature* **201**, 1204–1205.
5. Aue, W. A., Gehrke, C. W., Tindle, R. C., Stalling, D. L. and Ruyle, C. D. (1967). Application of the alkali-flame detector to nitrogen containing compounds. *J. Gas Chromat.* **5**, 381–382.
6. Crider, W. L. (1965). Hydrogen flame emission spectrophotometry in monitoring air for sulfur dioxide and sulfuric acid aerosol. *Anal. Chem.* **37**, 1770–1773.
7. Brody, S. S. and Chaney, J. E. (1966). Flame photometric detector: The application of a specific detector for phosphorus and sulfur compounds—sensitive to sub-nanogram quantities. *J. Gas Chromat.* **4**, 42–46.
8. Stevens, R. K., Mulik, J. D., O'Keeffe, A. E., and Krost, K. J. (1971). Gas chromatography of reactive sulfur gases in air at the ppb level. *Anal. Chem.* **43**, 827–831.
9. Prager, M. J. and Seitz, W. R. (1975). Flame emission photometer for determining phosphorus in air and natural water. *Anal. Chem.* **47**, 148–151.
10. Lovelock, J. E. and Lipsky, S. R. (1960). Electron affinity spectroscopy. A new method for identification of functional groups in chemical compounds separated by gas chromatography. *J. Amer. Chem. Soc.* **82**, 431–433.

* See "Note added in proof", p. 380, for clarification of the term correlation coefficient.

11. Lovelock, J. E., Maggs, R. J. and Adlard, E. R. (1971). Gas-phase coulometry by thermal electron attachment. *Anal. Chem.* **43**, 1962–1965.
12. Singh, B. H. and Lillian, D. (1974). Absolute determination of atmospheric halocarbons by gas phase coulometry. *Anal. Chem.* **46**, 1060–1063.
13. Steffenson, D. M. and Stedman, D. H. (1974). Optimization of operating parameters of chemiluminescence nitric oxide detectors. *Anal. Chem.* **46**, 1704–1708.
14. Clyne, M. A. A., Thrush, B. A. and Wayne, R. P. (1964). Kinetics of chemiluminescent reaction between nitric oxide and ozone. *Trans. Faraday Soc.* **60**, 359–370.
15. Clough, P. N. and Thrush, B. A. (1967). Mechanism of chemiluminescent reaction between nitric oxide and ozone. *Trans. Faraday Soc.* **63**, 915–925.
16. Fontijn, A., Sabadell, A. J. and Ronco, R. J. (1970). Homogeneous chemiluminescent measurement of nitric oxide with ozone. *Anal. Chem.* **42**, 575–579.
17. Magee, P. N. and Barnes, J. M. (1956). Production of malignant primary hepatic tumours in the rat by feeding dimethylnitrosamine. *Brit. J. Cancer* **10**, 114–122.
18. Fine, D. H., Rounbehler, D. P., Sawicki, E., Krost, K. J. and DeMarrais, G. A. (1976). N-nitroso compounds in the ambient community air of Baltimore, Maryland. *Anal. Lett.* **9**, 595–609.
19. Fine, D. H., Rufeh, F., Lieb, D. and Rounbehler, D. P. (1975). Description of non-volatile N-nitroso compounds. *Anal. Chem.* **47**, 1188–1191.
20. Gough, T. A. and Webb, K. S. (1973). A method for the detection of traces of nitrosamines using combined mass spectrometry and gas chromatography. *J. Chromat.* **79**, 57–63.
21. Stauff, J. and Jaeschke, W. (1975). A chemiluminescence technique for measuring atmospheric trace concentrations of sulphur dioxide. *Atmos. Environ.* **9**, 1038–1039.
22. McClenny, W. A., Martin, B. E., Baumgardner, R. E., Stevens, R. K. and O'Keeffe, A. E. (1976). Detection of vinyl chloride and related compounds by a gas chromatographic, chemiluminescence technique. *Environ. Sci. and Technol.* **10**, 810–813.
23. Nederbragt, G. W., Van der Horst, A. and Van Duijn, J. (1965). Rapid ozone determination near an accelerator. *Nature* **206**, 87.
24. Warren, G. J. and Babcock, G. (1970). Portable ethylene chemiluminescence ozone monitor. *Rev. Sci. Instrum.* **41**, 280–282.
25. Kummer, W. A., Pitts, J. N. and Steer, R. P. (1971). Chemiluminescent reactions of ozone with olefins and sulfides. *Environ. Sci. and Technol.* **5**, 1045–1047.
26. Hager, R. N. (1973). Derivative spectroscopy with emphasis on trace gas analysis. *Anal. Chem.* **45**, 1131A–1138A.
27. Stanley, C. W., Barney, J. E., Helton, M. R. and Yobs, A. R. (1971). Measurement of atmospheric levels of pesticides. *Environ. Sci. and Technol.* **5**, 430–435.
28. Singh, B. H., Lillian, D. and Appleby, A. (1975). Absolute determination of phosgene: Pulsed flow coulometry. *Anal. Chem.* **47**, 860–861.
29. Evans, K. P., Mathias, A., Mellor, N., Silvester, Ṙ. and Williams, A. E. (1975). Detection and estimation of bis-(chloromethyl)ether in air by gas chromatography–high resolution mass spectrometry. *Anal. Chem.* **47**, 821–824.

30. Chau, Y. K., Wong, P. T. S. and Goulden, P. D. (1975). G-C and AA method for the determination of dimethyl selenide and dimethyl diselenide *Anal. Chem.* **47**, 2279–2281.
31. LaHue, M. D., Axelrod, H. D. and Lodge, J. P. (1973). Measurement of atmospheric nitrous oxide using a molecular sieve 5A trap and gas chromatography. *Anal. Chem.* **43**, 1113–1115.
32. Lonneman, W. A. (1977). PAN measurement in dry and humid atmospheres. *Environ. Sci. and Technol.* **11**, 194–195.
33. Ross, W. D. and Sievers, R. E. (1968). Rapid ultra-trace determination of beryllium by gas-chromatography. *Talenta* **15**, 87–94.
34. Ross, W. D. and Sievers, R. E. (1972). Environmental air analysis for ultra-trace concentrations of beryllium by gas chromatography. *Environ. Sci. and Technol.* **6**, 155–158.
35. Belcher, R. and Blessel, K., Cardwell, T., Pravica, M., Stephen, W. I. and Uden, P. C. (1973). Volatile transition metal complexes of bis-acetylacetone-ethylenedi-imine and its fluorinated analogues. *J. Inorg. Nucl. Chem.* **35**, 1127–1144.
36. O'Keeffe, A. E. and Ortman, G. C. (1966). Primary standards for trace gas analysis. *Anal. Chem.* **38**, 760–763.
37. Middleton, D. R. (1976). Studies of atmospheric pollutants in urban districts. Ph.D. Thesis, University of Aston, Birmingham.
38. Stephens, E. R. (1964). Absorptivities for infrared determinations of peroxyacylnitrates. *Anal. Chem.* **36**, 928–929.
39. Hewitt, P. J. (1972). Instrumental neutron activation analysis of airborne contaminants using Ge–Li detectors. *Ann. Occup. Hyg.* **15**, 341–348.
40. Woldseth, R. (1973). "X-ray energy spectrometry." Kevex Corporation, Burlingame, California, 1st Edition.
41. Johnson, G. G. and White, E. W. (1970). "X-ray emission wavelengths and keV tables for non-diffractive analysis." ASTM Data Series DS 46.
42. Butler, J. D., Macmurdo, S. D. and Stewart, C. J. (1976). Characterization of aerosol particulates by scanning electron microscope and X-ray energy fluorescence analysis. *Int. J. Environ. Stud.* **9**, 93–103.
43. Slavin, W. (1968). "Atomic Absorption Spectroscopy". Interscience Publishers.
44. Massmann, H. (1968). Vergleich von Atomabsorption und Atomfluoreszenz in der Graphitkuvette. *Spectrochim. Acta.* **23B**, 215–226.
45. Matousek, J. P. and Brodie, K. G. (1973). Direct determination of lead airborne particulates by non-flame atomic absorption. *Anal. Chem.* **45**, 1606–1609.
46. Ranweiler, L. E. and Moyes, J. L. (1974). Atomic absorption procedure for analysis of metals in atmospheric particulate matter. *Environ. Sci. and Technol.* **8**, 152–156.
47. Commins, B. T. and Lawther, P. J. (1958). Volatility of 3,4-benzpyrene in relation to the collection of smoke samples. *Brit. J. Cancer* **12**, 351–354.
48. Commins, B. T. (1962). "Interim report on the study of techniques for determination of polycyclic aromatic hydrocarbons in air." Natl. Cancer Instit. Monograph No. 9, 225–233.
49. Murray, J. J., Pottie, R. F. and Pupp, C. (1974). The vapour pressures, enthalpies of sublimation of five polycyclic aromatic hydrocarbons. *Canad. J. Chem.* **52**, 557–563.

50. Murray, J. J., Pottie, R. F., Pupp, C. and Lao, R. C. (1974). Equilibrium vapour concentrations of some poly aromatic hydrocarbons, As_4O_6 and SeO_2 and the collection efficiencies of these air pollutants. *Atmos. Environ.* **8**, 915–925.
51. Thomas, J. F., Mukai, M. and Tebbens, B. D. (1968). Fate of airborne benzo(a)pyrene. *Environ. Sci. and Technol.* **2**, 33–39.
52. Brockhaus, A. (1974). Sampling of tetracyclic aromatic hydrocarbons under atmospheric conditions. *Atmos. Environ.* **8**, 521.
53. Gordon, R. J. (1974). Solvent selection in extraction of airborne particulate matter. *Atmos. Environ.* **8**, 189–191.
54. Grosjean, D. (1975). Solvent extraction and organic carbon determination in atmospheric particulate matter: The organic extraction-organic carbon analyzer (OE-OCA) technique. *Anal Chem.* **47**, 797–805.
55. Monkman, J. L., Moore, G. E. and Katz, M. (1962). Analysis of polycyclic hydrocarbons in particulate pollutants. *Amer. Ind. Hyg. Assoc. J.* **23**, 487–495.
56. Falk, H. L., Kotin, P. and Mehler, A. (1964). Polycyclic hydrocarbons as carcinogens for man. *Arch. Environ. Health* **8**, 721–730.
57. Falk, H. L. and Kotin, P. (1963). Atmospheric factors in the pathogenesis of the lung. *Adv. Cancer Res.* **7**, 475–514.
58. Hoffmann, D. and Wynder, E. L. (1968). Organic particulate pollutants. In "Air Pollution". A. C. Stern, Ed., Vol. 2, p. 191. Academic Press.
59. Cautreels, W. and van Cauwenberghe, K. (1976). Extraction of organic compounds from airborne particulate matter. *Water, Air and Soil Pollution* **6**, 103–110.
60. Rosen, A. A. and Middleton, F. M. (1955). Identification of petroleum refinery waste in surface waters. *Anal. Chem.* **27**, 790–794.
61. Hoffmann, D. and Wynder, E. L. (1962). "Analytical and biological studies on gasoline engine exhaust." Natl. Cancer Instit. Monograph No. 9, 91–110.
62. Candeli, A., Morozzi, G., Paolacci, A. and Zoccolillo, L. (1975). Analysis using thin layer and gas-liquid chromatography of polycyclic aromatic hydrocarbons in the exhaust products from a European car running on fuels containing a range of concentrations of these hydrocarbons. *Atmos. Environ.* **9**, 843–849.
63. Butler, J. D. and Crossley, P., (1979). An appraisal of relative airborne sub-urban concentrations of polycyclic aromatic hydrocarbons.
64. Jost, W. (1952). *Sci. Total Environ.* **11**, in press. "Diffusion in Solids, Liquids and Gases". Academic Press.
65. Sakabe, H. (1964). Air pollution in Japan. *Proc. Roy. Soc. Med.* **57**, 1005–1012.
66. Moore, G. E., Thomas, R. S. and Monkman, J. L. (1967). The routine determination of polycyclic hydrocarbons in airborne pollutants. *J. Chromat.* **26**, 456–464.
67. Cahnmann, H. J. (1957). Partially deactivated SiO_2-gel columns in chromatography: Chromatographic behaviour of benzo(a)pyrene. *Anal. Chem.* **29**, 1307–1311.
68. Karr, C., Childers, E. E. and Warner, W. C. (1963). Analysis of aromatic hydrocarbon samples by liquid chromatography with operating conditions analogous to that of gas chromatography. *Anal. Chem.* **35**, 1290.

69. Cleary, G. J. (1962). Discrete separation of PNH's in airborne particulates using very long Al_2O_3 columns. *J. Chromat.* **9**, 204–215.
70. Popl, M., Dolansky, V. and Mostecky, J. (1971). The separation of polynuclear hydrocarbons. *J. Chromat.* **59**, 329–334.
71. Sawicki, E., Meeker, J. E. and Morgan, M. J. (1965). Polynuclear aza compounds in automotive exhausts. *Arch. Environ. Health* **11**, 773–775.
72. Cooper, R. L. (1954). The detection of polycyclic hydrocarbons in town air. *The Analyst* **79**, 573–579.
73. Commins, B. T. (1958). A modified method for the determination of polycyclic hydrocarbons. *The Analyst* **83**, 386–389.
74. Sawicki, E., Meeker, J. E. and Morgan, M. J. (1965). The quantitative composition of air pollution source effluents in terms of aza heterocyclic compounds and polynuclear aromatic hydrocarbons. *Int. J. Air Water Pollut.* **9**, 291–298.
75. Sawicki, E., Elbert, W. C., Stanley, T. W., Hauser, T. R. and Fox, F. T. (1960). Separation and characterization of polynuclear aromatic hydrocarbons in urban airborne particulates. *Anal. Chem.* **32**, 810–815.
76. Dubois, L. and Monkman, J. L. (1965). A technique for accurately measuring BaP or BkF by fluorescence in air samples either separately or in a mixture. *Int. J. Air Water Pollut.* **9**, 131–133.
77. Little, J. N., Horgan, D. F. and Bombaugh, K. J. (1970). Performance characteristics of liquid chromatographic adsorbents. *J. Chromat. Sci.* **8**, 625–629.
78. Kirkland, J. J. (1969). High-speed liquid chromatography with controlled surface porosity supports. *J. Chromat. Sci.* **7**, 7–12.
79. Karger, B. L., Martin, M., Loheac, J. and Guichon, G. (1973). Separation of polynuclear hydrocarbon by liquid-solid chromatography using a 2,4,7-trifluorenone impregnated Corasil column. *Anal. Chem.* **45**, 496–500.
80. Vaughan, C. G., Wheals, B. B. and Whitehouse, M. J. (1973). The use of pressure assisted liquid chromatography in the separation of polynuclear hydrocarbons. *J. Chromat.* **78**, 203–210.
81. Ray, S., and Frei, R. W. (1972). Separation of polynuclear aza heterocycles by high speed liquid chromatography on a chemically bonded stationary phase. *J. Chromat.* **71**, 451–457.
82. Grant, D. W., Meiris, R. B. and Hollis, M. G. (1974). The use of Bentone-34 coated supports in column chromatography and their potential application to the field of organic pollution analysis. *J. Chromat.* **99**, 721–729.
83. Doran, T. and McTaggart, N. G. (1974). The combined use of high efficiency liquid and capillary gas chromatography for the determination of polynuclear hydrocarbons in automotive exhaust condensates and other hydrocarbon mixtures. *J. Chromat. Sci.* **12**, 715–721.
84. Dong, M., Ferrand, E. and Locke, D. C. (1976). High pressure liquid chromatographic method for routine analysis of major parent polycyclic aromatic hydrocarbons in suspended particulate matter. *Anal. Chem.* **48**, 368–371.
85. Krstulovic, A. M., Rosie, D. M. and Brown, P. R. (1976). Selective monitoring of polynuclear aromatic hydrocarbons by high pressure liquid chromatography with a variable wavelength detector. *Anal. Chem.* **48**, 1383–1385.

86. Sawicki, E., Stanley, T. W. and Elbert, W. C. (1964). Quenchofluorometric analysis for fluoranthenic hydrocarbons in the presence of other types of aromatic hydrocarbons. *Talanta* **11**, 1433–1441.
87. Sawicki, E., Johnson, H. and Kosinski, K. (1966). Chromatographic separation and spectral analysis of polynuclear aromatic amines and heterocyclic imines. *Microchem. J.* **10**, 72–102.
88. Sawicki, E., Stanley, T. W., Elbert, W. C., Meeker, J. E. and McPherson, S. (1967). Comparison of methods for the determination of benzo(a)pyrene in particulates from urban and other atmospheres. *Atmos. Environ.* **1**, 131–145.
89. Pierce, R. C. and Katz, M. (1975). Determination of atmospheric isomeric polycyclic arenes by thin-layer chromatography and fluorescence spectrophotometry. *Anal. Chem.* **47**, 1743–1748.
90. Brocco, D., Cimmino, A. and Possanzini, M. (1973). Determination of azaheterocyclic compounds in atmospheric dust by combination of thin-layer and gas chromatography. *J. Chromat.* **84**, 371–377.
91. Bender, D. F., Sawicki, E. and Wilson, R. M. (1964). Characterization of carbazole and polynuclear carbazoles in urban air and in air polluted by coal tar pitch fumes by thin-layer chromatography and spectrophotofluorometry. *Int. J. Air Water Pollution* **8**, 633–643.
92. Sawicki, E., Elbert, W. C. and Stanley, T. W. (1965). The fluorescence-quenching effect in thin-layer chromatography of polynuclear aromatic hydrocarbons and their aza analogs. *J. Chromat.* **17**, 120–126.
93. Sawicki, E., Elbert, W. C. and Stanley, T. W. (1965). Characterization of polynuclear aza heterocyclic hydrocarbons separated by column and thin-layer chromatography from air pollution source particulates. *J. Chromat.* **18**, 512–519.
94. Sawicki, E., Stanley, T. W., Elbert, W. C. and Morgan, M. J. (1965). Column and thin-layer chromatographic separation of polynuclear ring-carbonyl compounds. *Talanta* **12**, 605–616.
95. Sawicki, E., Stanley, T. W. and Elbert, W. C. (1965). Analysis of the urban atmosphere and air pollution source effluents for phenalen-1-one and 7 H-benze(de)anthracen-7-one. *Mikrochim. Acta*, 1110–1123.
96. Pierce, R. C. and Katz, M. (1976). Chromatographic isolation and spectral analysis of polycyclic quinones. Application to air pollution analysis. *Environ. Sci. and Technol.* **10**, 45–51.
97. Sawicki, E., Stanley, T. W., McPherson, S. and Morgan, M. J. (1966). Use of gas-liquid and thin-layer chromatography in characterizing air pollutants by fluorometry. *Talanta* **13**, 619–629.
98. Luly, A. M. and Sakodynsky, K. (1965). A paper chromatographic study of azaheterocyclic hydrocarbons using aqueous solvents. *J. Chromat.* **19**, 624–629.
99. Sawicki, E. and Pfaff, J. D. (1965). Analysis of aromatic compounds on paper and tlc by spectrophotophosphorimetry. *Anal. Chim. Acta* **32**, 521–534.
100. Lederer, M. and Roch, G. (1967). Paper chromatography of some azaheterocyclic hydrocarbons. *J. Chromat.* **31**, 618–627.
101. Hlucháň, E., Jeník, M. and Malý, E. (1974). Determination of airborne polynuclear hydrocarbons by paper chromatography. *J. Chromat.* **91**, 531–538.
102. Lane, D. A., Moe, H. K. and Katz, M. (1973). Analysis of polynuclear aromatic hydrocarbons, some heterocyclics, and aliphatics with single gas chromatography column. *Anal. Chem.* **45**, 1776–1778.

103. Zane, A. (1968). Separation of some polynuclear aromatic hydrocarbons by gas-solid chromatography on graphitized carbon black. *J. Chromat.* **38**, 130–133.
104. Fryčka, J. (1972). Evaluation of the separation of phenanthrene, anthracene and carbazole in pure tar products by gas-solid chromatography. *J. Chromat.* **65**, 341–344.
105. Janini, G. M., Johnston, K. and Zielinski, W. L. (1975). Use of nematic liquid crystal for gas-liquid chromatographic separation of polyaromatic hydrocarbons. *Anal. Chem.* **47**, 670–674.
106. Liberti, A., Cartoni, G. P. and Cantuti, V. J. (1964). Gas chromatographic determination of polynuclear hydrocarbons in dust. *J. Chromat.* **15**, 141–148.
107. Liberti, A., Cartoni, G. P., Cantuti, V. J. and Torri, A. G. (1965). Improved evaluation of polynuclear hydrocarbons in atmospheric dust by gas chromatography. *J. Chromat.*, **17**, 60–65.
108. Wilmshurst, J. R. (1965). Gas chromatographic analysis of polynuclear arenes. *J. Chromat.* **17**, 50–59.
109. Janini, G. M., Muschik, G. M. and Zielinski, W. L. (1976). N,N′-bis(p-butoxybenzylidene)-α,α'-bi-p-toluidine: Thermally stable liquid crystal for unique glc separation of polycyclic aromatic hydrocarbons. *Anal. Chem.* **48**, 809–813.
110. Lao, R. C., Thomas, R. S. and Monkman, J. L. (1975). Computerized gas chromatographic-mass spectrometric analysis of polynuclear hydrocarbons in environmental samples. *J. Chromat.* **112**, 681–700.
111. Grob, K. and Grob, G. (1971). Gas liquid chromatographic–mass spectrometric investigation of C_6—C_{20} organic compounds in urban atmospheres. *J. Chromat.* **62**, 1–13.
112. Cautreels, W. and van Cauwenberghe, K. (1976). The determination of organic compounds in airborne particulate matter by gas chromatography–mass spectrometry. *Atmos. Environ.* **10**, 447–457.
113. Pellizzari, E. D., Bunch, J. E., Berkley, R. E. and McRae, J. (1976). Determination of trace hazardous organic vapor pollutants in ambient atmospheres by gas chromatoglraphy–mass spectrometry-computer. *Anal. Chem.* **48**, 803–807.
114. Markey, S. P. (1970). Improved glass-frit interface for combined gas chromatography-mass spectrometry. *Anal. Chem.* **42**, 306–309.
115. McLafferty, F. W. (1966). Interpretation of Mass Spectra. W. A. Benjamin Inc.
116. Biemann, K. (1962). "Mass Spectrometry. Organic Chemical Application." McGraw Hill Inc.
117. Mass Spectra Tables, Serial No. 1020. Amer. Petroleum Instit. Res. Project 44. Carnegie Instit. Technol., Pittsburgh, Pa.
118. Sawicki, E. and Barry, R. (1959). New color tests for the larger polynuclear aromatic hydrocarbons. *Talanta* **2**, 128–134.
119. Sawicki, E., Stanley, T. W. and Hauser, T. R. (1958). Detection of hetero substituted aromatic derivatives and determination of aromatics in air. *Chemist Analyst.* **47**, 69–77.
120. Sawicki, E., Miller, R., Stanley, T. W. and Hauser, T. R. (1958). Detection of polynuclear hydrocarbons and phenols with benzal and piperonal chlorides. *Anal. Chem.* **30**, 1130–1133.
121. Sawicki, E., Stanley, T. W., Hauser, T. R. and Barry, R. (1959). Isatin tests for aromatic hydrocarbons and phenols. *Anal. Chem.* **31**, 1664–1667.

122. Sawicki, E., Noe, J. L. and Fox, F. T. (1971). Spot test detection and colorimetric determination of aniline, naphthylamine and anthramine derivatives with 4-azobenzene-diazonium fluoborate. *Talanta* **8**, 257–264.
123. Sawicki, E. and Elbert, W. C. (1959). Thermochromic detection of polynuclear compounds containing the fluorenic methylene group. *Chemist Analyst* **48**, 68–69.
124. Sawicki, E., Stanley, T. W. and Hauser, T. R. (1958). A thermochromic test for polycyclic p-quinones. *Anal. Chem.* **30**, 2005–2006.
125. Sawicki, E. and Stanley, T. W. (1960). Simple spot test for pyrene and its derivatives: Application to air pollution studies. *Chemist Analyst* **49**, 77–78.
126. Sanderson, H. P., Ferguson, M. B. and Katz, M. (1958). Evaluation of airborne particulates in atmospheric pollution studies. *Anal. Chem.* **30**, 1172–1180.
127. Dubois, L., Baker, C. J., Zdrojewski, A. and Monkman, J. L. (1970). Correlation of the blackness index of high volume air samples with the polycyclic hydrocarbon concentrations in urban air. *Pure and Appl. Chem.* **24**, 695–706.
128. Technical Note 1. "The measurement of airborne asbestos dust by the membrane filter method." Revised 1971. The Asbestos Research Council, 114, Park St., London, W1Y 4AB.
129. Markham, M. C. and Wosczyna, K. (1976). Determination of micro-quantities of chrysotile asbestos by dye absorption. *Environ. Sci. and Technol.* **10**, 930–931.
130. Schultze, H. F. (1968). Monitoring airborne radioactivity. *In* "Air Pollution", Vol. 2, A. C. Stern, p. 393. Academic Press, 1968.
131. "Methods of air sampling and analysis." Interscience Committee. American Public Health Association, 1972.
132. Audric, B. N. and Long, J. V. P. (1953). The background and ^{14}C detection efficiency of a liquid scintillation counter. *J. Sci. Instru.* **30**, 467–469.
133. Geyh, M. A., "Experienced gathered in the construction of low level counters." Proc. Int. Atom. Energ. Auth. Symp. Radioactive dating and methods of low level counting, Vienna 1967, pl. 575–589.
134. Loosli, H. H., Oeschger, H. and Wahlen, M. New attempts in low-level counting and a search for cosmic ray produced ^{39}Ar. Proc. Int. Atom. Ener. Auth. Symp., Vienna 1967, p. 593–601.
135. Butler, F. F. (1961). Determination of tritium in water and urine. *Anal. Chem.* **33**, 409–414.
136. Pratt, J. W. (1974). "Statistical and Mathematical Aspects of Pollution Problems". Marcel Dekker, Inc., New York.

5

Atmospheric Reactions

5.1 Introduction

Many of the important chemical reactions which occur in the atmosphere are of a photochemical nature. Before these are described some relevant features of photochemical processes will be briefly considered by way of background. There are two fundamental photochemical laws:

(1) a molecule must absorb incident light energy to become activated; and

(2) the absorption of one photon ($h\nu$) of radiation activates one molecule.

The efficiency of a photochemical process is determined by the quantum yield. Once activated in the primary stage a molecule can enter into reaction with other compounds to yield products. The quantum yield is defined as the ratio of the number of product molecules formed to the number of photons absorbed in the primary stage.

The amount of light absorbed in the first instance is governed by the well known Lambert–Beer law. This law can be expressed in slightly different ways, the one familiar to chemists is

$$\log_{10}\frac{I_0}{I} = ecl \qquad (5.1)$$

where I_0 and I are the incident and emergent light intensities after passage through a system of path length l and concentration c; e is a constant called the molar extinction coefficient for the substance when light of a specific wavelength, λ, has been used. For atmospheric work, however, the form

$$\log_{10}\frac{I}{I_0} = -\sigma nl \qquad (5.2)$$

where σ is the absorption cross-section and n is the number of molecules cm^{-3}, is sometimes preferred.

The photochemical initiation of a chemical reaction depends on the ability of the absorbed energy to excite and dissociate the absorbing molecule, so that active free radicals result. Furthermore, in recent years a realization has developed that various forms of ground state and excited state conditions of molecules and atoms can be involved in specific types of chemical and photochemical reactions. Such subtle differences are not disclosed by the usual chemical representation and because of their significance in atmospheric chemistry an explanation by way of introduction will be given.

Since oxygen plays such an important role in this field of chemistry oxygen will serve as a suitable illustration. Figure 5.1 shows the Franck–Condon diagram of potential energy *vs* internuclear separation of the

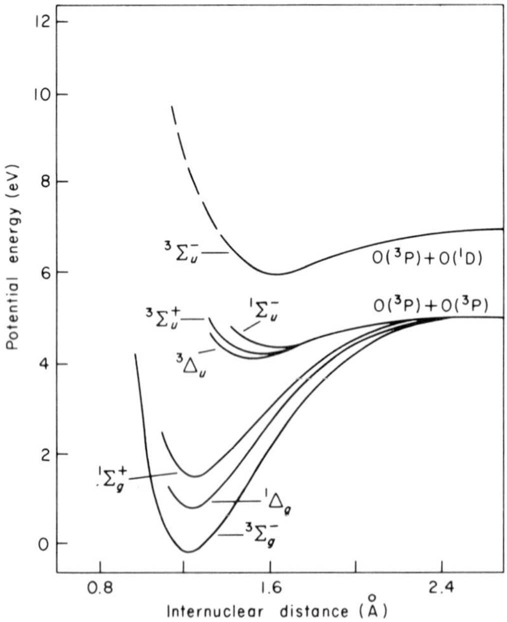

FIG. 5.1 Franck–Condon diagram of the potential energy curves of some states molecular oxygen

various states of molecular and atomic oxygen. The ground state condition of molecular oxygen shown in Fig. 5.2 is compounded from the combination of two oxygen atoms having the usual extranuclear electron distribution accommodated into the appropriate molecular orbitals i.e.

$$O[1s^2, 2s^2, 2p^4] + O[1s^2, 2s^2, 2p^4] \rightarrow O_2[KK(z\sigma)^2(y\sigma)^2(x\sigma)^2(w\pi)^4(v\pi)^2] \quad (5.3)$$

where KK replaces $(\sigma 1s)^2(\sigma^* 1s)^2$. This molecular orbital configuration can give rise to the three ground state conditions indicated in Fig. 5.1 by $^3\Sigma_g^-$,

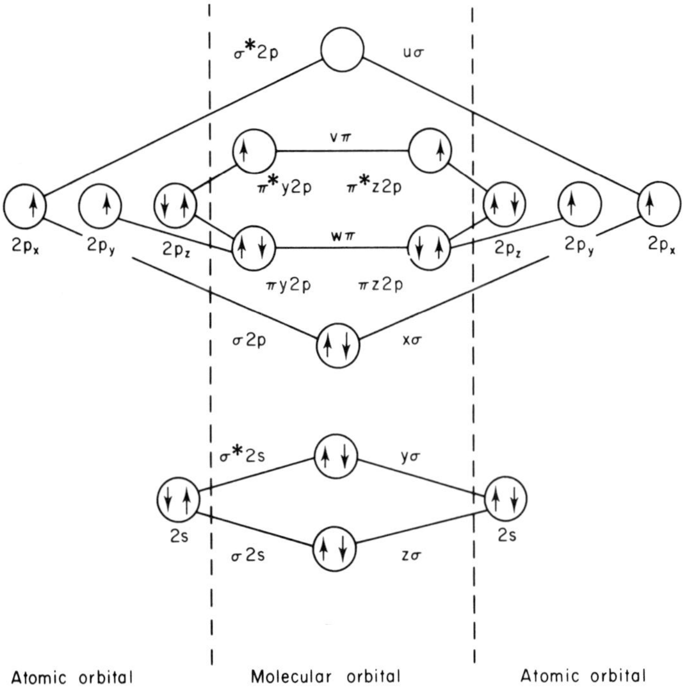

FIG. 5.2 Energy level diagram of oxygen showing the combination of atomic orbitals into molecular orbitals. (Reprinted with permission of Oxford University Press)

$^1\Delta_g$, and $^1\Sigma_g^+$. This occurs by a modification of the manner in which the electrons occupy the highest molecular orbitals [1] as shown in Table 5.1. It can be seen that these slightly elevated ground state energies associated with the $^1\Delta_g$ and $^1\Sigma_g^+$ states contravene Hund's rules of maximum multiplicity. In the one case although the electrons are singly occupying separate orbitals they have their spins anti-parallel and in the other instance they both occupy the same orbital leaving the other unoccupied.

Figure 5.1 also indicates the relationship between energies of oxygen atoms which are produced upon dissociation of molecular oxygen. These are distinguished by the $O(^3P)$ atom which is lower in energy than the metastable $O(^1D)$ atom.

The ability of light to dissociate molecular oxygen depends upon the wavelength of the incident radiation. These wavelengths and the designation of the product oxygen atoms or excited molecular oxygen produced by light irradiation of molecular oxygen and ozone are given in Table 5.2. The table has been compiled from data given by Volman [2].

TABLE 5.1 *Influence of electron occupancy in the vπ molecular orbital on the ground state of molecular oxygen* [1]

Electronic states of the oxygen molecule[a]	Occupancy of highest orbitals ($v\pi$)	Energy above ground state (kJ)
Second excited state ($^1\Sigma_g^+$)	(↑) (↓)	157
First excited state ($^1\Delta_g$)	(↑↓) ( )	94
Ground state ($^3\Sigma_g^-$)	(↑) (↑)	

[a] In this chapter the designation of the ground, first excited and second excited states of molecular oxygen will be abbreviated to $^3\Sigma_g^-$, $^1\Delta_g$ and $^1\Sigma_g^+$, respectively.

TABLE 5.2 *Photodissociation reactions of molecular oxygen and ozone*

Wavelength (nm)	Reaction	Equation
$\lambda < 175$	$O_2(^3\Sigma_g^-) + h\nu \rightarrow O(^1D) + O(^3P)$	(5.4)
$175 < \lambda < 200$	$O_2(^3\Sigma_g^-) + h\nu \rightarrow O_2(^3\Sigma_u^-)$	(5.5)
$200 < \lambda < 242$	$O_2(^3\Sigma_g^-) + h\nu \rightarrow O(^3P) + O(^3P)$	(5.6)
$242 < \lambda < 279$	$O_2(^3\Sigma_g^-) + h\nu \rightarrow O_2(^3\Sigma_u^+)$	(5.7)
$\lambda < 1140$	$O_3 + h\nu \rightarrow O_2(^3\Sigma_g^- + O(^3P)$	(5.8)
$\lambda < 308$	$O_3 + h\nu \rightarrow O_2(^1\Delta_g) + O(^1D$	(5.9)
$\lambda < 266$	$O_3 + h\nu \rightarrow O_2(^1\Sigma_g^+) + O(^1D)$	(5.10)

5.2 Natural ozone occurrence

Ultraviolet radiation from the sun interacts with oxygen in the stratosphere to generate atomic oxygen by the reaction:

$$O_2 + h\nu \rightarrow O(^3P) + O(^3P) \qquad (5.6)$$

$$\lambda < 242 \text{ nm}$$

In the presence of a third body, M, molecular and atomic oxygen combine to form ozone:

$$O_2 + O(^3P) + M \rightarrow O_3 + M \qquad \Delta H^0 = -100\ \text{kJ} \qquad (5.11)$$

In the stratosphere at altitudes between 10 and 60 km the third body is usually nitrogen. Solar radiation is also responsible for the decomposition of ozone:

$$O_3 + h\nu \rightarrow O_2 + O(^3P) \qquad (5.8)$$

$$\lambda < 1140\ \text{nm}$$

and

$$O(^3P) + O_3 \rightarrow O_2(^1\Sigma_g^+) + O_2(^3\Sigma_g^-) \qquad (5.12)$$

$$\Delta H^0 = -390\ \text{kJ}$$

This sequence of reactions accounts for the natural existence of ozone in the upper atmosphere. The actual concentration at any time depends upon the intensity of solar radiation and altitude. Unpolluted air at sea level contains 1–5 parts 10^{-8}; stratospheric concentrations can be a thousand times this amount. Maximum concentrations occur at altitudes of 25–35 km (Fig. 5.3) [3]. It will be noticed that reactions (5.11) and (5.12) are

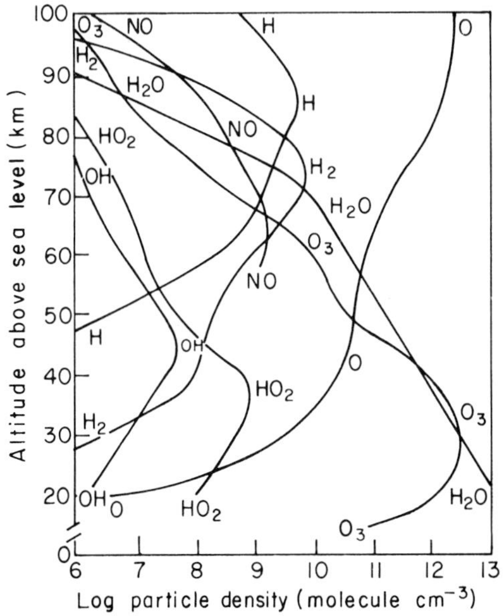

FIG. 5.3 Daytime equilibrium concentration–altitude profiles of H_2O, H_2, O_3, OH, HO_2, H, O, and NO [3]

exothermic, so that in the complete reaction sequence energy is continually being dissipated. This accounts for the observed increase of temperature with altitude in the stratosphere that is shown in Fig. 5.4.

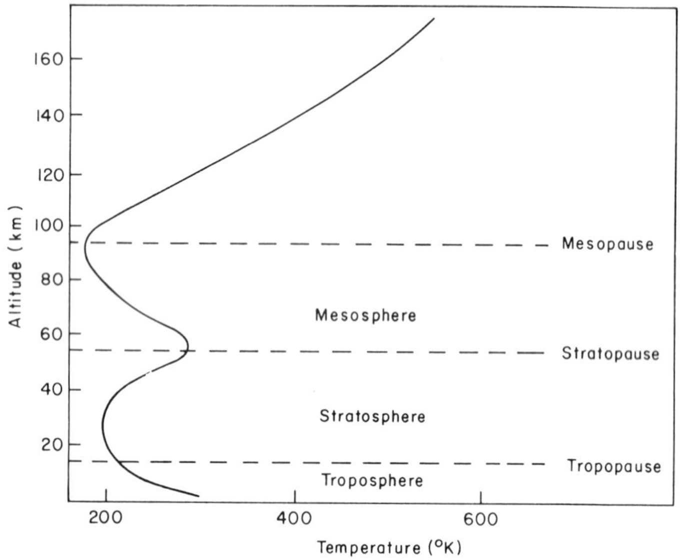

FIG. 5.4 Atmospheric temperature changes with altitude

The other feature of stratospheric ozone just described is that it is vitally important to life on earth. Stratospheric ozone absorbs potentially harmful u.v. radiation from the sun in reactions (5.8) to (5.10). This screening effect of ozone ensures that the lower limit of sunlight which penetrates to the earth's surface is about 290 nm or 420 kJ mol^{-1}. This energy is sufficient to dissociate ozone and nitrogen dioxide but, unlike the situation in the stratosphere, too low to dissociate oxygen.

In contrast to stratospheric ozone reactions the most important reactions in the troposphere between 0 and 10 km are:

$$NO_2 + h\nu \rightarrow NO + O(^3P) \qquad (5.13)$$

$$\lambda < 430 \text{ nm}$$

the exothermic reaction

$$O(^3P) + O_2 + M \rightarrow M + O_3 \qquad \Delta H^0 = -100 \text{ kJ} \qquad (5.11)$$

and

$$O_3 + NO \rightarrow NO_2 + O_2 \qquad (5.14)$$

Equilibrium ozone concentrations in the atmosphere under these conditions will be determined in a matter of minutes by the nitrogen dioxide concentration. It is evident from the reaction series (5.13), (5.11) and (5.14) that any process that is capable of increasing the atmospheric nitrogen dioxide concentration could cause the production of more ozone. In a polluted atmosphere nitric oxide can provide the source of nitrogen dioxide.

5.3 Tropospheric chemical reactions

Homogeneous reactions between molecules and radicals that take place in the troposphere can be very complex. Some of the first recognized manifestations of atmospheric chemistry were the Los Angeles "photochemical smogs" that developed during the 1940s. Besides atmospheric pollutants, the other prime requirements needed for formation of photochemical smog are low windspeeds, two-eighths cloud cover or less, and temperatures in excess of 18°C. Under these circumstances a series of chemical reactions can occur between oxides of nitrogen, ozone, hydrocarbons and sulphur dioxide. A great deal of research has been done to try to understand the fundamental chemistry involved. Much has been achieved and in the process of this achievement a lot has been learnt which is applicable to other branches of environmental chemistry. The widespread significance of the OH radical already mentioned in Chapter 2 serves to emphasize this point.

In this section (5.3) some reactions that contribute to photochemical smog are considered, but the coverage is not intended to be limited to this topic alone. For a detailed review of the mechanism of photochemical smog formation the paper by Demerjian *et al.* [4] should be consulted.

5.3.1 Participation of nitrogen oxides

Nitric oxide is synthesized from the air in thunderstorms when energy is discharged in the lightning flash. Naturally occurring nitrates present in the earth's crust originate from the severe electrical storms which raged in evolutionary times. Nitric oxide of interest in photochemical reactions, however, comes from the combustion of fossil fuel. The mechanism of formation has already been discussed in section 2.3.2.2. These reactions are again summarized by equations (5.15) to (5.18):

$$CO + OH \rightarrow CO_2 + H \qquad (5.15)$$

$$H + O_2 \rightarrow OH + O \qquad (5.16)$$

$$O+N_2 \rightarrow NO+N \tag{5.17}$$

$$N+O_2 \rightarrow NO+O \tag{5.18}$$

Combustion chamber concentrations of nitric oxide from petrol and diesel engine sources vary between 50 and 5000 parts 10^{-6}. Once released into the atmosphere from the exhaust pipe, nitric oxide is diluted by air. The rate of oxidation to nitrogen dioxide is a slow process at ambient concentrations of around 1 part 10^{-8} or less, since kinetically the reaction depends on the square of the nitric oxide concentration:

$$2NO+O_2 \rightarrow 2NO_2 \tag{5.19}$$

In practice the most important reaction for the oxidation of nitric oxide is believed to involve the hydroperoxyl or peroxyalkyl radical:

$$\underset{(a)\quad(b)}{NO+HO_2(RO_2)} \rightarrow \underset{(a)\quad(b)}{NO_2+OH(OR)} \tag{5.20a,b}$$

Hence any process which converts OH to HO_2 or RO to RO_2 will participate in a free radical chain reaction, the overall effect of which will be the production of ozone through the sequence of reactions given by (5.11), (5.13) and (5.14). Table 5.3 shows that nitrogen dioxide is also present and by analogy with equation (5.14), viz.

$$O_3+NO \rightarrow NO_2+O_2 \tag{5.14}$$

ozone can react with nitrogen dioxide to give nitrogen trioxide:

$$O_3+NO_2 \rightarrow NO_3+O_2 \tag{5.21}$$

TABLE 5.3 *Typical concentrations of the main pollutants in photochemical smog* [5]

Pollutant	Concentration (parts 10^{-8})
Carbon monoxide	200–2000
Nitric oxide	1–15
Nitrogen dioxide	5–20
Ozone	2–20
Total hydrocarbons (excluding methane)	20–50
Alkenes	2–6
Aromatics	10–30
Aldehydes	5–25
Peroxyacylnitrates PAN	1–4

and in addition, NO_3 can be formed by reaction with atomic oxygen [6]:

$$O(^3P)+NO_2+M \rightarrow NO_3+M \tag{5.22}$$

These NO_3 radicals will mainly be removed in about equal amounts by reaction with NO and NO_2.

$$NO_3+NO_2 \rightarrow N_2O_5 \tag{5.23}$$

$$NO_3+NO \rightarrow 2NO_2 \tag{5.24}$$

In the presence of moisture nitrous and nitric acids are formed as shown by the equations (5.25) and (5.26):

$$N_2O_5+H_2O \rightarrow 2HNO_3 \tag{5.25}$$

$$NO_2+NO+H_2O \rightarrow 2HNO_2 \tag{5.26}$$

Nitrous acid readily photolyses in sunlight at wavelengths less than 400 nm to give nitric oxide and the hydroxyl radical:

$$HONO+h\nu \rightarrow OH+NO \tag{5.27}$$

$$\lambda < 400 \text{ nm}$$

5.3.2 Participation of carbon monoxide

The global concentration of carbon monoxide in the upper troposphere defined in terms of the mole fraction, or mixing ratio as it is called in the atmospheric sciences, is about 0·1–0·15 parts 10^{-6} rising to 0·2 parts 10^{-6} in air over the north Atlantic [7]. In a polluted atmosphere, on the other hand, as shown in Table 5.3, carbon monoxide is the principal component present during a photochemical smog episode. This carbon monoxide brings about the oxidation of OH to HO_2 in the following manner:

$$OH+CO \rightarrow CO_2+H\ [8] \tag{5.15}$$

$$H+O_2+M \rightarrow HO_2+M \tag{5.28}$$

As discussed in section 2.2.1 in connection with the carbon cycle, the oxidation of carbon monoxide by the OH radical accounts for the principal removal route of carbon monoxide from the atmosphere. On this basis the lifetime of carbon monoxide in the atmosphere will be about 0·1 year.

Even in unpolluted air there are several reactions which produce OH and HO_2 radicals. Photolysis according to equations (5.9) or (5.10) gives the metastable $O(^1D)$ [9]:

$$O_3+h\nu \rightarrow O_2(^1\Delta_g)+O(^1D) \tag{5.9}$$

$$\lambda < 308 \text{ nm}$$

$$O_3 + h\nu \rightarrow O_2(^1\Sigma_g^+) + O(^1D) \quad (5.1($$

$$\lambda < 266\ \text{nm}$$

This atomic oxygen reacts with water vapour to yield OH radicals:

$$O(^1D) + H_2O \rightarrow 2OH \quad (5.2\text{\S}$$

This reaction of $O(^1D)$ with water vapour represents the main source (OH radicals, although photolysis of nitrous acid, equation (5.27), als contributes to give an estimated OH concentration at sea level of 2–3 × 1(molecules cm^{-3}, with an average lifetime of 3–4 days.

The hydroperoxyl radical HO_2 is formed from the photolysis of forma dehyde in the following manner [10]:

$$HCHO + h\nu \xrightarrow{k=5{\cdot}2\times10^{-3}\,min^{-1}} H_2 + CO \quad (5.30a$$

$$HCHO + h\nu \xrightarrow{k=2{\cdot}0\times10^{-3}\,min^{-1}} H + HCO \quad (5.30b$$

$$\lambda < 370\ \text{nm}$$

The radicals produced from reaction (5.30b) interact with molecular oxy gen to give the hydroperoxyl radical:

$$HCO + O_2 \rightarrow HO_2 + CO \quad (5.31$$

$$H + O_2 + M \rightarrow HO_2 + M \quad (5.28$$

Besides the photodissociation of formaldehyde leading to the HO_2 radica the OH radical itself can produce the formyl radical from formaldehyd [11]:

$$HCHO + OH \rightarrow HCO + H_2O \quad (5.32$$

Conversion to HO_2 then proceeds through reactions (5.31) and (5.28) a before. Other possibilities that have been investigated [12, 13] are:

$$OH + H_2 \rightarrow H + H_2O \quad (5.33$$

followed by reaction (5.28), and

$$OH + H_2O_2 \rightarrow HO_2 + H_2O \quad (5.34$$

but these are less significant. The origin of the hydrogen peroxide indicatec in equation (5.34) can be either through combination of hydroxyl anc hydroperoxyl radicals as shown in equations (5.35) and (5.36):

$$M + HO + HO \rightarrow H_2O_2 + M \quad (5.35$$

$$HO_2 + HO_2 \rightarrow H_2O_2 + O_2 \quad (5.36$$

or a hydrogen abstraction reaction following an attack of an hydrocarbon species by the hydroperoxyl radical. The reaction shown in equation (5.37) between the hydroperoxyl radical and acetaldehyde serves as a typical example:

$$HO_2 + CH_3CHO \rightarrow H_2O_2 + CH_3CO \qquad (5.37)$$

5.3.3 Participation of hydrocarbons

5.3.3.1 Hydrocarbon reactivity series

Most of the information on the interaction between hydrocarbons, oxides of nitrogen and sunlight comes from either:

(1) laboratory environmental chamber studies, in which air, nitric oxide and hydrocarbons are introduced into a chamber which is then irradiated with u.v. light and the resulting products and product distributions determined; or
(2) Computer predictions of the concentrations of reaction products which are based upon the reaction rate constants of all processes assumed to be participating.

Simulations from these experiments have enabled detailed mechanisms to be worked out and the relative importance of different radicals and different hydrocarbon species to be evaluated. On this sort of evidence Altshuller and Bufalini [14] have summarized the findings for hydrocarbon reactivity in such systems. The criteria applied for evaluating hydrocarbon reactivity varies. These assessments include the following considerations:

(1) efficiency of conversion of nitric oxide to nitrogen dioxide;
(2) rate of hydrocarbon consumption;
(3) rate of ozone formation;
(4) extent of aerosol formation;
(5) eye irritation potential; and
(6) plant damage capability.

As a consequence of these different standards several reactivity series have been constructed: e.g. in order of decreasing activity, disubstituted internal alkenes > cyclopentenes > monosubstituted internal alkenes > unsubstituted internal alkenes = cyclohexenes > tri- and tetra-alkylbenzenes = terminal alkenes > C_4 + paraffins and monoalkylbenzenes > propane > benzene > ethane > methane [15].

Recently, a rather different approach for assessing hydrocarbon reactivity has been proposed by Darnall *et al.* [16]. This is based on the ability of hydrocarbons to react with the OH radical. This suggestion arises from the fact that Niki *et al.* [17] showed that the reactivity of a number of hydrocarbons, as measured by the rate of conversion of NO to NO_2, correlates significantly better with OH rate constants than with either

$O(^3P)$ or O_3 rate constants. The basic assumption of the OH radical reactivity scale implies that competition by ozone, $O(^3P)$ and HO_2 radicals for hydrocarbons all give reaction rates several orders of magnitude slower than that of the OH radical.

When hydrocarbons are assessed in this way they fall into five classes. Methane, the least reactive hydrocarbon, is the sole member of Class I. Class II, defined by reactivities in the range 10–100, includes acetylene and ethane. Class III, range 100–1000, contains benzene, propane, *n*-butane, *iso*-pentane, methylethylketone, 2-methylpentane, toluene, *n*-propylbenzene, cumene, ethene, *n*-hexane, 3-methylpentane, ethylbenzene and *p*-xylene. Class IV, range 1000–10 000 comprises *p*-ethyltoluene, *o*-ethyltoluene, *o*-xylene, methylisobutylketone, *m*-ethyltoluene, *m*-xylene; 1,2,3-trimethylbenzene, propene; 1,2,4-trimethylbenzene; 1,3,5-trimethylbenzene, *cis*-but-2-ene, β-pinene and 1,3-butadiene. Finally, the most reactive compounds whose reactivities are greater than 10 000 are placed in Class V, e.g. 2-methylbut-2-ene, 2,3-dimethylbut-2-ene and *d*-limonene.

An advantage of this scale that should be recognized is that it can discriminate between ethane and propane. Furthermore, a number of aromatic hydrocarbons may be distinguished which require a longer period of time to react but can, nevertheless, contribute significantly to ozone formation during longer irradiation periods. These factors will have important consequences when pollutants are transported downwind from urban sources to more remote areas.

An empirical treatment which is only applicable to alkenes for assessing their activity in photochemial smog reactions has been described by Yeung and Phillips [18]. The method is based on the ability of alkenes to bring about the oxidation of NO to NO_2. Using experimental rate data obtained by Glasson and Tuesday [19] for NO_2 formation in the presence of various alkenes, a characteristic rate factor has been derived. This characteristic rate factor can be related to the structure of the alkene as shown in Table 5.4. The NO_2 formation rates expressed in parts 10^{-9} min^{-1} given in the table are derived from the expression,

$$\text{Rate of } NO_2 \text{ formation } (r_{NO}) = \frac{[NO]_0}{2T_{1/2}} \quad (5.38)$$

where $[NO]_0$ is the initial NO concentration (parts 10^{-9}) and $T_{1/2}$ is the time in minutes required to produce 190 parts 10^{-9} from an initial mixture of 1·0 parts 10^{-6} hydrocarbon, 0·38 parts 10^{-6} NO and 0·02 parts 10^{-} NO_2. To illustrate the calculation let us predict the rate of oxidation of NC in the presence of (a) *cis*-pent-2-ene, (b) 2-methylbut-2-ene, and (c) 2

methylpent-2-ene, from the information given in Table 5.4.
(a) *Cis*-pent-2-ene

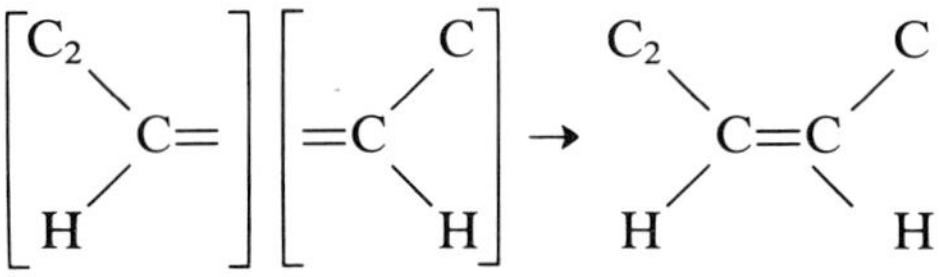

cis-pent-2-ene

$(2.23 \text{ parts } 10^{-9} \text{ min}^{-1})^{1/2}(2.66 \text{ parts } 10^{-9} \text{ min}^{-1})^{1/2} = 5{\cdot}9 \text{ part } 10^{-9} \text{ min}^{-1}$
(observed rate $5{\cdot}4$ parts 10^{-9} min^{-1})

TABLE 5.4 *Structural relationships in the activity of alkenes (compiled from reference* 18)

Alkene	NO_2 formation rate (parts 10^{-9} min^{-1})	Characteristic rate factor for structure shown (parts 10^{-9} min^{-1})$^{1/2}$		
ethene	1·7	1·3		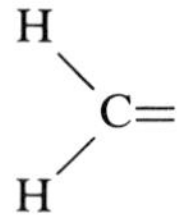
cis-but-2-ene	7·1	2·66	*cis*	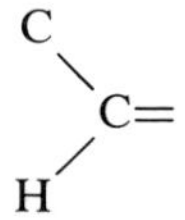
trans-but-2-ene	11·2	3·35	*trans*	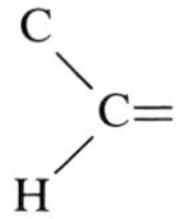
cis-hex-3-ene	5·0	2·23	*cis*	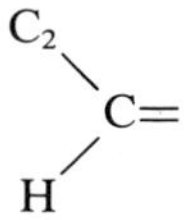
trans-hex-3-ene	5·9	2·43	*trans*	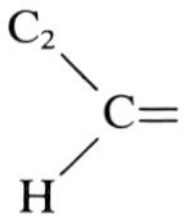
2,3-dimethylbut-2-ene	59	7·68		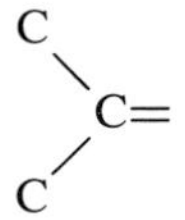

(b) 2-Methylbut-2-ene

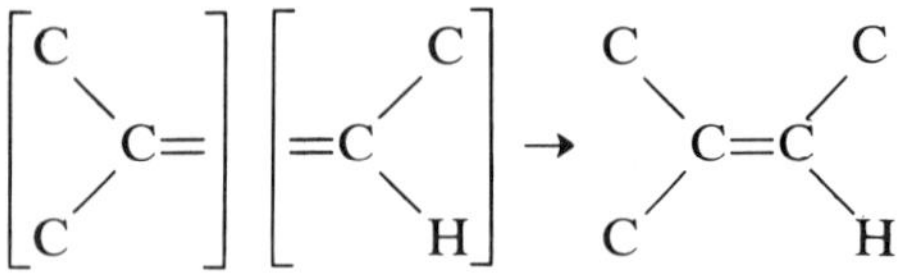

2-methylbut-2-ene

(7.68 parts 10^{-9} $min^{-1})^{1/2}$(2·66 parts 10^{-9} $min^{-1})^{1/2} = 20·4$ parts 10^{-9} min^{-1} (observed rate 18·4 parts 10^{-9} min^{-1})

(c) 2-Methylpent-2-ene

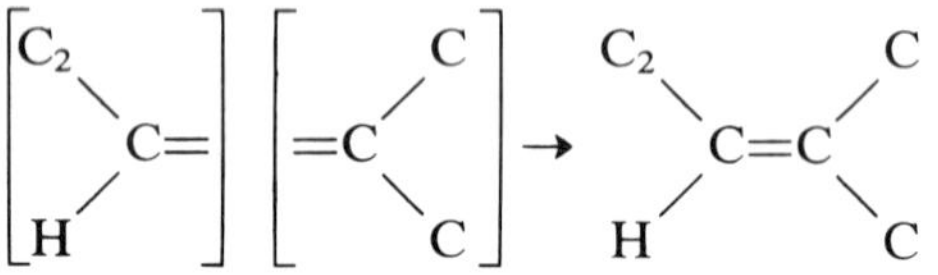

2-methylpent-2-ene

(2·23 parts 10^{-9} $min^{-1})^{1/2}$(7·68 parts 10^{-9} $min^{-1})^{1/2} = 17·1$ parts 10^{-9} min^{-1} (observed rate 16 parts 10^{-9} min^{-1})

Activities of alkenes that were investigated ranged over two orders of magnitude. Statistically, the correlation coefficient (p) at the 95% confidence limit between the measured and predicted rates were in the range $0·993 < p < 0·999$. As shown in Fig. 5.5, the rate of oxidation depends on the extent of substitution of alkene and upon the chain length of the substituents.

Undoubtedly, the model is reasonably self-consistent within the framework of the original conditions stipulated for alkene, NO and NO_2. In this respect, it follows the well known procedures used in estimating thermodynamic properties of compounds, for instance, which are based on chemical structure and configuration [20]. Whether the reliability of the method extends beyond the original conditions is open to question. As pointed out by Bufalini [21], the rate of oxidation of NO varies with the ratio of alkene to NO_x in the system and different ratios may well invalidate the predictions.

5.3.3.2 Reactions of methane

As discussed in the previous section, methane is the least active hydrocarbon and the principal atmospheric removal mechanism already considered in section 2.2.1 involves initial attack by an OH radical shown in equation (5.39). This reaction is some 15 times slower than the corresponding reaction between CO and OH, equation (5.15):

$$OH + CH_4 \rightarrow CH_3 + H_2O \quad (5.39)$$

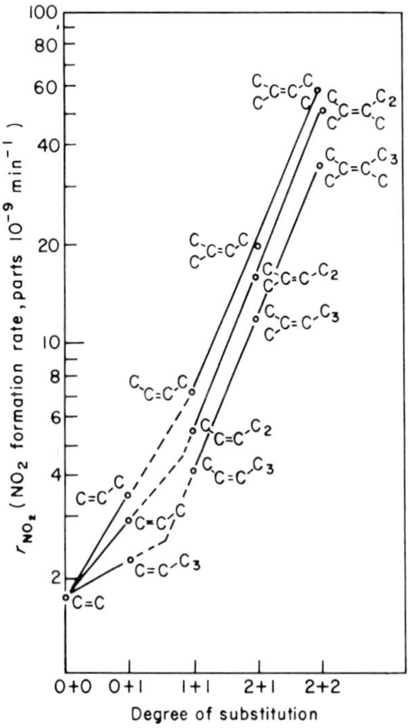

FIG. 5.5 Dependence of the rate of formation of NO_2 on the extent of substitution of the olefins [18]

According to Levy [22] this is followed by

$$CH_3 + O_2 + M \rightarrow CH_3OO + M \quad (5.40)$$

and then

$$CH_3OO + NO \rightarrow CH_3O + NO_2 \quad (5.41)$$

or

$$CH_3OO + CH_3OO \rightarrow 2CH_3O + O_2 \quad (5.42)$$

and finally,

$$CH_3O + O_2 \rightarrow HO_2 + HCHO \quad (5.43)$$

It should be noted that the hydroperoxyl radical HO_2 will be available for reaction with nitric oxide by reaction (5.20a):

$$NO + HO_2 \rightarrow NO_2 + OH \qquad (5{\cdot}20a)$$

and hence this sequence of reactions provides another route for the oxidation of nitric oxide without the participation of ozone by reaction (5.14)

$$O_3 + NO \rightarrow NO_2 + O_2 \qquad (5.14)$$

and this means that the concentration of ozone in the atmosphere will increase. A diagrammatic representation of the atmospheric oxidation of methane is shown in Fig. 5.6.

5.3.3.3 Reaction with alkenes

The olefinic $>C{=}C<$ double bond, of course, represents a weakness in this type of hydrocarbon. This is the location in an alkene where oxygen atoms and ozone attack. As an illustration equation (5.44) shows the result of interaction between propene and an oxygen atom. The unstable intermediate breaks up according to equations (5.45) and (5.46) into alkyl radical and oxygenated radicals:

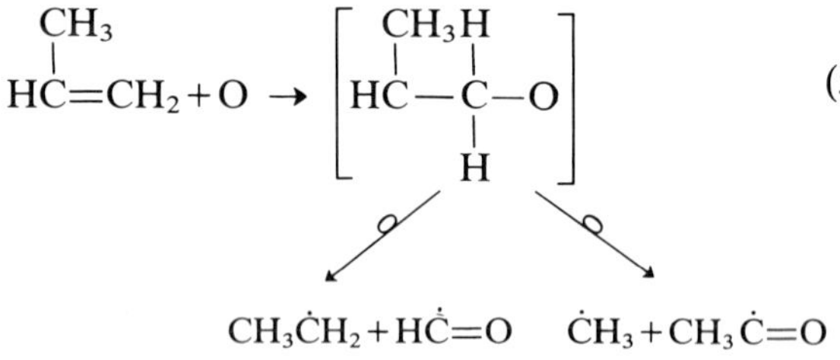

(5.44)

main products (5.45) minor products (5.46)

These free radicals are very reactive chemically and immediately become involved in other processes.

Initial gas phase reaction between ozone and alkenes is thought to proceed in two stages. The first stage produces an initial complex [reaction (5.47)] which then breaks up into an aldehyde or ketone and a peroxy biradical, one form of which is the zwitterion shown in equation (5.48):

$$O_3 + RCH{=}CHR \rightarrow [\text{initial complex}] \qquad (5.47)$$

$$[\text{initial complex}] \rightarrow RCHO + R\dot{C}HO\dot{O} \rightleftharpoons R\overset{+}{C}HOO^{-} \qquad (5.48)$$

By analogy with similar reactions in the liquid phase [23] the initial complex may be in the form of a 4-membered σ-complex,

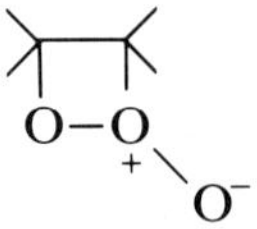

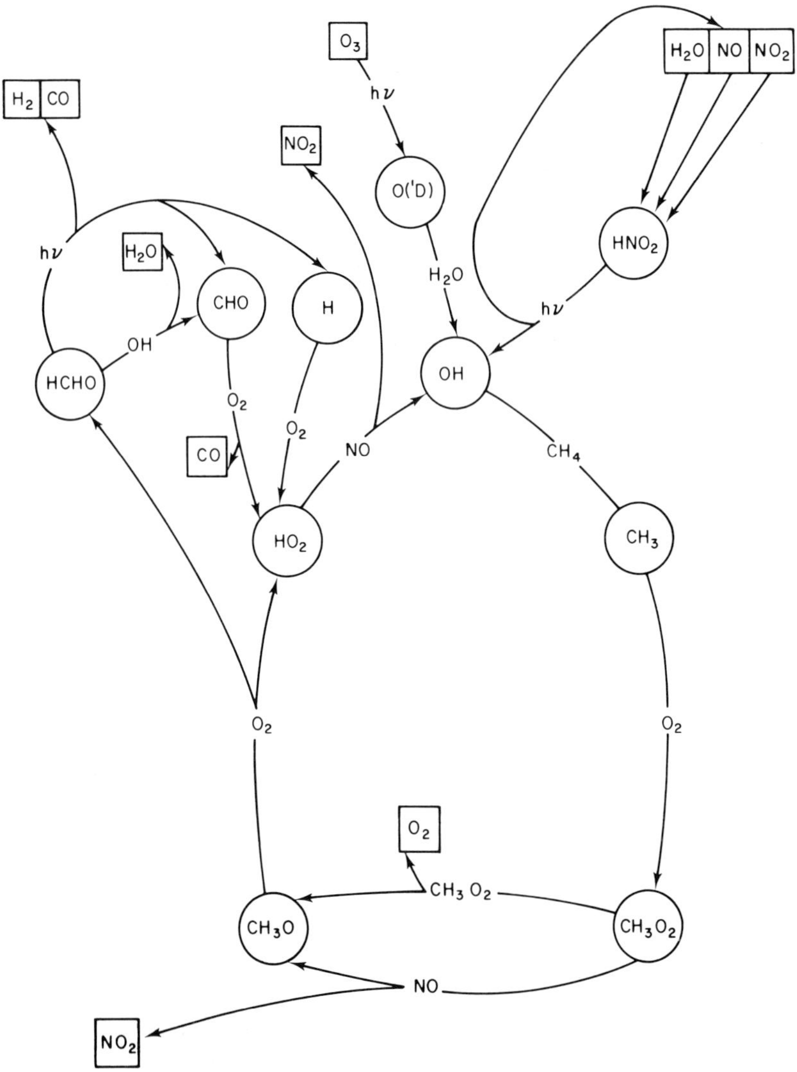

FIG. 5.6 Simplified scheme summarizing the atmospheric chemistry of methane

or a 5-membered π-complex

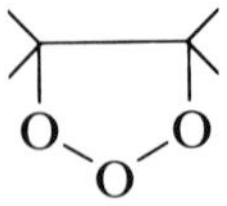

The 4-membered ring system is termed the molozonide, which in the presence of excess aldehyde can lead to oxidation of the aldehyde to an acid and cleavage products from the alkene. The sequence of reactions is indicated below:

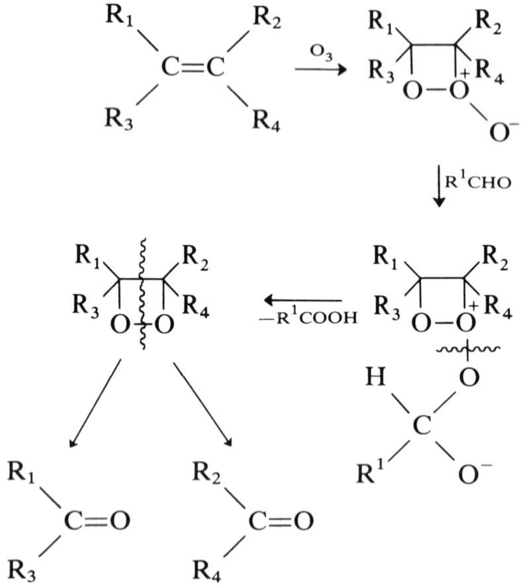

The biradical or zwitterion shown in equation (5.48) can undergo further reaction with other radical species, e.g. NO, or decompose, as indicated by equations (5.49a) and (5.49b):

$$RCHOO \rightarrow RH + CO_2 \quad (5.49a)$$

$$RCHOO \rightarrow ROH + CO \quad (5.49b)$$

Although these currently held views are based on experimental studies in the liquid phase, there is every reason to believe that similar transformations occur in the gas phase and for that matter at catalytic surfaces.

In addition to these reactions of alkenes with atomic oxygen and ozone, Winer and Bayes [24] demonstrated in 1966 that the first excited state of molecular oxygen, $O_2(^1\Delta_g)$, could react with 2,3-dimethylbut-2-ene tetramethylethylene, i.e. TME) according to equation (5.50), yielding the peroxide 2,3-dimethyl-2-hydroperoxybut-1-ene:

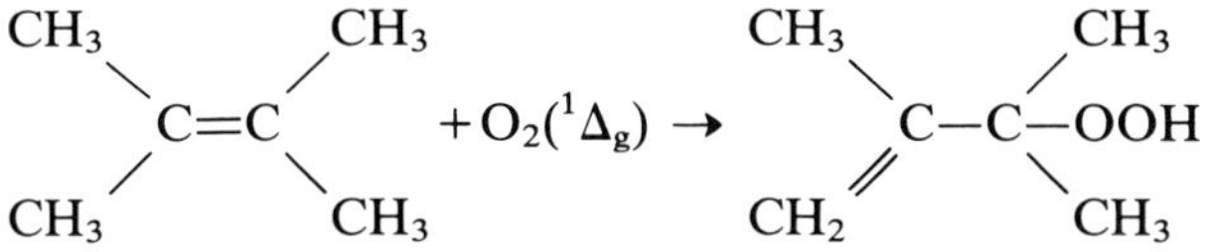

(5.50)

This important concept has led to a new interest being shown in the role of singlet oxygen in atmospheric photochemistry.

As discussed by Pitts *et al.* [25] collisional deactivation of $O_2(^1\Delta_g)$ by O_2 or N_2 is an inefficient process. Thus, the relatively long life of $O_2(^1\Delta_g)$ would permit these excited molecules, if they existed in polluted air, to encounter an alkene molecule. The question of origin of $O_2(^1\Delta_g)$, therefore, arises if serious consideration is to be given to its involvement in urban atmospheric and tropospheric chemistry. The production through direct absorption of radiation by way of reactions (5.51) and (5.52) was considered by Leighton [26] but was rejected as untenable:

$$O_2(^3\Sigma_g^-) + h\nu \rightarrow O_2(^1\Sigma_g^+) \qquad \lambda = 762 \text{ nm} \tag{5.51}$$

$$O_2(^3\Sigma_g^-) + h\nu \rightarrow O_2(^1\Delta_g) \qquad \lambda = 1269 \text{ nm} \tag{5.52}$$

It is now appreciated that there are other possibilities, these include the following sources of origin:

(1) direct photolysis of ozone [27],
(2) as a by-product of energy transfer processes [25, 28].

Equation (5.9) given in Table 5.2, viz.

$$O_3 + h\nu \rightarrow O_2(^1\Delta_g \text{ or } ^1\Sigma_g^+) + O(^1D) \qquad \lambda = 308 \text{ nm} \tag{5.9}$$

suggests one source of $O_2(^1\Delta_g)$ that has already been mentioned. It is possible that equation (5.9) is an oversimplification because in practice in the presence of oxygen, $O_2(^3\Sigma_g^-)$, atomic oxygen $O(^1D)$ may well be the critical factor [29] through reaction (5.53):

$$O(^1D) + O_2(^3\Sigma_g^-) \rightarrow O_2(^1\Delta_g \text{ or } ^1\Sigma_g^+) + O(^3P) \tag{5.53}$$

The proposals of Pitts *et al.* [25] neatly by-pass the problems associated with weak direct absorption of radiation according to equations (5.51) and (5.52). They are based on the fact that a high yield of $O_2(^1\Delta_g)$ can be obtained when solar radiation is absorbed initially by an organic molecule which then on collision with oxygen transfers some electronic energy to molecular oxygen. The overall mechanism of this energy transfer can be expressed by equations (5.54), (5.55) and (5.56):

$$D(S_0)+h\nu \rightarrow D(S_1) \tag{5.54}$$

$$D(S_1) \xrightarrow[\text{crossing}]{\text{intersystem}} D(T_1) \tag{5.55}$$

$$D(T_1)+O_2(^3\Sigma_g^-) \rightarrow D(S_0)+O_2(^1\Delta_g \text{ or } ^1\Sigma_g^+) \tag{5.56}$$

where D denotes the donor organic molecule that absorbs sunlight and the symbols S_0, S_1 and T_1 stand for the ground state, first excited state and triplet state of this molecule, respectively. These energy transitions are shown schematically in Fig. 5.7. Hence, in many cases, it is the triplet state of the donor molecule that transfers energy to oxygen. Kummler and Bortner [30] were the first to demonstrate that excited benzaldehyde could act in this capacity as a donor molecule which when in collision with oxygen generated $O_2(^1\Delta g)$. Further proof of energy transfer reactions was obtained by Steer *et al.* [31] who showed that irradiation of a mixture of oxygen, benzene and tetramethylethylene gave 2,3-dimethyl-2-hydroperoxybut-1-ene. In this instance benzene is acting as the donor.

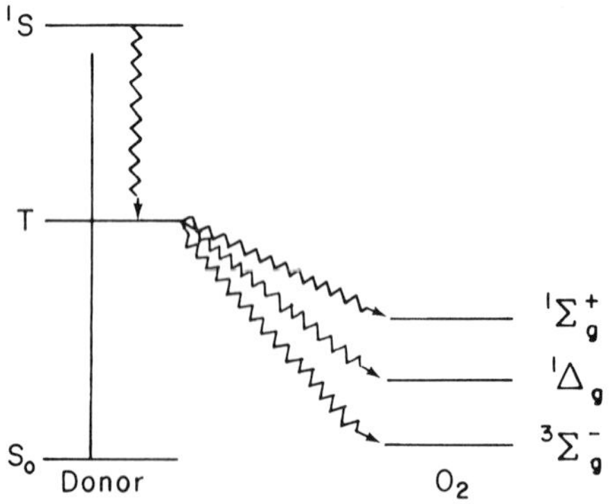

FIG. 5.7 Schematic representation of energy transfer as envisaged by Pitts *et al.* [25] from a donor organic molecule to generate excited states of molecular oxygen

Polluted air contains aldehydes, ketones and aromatic hydrocarbons, all of which may act as triplet sensitizers to produce the potent oxidizing agent $O_2(^1\Delta_g)$. These ingredients add yet another dimension to the complexity of the reactions that are possible, especially in urban environments in strong sunlight.

Early investigations into the role of alkenes in photochemical smog reactions envisage schemes of the sort shown in Fig. 5.8. These formulations assume that initial reaction between either oxygen atoms or ozone generate biradicals, which subsequently follow a decomposition–oxidation sequence leading to a stable molecule and a monoradical. These monoradicals with the odd electron centred on the carbon atom are rapidly oxidized to peroxyalkyl radicals (ROO) which in turn oxidize NO to NO_2 through reaction (5.20). During this oxidation step the peroxy radicals are themselves converted to alkoxyl radicals (RO) which ultimately end up as

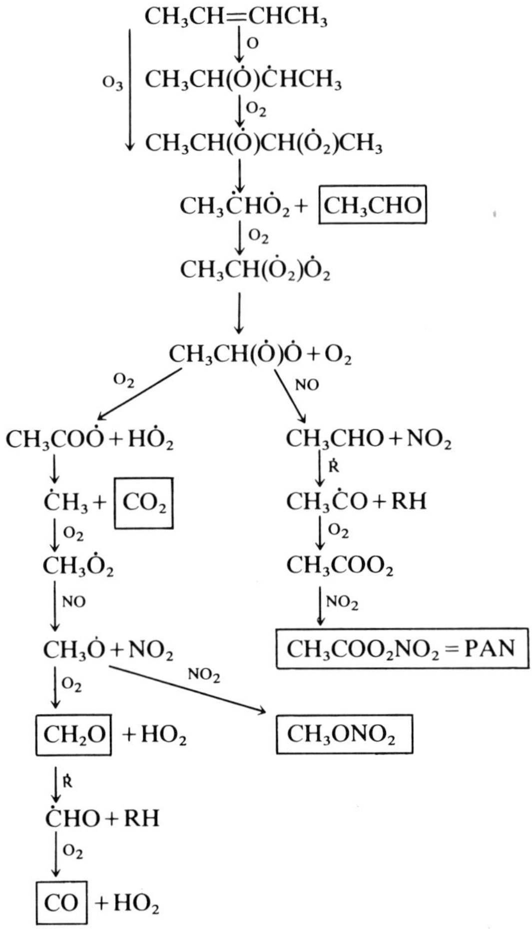

FIG. 5.8 Oxidative degradation scheme for an alkene initiated by oxygen atom or ozone attack at the double bond [5]

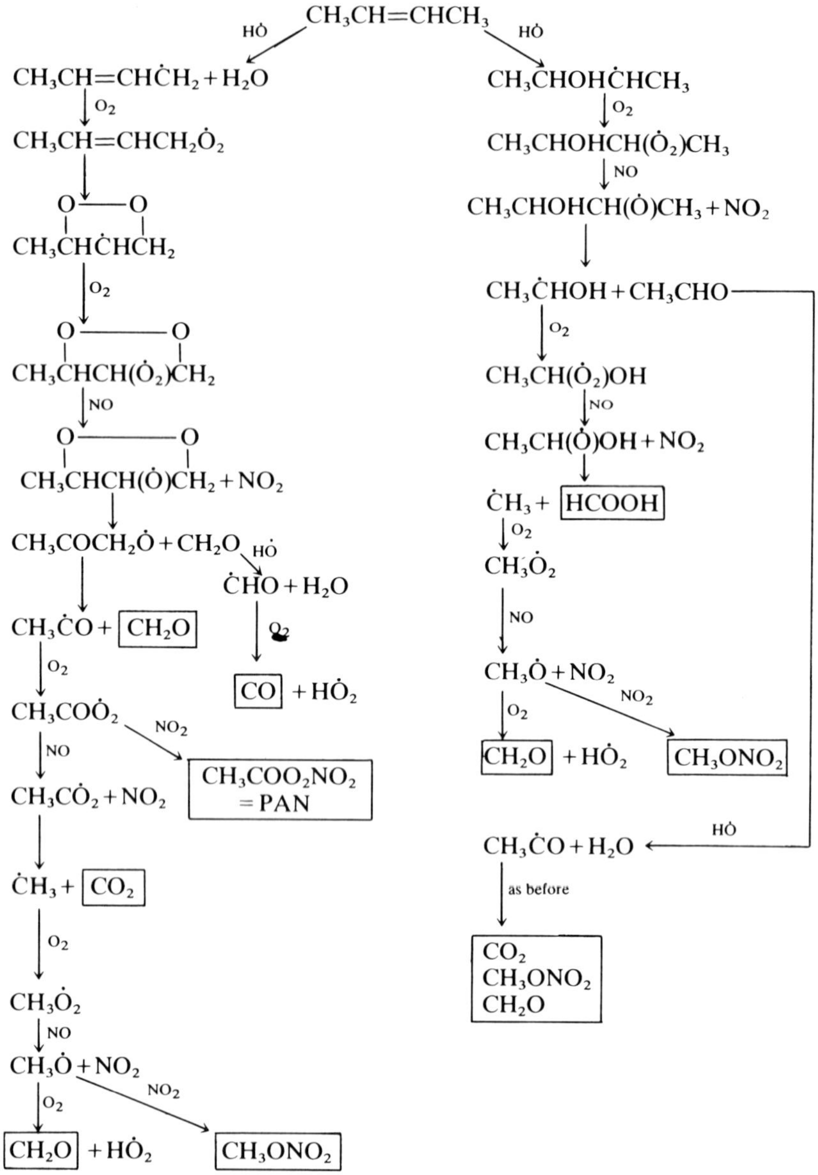

FIG. 5.9 Oxidative degradation scheme for an alkene initiated by hydroxyl radical attack at the double bond [5]

stable carbonyl products and another free radical. Eventually, the hydrocarbon is oxidized to products containing fewer carbon atoms than the original hydrocarbon.

This scheme is inadequate because it cannot predict the disappearance of hydrocarbon from the system at a fast enough rate. In keeping with the discussion in section 5.3.3.1 it is now believed that a large contribution to hydrocarbon removal is initiated by a free radical chain reaction involving the OH radical. This modified scheme is shown in Fig. 5.9. The OH radical reacts with the alkene by either hydrogen atom abstraction or direct addition across the double bond. These are the left- and right-hand routes, respectively, of the scheme shown in Fig. 5.9. The free radical species formed then undergoes similar transitions to oxidized products as already discussed.

A comparison of the experimental results found for the irradiation of *trans*-but-2-ene with NO, NO_2 and air with a computer simulation based on rate data for the OH radical initiated reaction has been carried out by Kerr *et al.* [5]. Their results, which involve a consideration of over 200 reactions are shown in Fig. 5.11. Generally, good agreement between the experimental and calculated curves has been achieved. The largest discrepancy is for carbon monoxide where the model predicts lower concentrations than are found experimentally (Fig. 5.10).

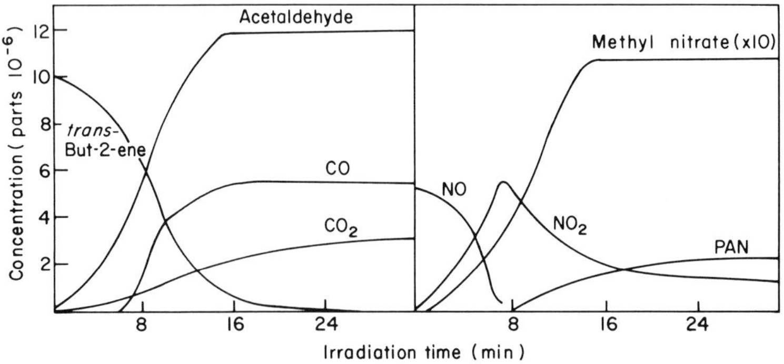

FIG. 5.10 Variation in concentration of reactants and products during the irradiation of *trans*-but-2-ene, NO, NO_2 and air [5]

The general form of these reactant–product–time profiles should be noted and compared with those obtained from experiments in which dilute motor vehicle exhaust gas has been irradiated (Fig. 5.12), and Fig. 5.13 which reconstructs the events during a photochemical smog episode. Between 06·00 and 08·30 when peak traffic flows occur hydrocarbons and oxides of nitrogen are discharged into the atmosphere. Initially, at 06.00

NO concentrations are greater than those of NO_2. As the temperature rises during the morning and sunlight initiates photochemical reactions NO becomes oxidized to NO_2. The participation of alkoxy and peroxyalkyl radicals derived from hydrocarbons in the polluted atmosphere causes the overall hydrocarbon concentration to fall. Since the oxidation of NO to NO_2 has been accomplished largely by these radicals through reaction (5.20) without involving ozone, the ozone concentration gradually increases during the course of the day reaching a maximum between 12.00 and 15.00 (Fig. 5.13). On occasions concentrations of ozone as high as 60 parts 10^{-8} have been recorded in Los Angeles. The increase in ozone in the

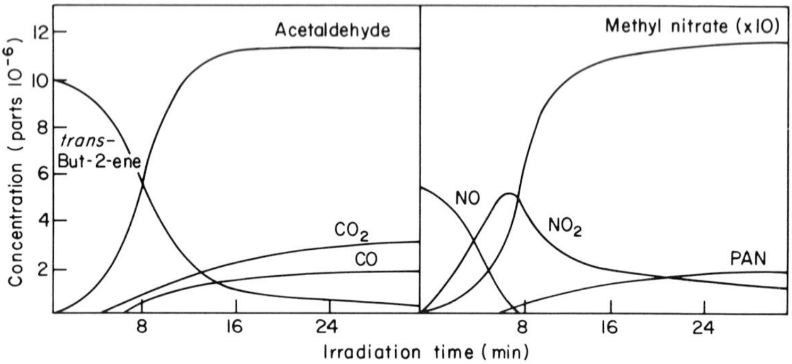

FIG. 5.11 Computer-simulated product and reactant concentrations during the irradiaton of a mixture of *trans*-but-2-ene, NO, NO_2 and air [5]

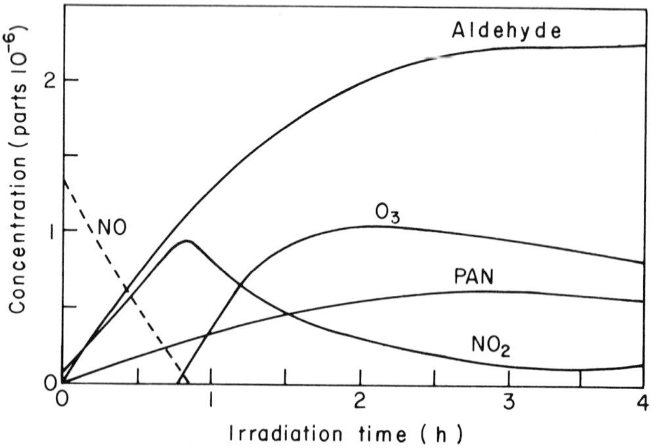

FIG. 5.12 Concentration changes in oxides of nitrogen and certain products on irradiation of dilute automobile exhaust in air [26]

early afternoon is accompanied by an increase in aldehydes and peroxyacylnitrates, PAN. This shows that these compounds originate from photochemical reaction rather than being emitted directly to atmosphere from vehicles. The nitrates shown in Figs 5.9, 5.10 and 5.11 are formed by

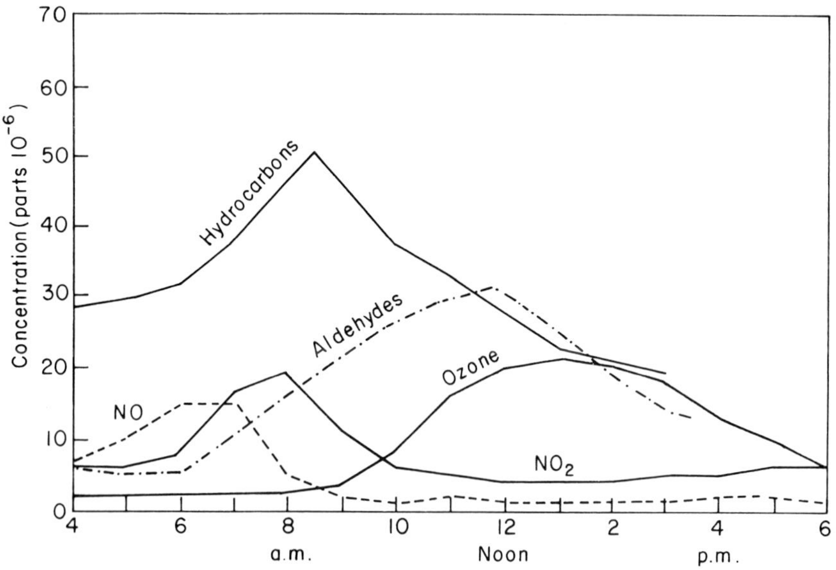

FIG. 5.13 Average concentrations during days of eye irritation in downtown Los Angeles. Hydrocarbons, aldehydes, and ozone for 1953–54. Nitric oxide and nitrogen dioxide for 1958. From data of the Los Angeles County Air Pollution Control District [26]

reaction of NO_2 with alkoxyl and peroxyalkyl radicals according to equations (5.57) and (5.58), e.g.

$$CH_3O + NO_2 \rightarrow CH_3ONO_2 \quad (5.57)$$

and

$$CH_3\overset{\overset{\displaystyle O}{\|}}{C}O_2 + NO_2 \rightarrow CH_3\overset{\overset{\displaystyle O}{\|}}{C}O_2NO_2 \quad (5.58)$$

During daylight reaction (5.20b):

$$NO + RO_2 \rightarrow NO_2 + RO \quad (5.20b)$$

is more important than reaction (5.20a),

$$NO + HO_2 \rightarrow NO_2 + OH \quad (5.20a)$$

In the morning reaction (5.24):

$$NO_3 + NO \rightarrow 2NO_2 \qquad (5.24)$$

is of little consequence but at the end of the day when photolysis ceases the emphasis changes. The concentrations of RO_2 and HO_2 are mainly derived from the photodissociation of aldehydes, as their concentrations decrease at sundown residual ozone reacts with NO_2 by reaction (5.21):

$$NO_2 + O_3 \rightarrow NO_3 + O_2 \qquad (5.21)$$

Thus, at night reaction (5.24) provides a route for the oxidation of NO to NO_2 [32].

Ethylene itself does not give nitrogen-containing organic products; with this system nitrogen oxides are converted by ozone in the presence of water vapour to nitric acid through reactions (5.21), (5.23) and (5.25).

5.3.4 Participation of sulphur dioxide

Our understanding of the involvement of sulphur dioxide in atmospheric reactions is less advanced than those discussed in the previous section for the nitrogen oxides. The literature on the subject abounds with rival viewpoints expounding the validity and importance, or otherwise, of photochemical oxidative processes over homogeneous and heterogeneous catalytic reactions. There is no doubt that each have their merits and in practice it is environmental factors that exert some influence [33]. For example, the principal oxidative mechanism within a stack gas plume is most probably catalytic rather than photochemical. In the general urban scene, on the other hand, climate permitting, photochemical processes play a predominant role. In the absence of favourable photochemical conditions or away from badly polluted atmospheres the significance of the free radical OH and HO_2 oxidative routes must not be overlooked. According to Roberts and Friedlander [34] these routes can be summarized as follows:

(1) Liquid phase oxidation of SO_2 catalysed by metals [35].
(2) Catalysed oxidation of SO_2 adsorbed on soot particles [36].
(3) Homogeneous gas phase oxidation by OH radicals [37].
(4) Homogeneous gas phase oxidation of SO_2 by an intermediate of the ozone–NO_x–hydrocarbon photochemical reaction [38].
(5) Homogeneous gas phase oxidation by other intermediate species in (4), e.g. HO_2, RO_2 and RO.

In the absence of particulate matter, computer-simulation studies for an alkene–NO_x–SO_2 system indicate that routes (3) and (4) are significant [4].

Measurement of the oxidation rate for SO_2 within a smoke plume have been shown by Gartrell *et al.* [39] to be strongly influenced by the relative humidity. Little oxidation occurs for relative humdities below 70% but at higher values conversions of 22% after 12 min and up to 32% in 96 min were observed. These oxidation rates are too fast to be attributed to photochemical processes and are more consistent with those expected for metal catalysed liquid phase reactions. Plumes from power stations using solid fuel as the energy source will contain an abundance of iron and other heavy metals in the fly-ash. When dissolved in liquid droplets within the plume these metals can act as catalysts for the conversion of sulphur dioxide into sulphuric acid:

$$2SO_2 + H_2O + O_2 \rightarrow 2H_2SO_4 \qquad (5.59)$$

Manganese and iron salts are well known examples of catalysts which have this capability. Foster [35] has analysed these systems in some detail in the following manner:

If G parts 10^{-6} is the concentration of SO_2 in undiluted stack gas, then the SO_2 content at a point within the plume will be given by $GD \times 10^{-6}/22{\cdot}4$ mol l^{-1}_{plume}, where D is the effluent dilution factor at this point. If some fraction f_0 of the SO_2 has been converted to sulphate, then

$$SV = \frac{f_0 GD \times 10^{-6}}{22{\cdot}4} \qquad (5.60)$$

where V represents the water droplet content of the plume at this point, l_{H_2O}/l_{plume}, and S is the sulphate concentration made up from metal sulphates and sulphuric acid, mol $l^{-1}_{H_2O}$.

The total dust content of the plume will contain a fraction f_i of catalytic species i which have dissolved in the droplet, so that the weight of soluble oxide, i, in the plume at this point will be f_iDW g l^{-1}_{plume}, where W is the original effluent dust burden (g l^{-1}). For an oxide species of molecular weight, M_i, which supplies, n_i, catalytic ions from each molecule, the catalyst concentration within the droplet will be:

$$Ci = \frac{n_i f_i DW}{M_i V} \text{ mol } l^{-1}_{H_2O} \qquad (5.61)$$

The kinetics of oxidation of SO_2 is known to be second-order with respect to manganese salts, therefore:

$$\text{Rate of oxidation by } Mn_3O_4 = k_i C_i^2 V \text{ mol } l^{-1}_{plume} \text{ min}^{-1} \qquad (5.62)$$

where k_i is the reaction rate constant, l_{H_2O} mol^{-1} min^{-1}.

Using equation (5.60), this rate of oxidation can be expressed in terms of the percentage of total sulphur present in the plume:

$$\text{Rate of } SO_2 \text{ oxidation by } Mn_3O_4 = \frac{22{\cdot}4k_iC_i^2V \times 10^2}{GD \times 10^{-6}} \% \text{ min}^{-1} \qquad (5.63)$$

Substituting for C_i from equation (5.61) and for V from equation (5.61) into equation (5.63) gives:

$$\text{Rate of } SO_2 \text{ oxidation by } Mn_3O_4 = \frac{10^{16}Sk_i}{}\left(\frac{Wn_if_i}{GM_i}\right)^2 \% \text{ min}^{-1} \qquad (5.64)$$

Taking values of $W = 2\times10^{-3}$; $G = 2{\cdot}5\times10^3$; $D = 10^{-3}$; $f_0 = 10^{-1}$; $S =$ 1·0; molecular weight of Mn_3O_4 229; $n_i = 3$ and $f_i = 2\times10^{-4}$; the rate of SO_2 oxidation from equation (5.64) comes to $4{\cdot}4\times10^{-7}\ k_i$ % min^{-1}.

In the selection of the values used in equation (5.64) the estimation of the catalytic content of the droplet, f_i, is the least certain. This is because of insufficient information of the chemical constitution and solubilities of dust particles in the plume. It has been assumed that 10% of the metal oxide in the particulate matter is dissolved in the droplet.

In the case of the iron catalyst, Foster considers that the overall rate of SO_2 oxidation within the plume is first-order with respect to dissolved SO_2 and inversely proportional to the pH. Using this information the rate of oxidation due to the iron catalyst can be written:

$$\text{Rate of } SO_2 \text{ oxidation by } Fe_2O_3 = \frac{k_i'C_i[SO_2]V}{[H^+]} \text{ mol l}^{-1}_{\text{plume}} \text{ min}^{-1} \qquad (5.65)$$

where k_i' min^{-1} is the first-order reaction rate constant, C_i is the concentration of the Fe^{3+} ion, $[SO_2]$ the concentration of SO_2 and $[H^+]$ the concentration of hydrogen ions in the droplet. The solution of SO_2 in the aqueous phase will depend on Henry's law, so that it is directly related to the partial pressure of undissolved SO_2 in the plume. This will be

$$[SO_2] = 10^{-6}GD(1-f_0)K_H \qquad (5.66)$$

where K_H is the solubility constant, 1·5 mol l^{-1} atm^{-1}. Substituting equations (5.66) and (5.61) into (5.65) gives

Rate of SO_2 oxidation by Fe_2O_3

$$= \frac{2{\cdot}24\times10^3DWK_H(1-f_0)k_i'n_if_i}{[H^+]M_i} \% \text{ min}^{-1} \qquad (5.67)$$

Using the same values as before except that $M_i = 160$: $n_i = 2$ and $f_i = 10^{-2}$ gives an SO_2 oxidation rate of $7{\cdot}6\times10^{-7}\ k_i'/[H^+]$% min^{-1}.

In order to complete the estimate of SO_2 oxidation rate some knowledge of the pH of the droplet is necessary. The acidity is known to affect the oxidation rate in both instances, and in the iron case the $[H^+]$ is actually included in the rate expression. In reality the acidity of the droplet changes with time because of the neutralizing influence of alkaline components, notably calcium oxide (Table 2.11, section 2.3.1.1), which is present in the fly-ash. On this evidence, during the initial neutralizing stage the pH of the droplet is assumed to be in the range 5–6. An estimate of the reaction rate constant, k_i, applicable to the manganese catalysed reaction is put at about $2 \times 10^5 \ \mathrm{l\,mol\,min^{-1}}$ and around $2 \ \mathrm{min^{-1}}$ for k_i' of the iron-catalysed reaction. When these values are used the rate of oxidation of SO_2 comes to about 0·1 and between 0·15 and 1·5% $\mathrm{min^{-1}}$ for the Mn- and Fe-catalyzed reactions, respectively. This, as Foster points out, is the correct order of magnitude for oxidation rates within aerosol plumes of this type. Regarding the actual mechanism of the oxidation of sulphur dioxide within the droplet, this has recently been discussed by Barrie and Georgii [40]. They consider the oxidation to be governed by two chain reactions, one involving SO_3^- and the other HO_2 that are proceeding simultaneously:

1. SO_3^- chain carrier

$$SO_3^- + O_2 \rightarrow SO_5^- \tag{5.68}$$

$$SO_5^- + SO_3^{2-} \rightarrow SO_4^{2-} + SO_4^- \tag{5.69}$$

$$SO_4^- + SO_3^{2-} \rightarrow SO_3^- + SO_4^{2-} \tag{5.70}$$

The SO_5^- anion indicated in equation (5.68) will require sp^3d^2 hybridization giving a square pyramid configuration:

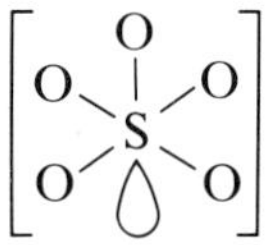

2. HO_2 chain carrier

$$HO_2 + SO_3^{2-} \rightarrow SO_4^{2-} + OH \tag{5.71}$$

$$OH + SO_3^{2-} \rightarrow SO_4^{2-} + H \tag{5.72}$$

$$H + O_2 \rightarrow HO_2 \tag{5.73}$$

When a metal such as iron is catalysing these reactions, the catalyst participates through the following reaction sequence:

$$Fe^{3+} + 3SO_3^{2-} \rightarrow [Fe(SO_3)_3]^{3-} \tag{5.74}$$

$$[Fe(SO_3)_3]^{3-} + O_2 \rightarrow [Fe(SO_3)_3]^{2-} + O_2^- \quad (5.75)$$

$$O_2^- + H^+ \rightarrow HO_2 \quad (5.76)$$

$$[Fe(SO_3)_3]^{2-} + SO_3^{2-} \rightarrow [Fe(SO_3)_3]^{3-} + SO_3^- \quad (5.77)$$

Turning our attention now to photochemical processes Calvert and co-workers [41] have studied the possible involvement of the excited states of SO_2 in the troposphere. SO_2 absorbs radiation between 290 and 340 nm to give a singlet excited species $^1SO_2$:

$$SO_2 + h\nu \rightarrow {}^1SO_2 \quad (5.78)$$
$$\lambda = 290 - 340 \text{ nm}$$

and between 340 and 400 nm to give the triplet state, $^3SO_2$:

$$SO_2 + h\nu \rightarrow {}^3SO_2 \quad (5.79)$$
$$\lambda = 340\text{–}400 \text{ nm}$$

In the lower atmosphere $^1SO_2$ collisional deactivation with other SO_2 molecules constitutes the principal removal mechanism for the singlet condition. This occurs by way of reactions shown in equations (5.80a) and (5.80b):

$$^1SO_2 + SO_2 \xrightarrow{91\%} SO_2 + SO_2 \quad (5.80a)$$

$$^1SO_2 + SO_2 \xrightarrow{9\%} {}^3SO_2 + SO_2 \quad (5.80b)$$

Sulphur trioxide is also formed with a quantum yield of about 0·04 to 0·05 and this is thought to arise [42] through the reaction

$$^3SO_2 + SO_2 \rightarrow SO_3 + SO \qquad \Delta H^0 = -109 \text{ kJ} \quad (5.81)$$

According to Calvert and co-workers [41] the triplet state of sulphur dioxide is the major, possibly exclusive state in the photochemistry of SO_2.

When NO_x is introduced into the photochemical SO_2–air system, reaction between atomic oxygen $O(^3P)$ as shown in equation (5.82) becomes feasible; although the limiting factor will be the low density of $O(^3P)$ in the troposphere:

$$O(^3P) + SO_2 + M \rightarrow SO_3 + M \quad (5.82)$$

Direct interaction between various nitrogen oxides with SO_2 shown in equations (5.83), (5.84) and (5.85) are insignificant [33]

$$SO_2+NO_2 \rightarrow SO_3+NO \quad (5.83)$$

$$SO_2+NO_3 \rightarrow SO_3+NO_2 \quad (5.84)$$

$$SO_2+N_2O_5 \rightarrow SO_3+N_2O_4 \quad (5.85)$$

as indeed is the direct oxidation of SO_2 by ozone [43]

$$SO_2+O_3 \rightarrow SO_3+O_2 \quad (5.86)$$

Once hydrocarbon is added to the SO_2–NO_x–air system the rate of oxidation of SO_2 increases. This observation was first reported in 1958 by Schuck *et al.* [44] when they showed that irradiation of motor vehicle exhaust gas into which SO_2 had been introduced gave a marked rapid increase in aerosol formation. It is now believed that as in other atmospheric and atmospheric photochemical reactions the radicals OH and HO_2 are primarily responsible for this behaviour [37]:

$$SO_2+HO_2 \rightarrow SO_3+OH \quad (5.87)$$

$$SO_2+OH+M \rightarrow HSO_3+M \quad (5.88)$$

The principal reaction of HO_2 in the troposphere is the conversion of NO to NO_2 by reaction (5.20a), so in fact although reaction (5.87) will occur it seems only to be of minor importance in urban atmospheres. In the lower stratosphere, however, reaction (5.87) could be more significant than reaction (5.82). This can be seen by inspection of Fig. 5.3 where the HO_2 density below altitudes of about 20 km is greater than that of atomic oxygen.

Another feature of reaction (5.87) that should be noted concerns the inhibition of SO_2 oxidation by NO discovered in environmental chamber studies by Smith and Urone [45]. Increasing concentrations of NO in an irradiated mixture of alkene–SO_2–NO_x–H_2O were shown to decrease the rate of oxidation to SO_3. Upon introduction of NO competition for the HO_2 radical between SO_2 and NO occurs, equations (5.87) and (5.20a). Since NO is a more efficient free radical scavenger than SO_2 reaction (5.20a) is favoured more than (5.87).

Reaction (5.88) is thought to be of major importance in both the troposphere and stratosphere. Using the value of $1{\cdot}5\times10^{-31}$ cm^3 molecule^{-1} s^{-1} for the reaction rate constant of reaction (5.88) found by Payne *et al.* [46] and an average number density of $3{\cdot}7\times10^6$ for the OH radical, Levy [47] estimates the residence time of 1–2 days for SO_2 in the lower atmosphere. This assumes that all the HSO_3 is quickly converted to sulphate. In these circumstances reaction (5.88) will be the main path for

SO_2 removal. In the presence of water vapour sulphur trioxide is rapidly converted to sulphuric acid, although even in this case there is some uncertainty about whether the process is surface induced:

$$SO_3 + H_2O \rightarrow H_2SO_4 \quad (5.89)$$

There is some information on the nature of products formed by SO_2 with alkanes when SO_2-alkane mixtures are irradiated [48]. For the system in which the excited triplet condition $^3SO_2$ is involved, there are basically two mechanistic routes, either insertion shown in reaction (5.90) or abstraction (5.91) for the formation of sulphinic acids and derivatives:

$$\text{insertion} \quad {}^3SO_2 + RH \rightarrow RSO_2H \quad (5.90)$$

$$\text{abstraction} \quad {}^3SO_2 + RH \rightarrow SO_2H + R \quad (5.91)$$

Of the alternatives, the authors tend to favour reaction (5.90). Although these direct SO_2-alkane reactions occur they proceed slowly and they cannot on their own account for the increased rate of sulphur dioxide disappearance when hydrocarbons are present in NO_x–SO_2–air systems.

The quenching of $^3SO_2$ by alkenes has also been studied by Calvert and co-workers [49] and in this instance a biradical intermediate is suggested as a route to sulphur containing hydrocarbon derivatives:

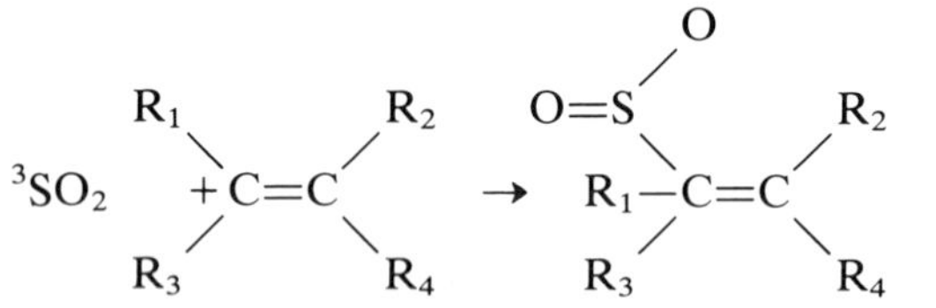

(5.92)

These postulates must await confirmation. In any event their participation in the troposphere will be minimal.

For conditions more appropriate to the troposphere Calvert and McQuigg [37] show that the OH and HO_2 radical reactions can account for at least 90% of the maximum oxidation rate of SO_2 of 4·5% h^{-1} in a *trans*-but-2-ene NO_x–SO_2–air system. The remaining 10% is made up by reaction of SO_2 with oxygenated hydrocarbon radicals such as RO, RCHOO and $O(^3P)$. The RCHOO radical, its zwitterion equivalent shown in equation (5.48) can react with SO_2 to give carbonyl products:

$$RCH{=}CHR + O_3 \rightarrow \text{[initial complex]} \quad (5.47)$$

$$\text{[initial complex]} + SO_2 \rightarrow 2RCHO + SO_3 \quad (5.93)$$

$$\text{[initial complex]} \rightarrow RCHO + RCHOO \quad (5.48)$$

$$RCHOO + SO_2 \rightarrow RCHO + SO_3 \quad (5.94)$$

The fact that yields of carbonyl compounds tend to increase in the presence of SO_2 lends support to reactions (5.93) and (5.94). This is also consistent with reduced ozone formation in these circumstances as the oxygenated hydrocarbons would otherwise have the opportunity to regenerate ozone through reaction (5.95) [50].

$$R\overset{+}{C}HOO^{-} + O_2 \rightarrow RCHO + O_3 \tag{5.95}$$

The reactions of SO_2 discussed in this section are summarized in Fig. 5.14.

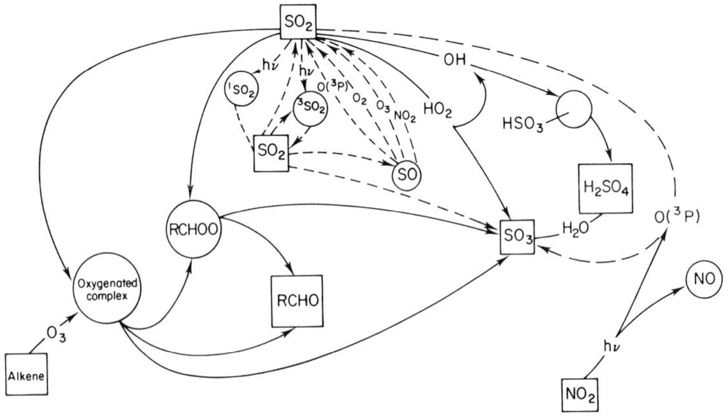

FIG. 5.14 Involvement of SO_2 in atmospheric chemical reactions

5.3.5 Photochemical oxidation of chlorinated alkenes

Unlike the fully halogenated compounds trichlorofluoromethane and dichlorodifluoromethane which are chemically unreactive, vinyl chloride, 1,1- and 1,2-dichloroethylene; tri- and tetra-chloroethylene may be photo-oxidized in the troposphere. Environmental chamber studies in which mixtures of halocarbon, air, and either NO_2 or NO_x have been carried out by Gay *et al.* [51]. They show that the reactivities of these compounds fall in decreasing order: 1,1-dichloroethylene > 1,2-dichloroethylene = trichloroethylene > ethylene > vinyl chloride ≫ tetrachloroethylene. Identification of reaction products was obtained from two separate series of experiments. For one series a long-path infrared photochemical reaction chamber was used and in the other wet chemical and gas chromatographic techniques were employed. Their results are summarized in Table 5.5.

Ozone was produced in all cases, showing that biradicals or peroxyalkyl radicals, RO_2, must react with NO to give NO_2 according to equation

TABLE 5.5 *Products formed after irradiation of halocarbons with NO or NO_x*

COMPOUND	FORMULA	Carbon monoxide CO	Hydrochloric acid HCl	Nitric acid HNO_3	Ozone O_3	Formic acid HCOOH	Formaldehyde HCHO	Formylchloride HCOCl	Monochloroacetylchloride $H_2CClCOCl$	Dichloroacetylchloride $HCCl_2COCl$	Trichloroacetylchloride CCl_3COCl	Phosgene $COCl_2$	Chlorinated PAN
vinyl chloride	$ClHC{=}CH_2$	+	+	t	+	+	+	t					?
1,1-dichloroethylene	$Cl_2C{=}CH_2$	+	+	+	+	+	+		+			+	
1,2-dichloroethylene	$ClHC{=}CHCl$	+	+	+	+	+							?
trichloroethylene	$Cl_2C{=}CHCl$	+	+	+	+	+				+		+	
tetrachloroethylene	$Cl_2C{=}CCl_2$	+	+	?	+	+					+	+	

t denotes trace.

(5.20b). This allows ozone in the system to build up in the familiar way. Some suggestions about the mechanism of formation of the products have been made and these are briefly summarized.

In the case of vinyl chloride either an initial oxygen atom attack as shown in equation (5.96) takes place or a radical reaction (5.97) occurs at the double bond:

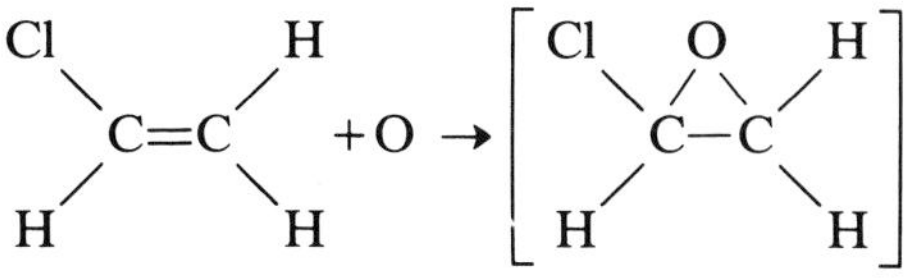

(5.96)

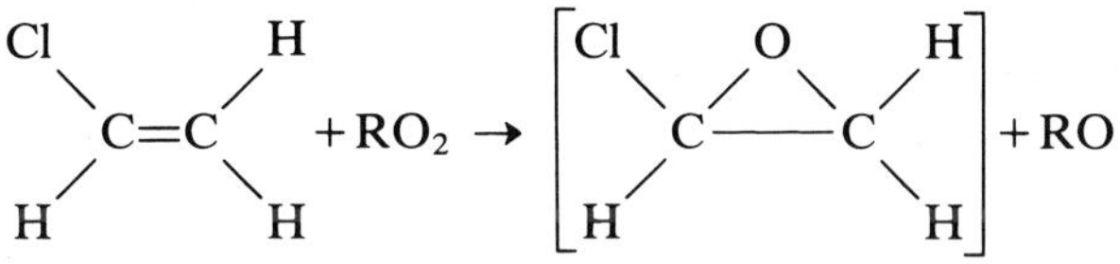

(5.97)

This epoxide intermediate then rearranges to give monochloroacetaldehyde:

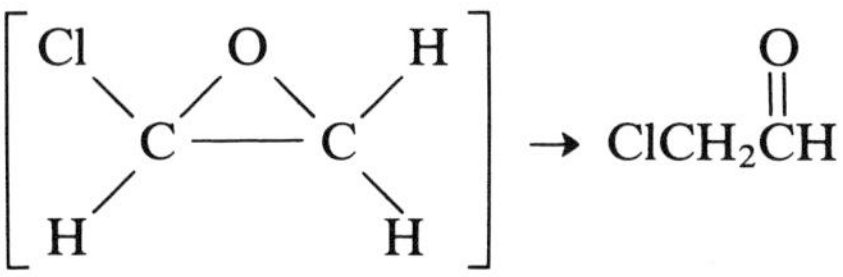

(5.98)

The chlorinated aldehyde can act as the precursor to chlorinated PAN through the following series of reactions:

$$\mathrm{ClCH_2\overset{\overset{O}{\|}}{C}H + OH \rightarrow ClCH_2\overset{\overset{O}{\|}}{C} + H_2O} \tag{5.99}$$

$$\mathrm{ClCH_2\overset{\overset{O}{\|}}{C} + O_2 \rightarrow ClCH_2\overset{\overset{O}{\|}}{C}{-}O{-}O} \tag{5.100}$$

$$\mathrm{ClCH_2\overset{\overset{O}{\|}}{C}{-}O{-}O + NO_2 \rightarrow ClCH_2\overset{\overset{O}{\|}}{C}{-}O{-}O{-}NO_2} \tag{5.101}$$

When two chlorine atoms are located on the same carbon atom as in 1,1-dichloroethylene the major product is chloroacetyl chloride and some phosgene is produced as well. Again, there are two possibilities: either ozone attack across the double bond followed by a break up of the molecule as indicated in equation (5.102) to yield phosgene, or by analogy with reaction (5.98) epoxide formation followed by rearrangement to give monochloroacetylchloride, equation (5.103).

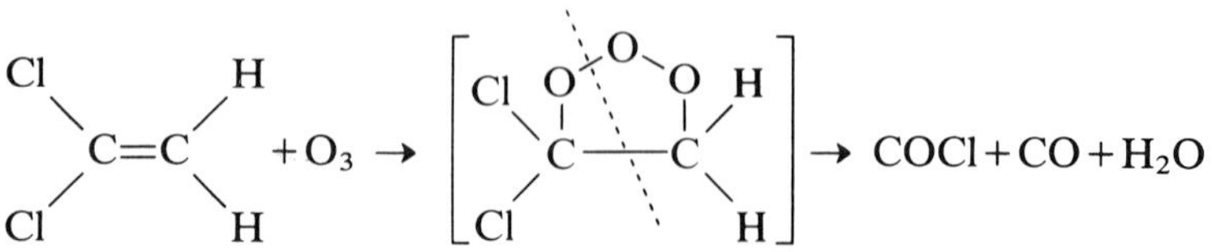
(5.102)

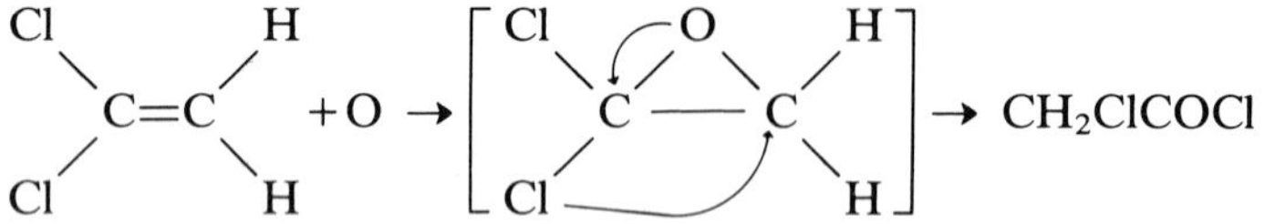
(5.103)

From the evidence presented in Table 5.5 where trichloroethylene gives dichloroacetylchloride together with phosgene and tetrachloroethylene gives trichloroacetylchloride plus phosgene similar mechanistic sequences are believed to be obeyed.

This behaviour may be contrasted with that of 1,2-dichloroethylene and for that matter vinyl chloride where acetyl chlorides are not formed. In these cases, after formation of the epoxy intermediate either by RO_2 or O atom attack across the double bond, chlorine atom transfer leads to dichloroacetaldehyde, equation (5.104).

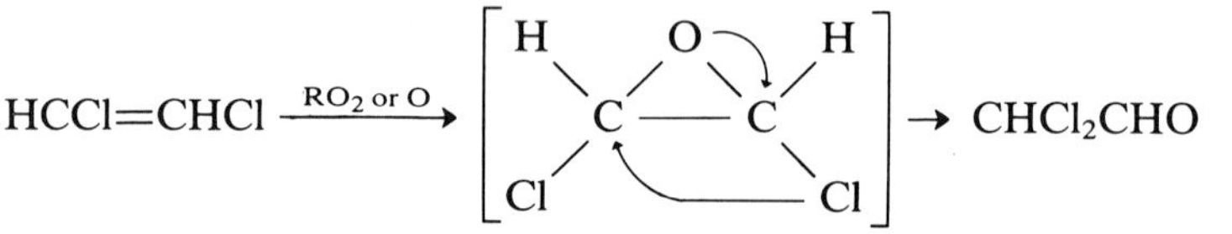
(5.104)

Chlorinated-PAN products may be formed through the sequence of reactions (5.99) to (5.101). Support for these proposals was obtained by Bufalini and co-workers from the fact that yields of HCl and CO were enhanced when vinyl chloride and 1,2-dichloroethylene were the reactants. Some evidence for the formation of PAN-type compounds was also found from infrared spectral absorptions at 800 cm^{-1}, 1300 cm^{-1} and 1750 cm^{-1}.

5.3.6 Inhibition of photochemical smog

We have seen from the chemistry described in this section 5.3 that radicals play a predominant role in atmospheric reactions. If a way could be found of slowing the conversion of NO to NO_2 by removal of the chain carrying OH radicals, in such a manner that the chains are not regenerated, it might be possible to exert some control of photochemical smog. Initial experiments with this objective in mind were conducted in smog chambers using phenol, benzaldehyde and aniline [52, 53]. These compounds were chosen because they have easily extractable H-atoms and do not have H-atoms attached to the α-carbon atom so that chain regeneration is inhibited. After initial abstraction of the H-atom as shown in reaction (5.105) with benzaldehyde as the example, the usual course of reaction leads to PAN-type products which in this case is the extremely lachrymatory peroxybenzylnitrate:

$$\phi CHO + OH \rightarrow \phi\overset{\overset{\displaystyle O}{\|}}{C} + H_2 \qquad (5.105)$$

$$\phi\overset{\overset{\displaystyle O}{\|}}{C} + O_2 \rightarrow \phi\overset{\overset{\displaystyle O}{\|}}{C}—O—O \qquad (5.106)$$

$$\phi\overset{\overset{\displaystyle O}{\|}}{C}—O—O + NO_2 \rightarrow \phi\overset{\overset{\displaystyle O}{\|}}{C}—O—O—NO_2 \qquad (5.107)$$

This approach has been developed by Heicklen and co-workers, who suggest that diethylhydroxylamine (DEHA) could be released into the atmosphere as a radical scavenger to prevent photochemical smog episodes [54, 55]. Smog chamber experiments indicate that when DEHA is introduced into a C_2H_4–NO–O_2 mixture at a concentration ratio of $[C_2H_4]/[DEHA]$ of 4 the initial rate of removal of ethylene is reduced by a factor of 5 and the initial conversion of NO to NO_2 is reduced by a factor of 20 compared with a mixture of similar composition except for DEHA. The reaction products in the presence of DEHA are CH_3CHO, $C_2H_5NO_2$, $C_2H_5ONO_2$, C_2H_5OH, HONO, NO_2, N_2O and CO_2 with trace amounts of CH_2O, CO, H_2O and HNO_3.

Mechanistically it is probable that initial reaction occurs between DEHA and a radical R, e.g. OH, $O(^3P)$:

$$R + (C_2H_5)_2NOH \rightarrow RH + (C_2H_5)_2NO \qquad (5.108)$$

The sequence of reactions that follow are speculative, but further reaction of the radical $(C_2H_5)_2NO$ with NO to yield a dimer intermediate has been proposed:

$$(C_2H_5)_2NO + NO \rightarrow (C_2H_5)_2N(O)NO \qquad (5.109)$$

The formation of the principal products may arise through the reaction of this intermediate with oxygen by way of a molecular process:

$$(C_2H_5)_2N(O)NO + O_2 \rightarrow 2CH_3CHO + N_2O + H_2O \qquad (5.110)$$

or a radical process:

$$(C_2H_5)_2N(O)NO + O_2 \rightarrow C_2H_5O_2 + C_2H_5O + N_2O \qquad (5.111)$$

although the latter cannot be a major pathway because reaction (5.111) generates rather than removes radicals. Nevertheless, Heicklen and his collaborators attribute some significance to (5.111) as a source of C_2H_5O which can be converted to ethanol by a hydrogen abstraction reaction with DEHA:

$$C_2H_5O + (C_2H_5)_2NOH \rightarrow C_2H_5OH + (C_2H_5)_2NO \qquad (5.112)$$

The suppression of photochemical smog by DEHA advocated by Heicklen poses important ethical questions for environmentalists. The principle of the injection of chemicals into the atmosphere to combat the effect of those already put there from anthropogenic sources raises moral issues that must be resolved. On these grounds alone debate is liable to be prolonged. Opposition to the scheme comes from respected scientists who fear the possible mutagenicity of DEHA. Structurally DEHA is very similar to the notorious carcinogen diethylnitrosamine and, therefore, must be regarded as potentially hazardous. As discussed by Maugh [56], informed opinion in the USA agrees that extensive tests on mutagenic properties of DEHA must be undertaken before even trial experiments in the atmosphere are contemplated.

5.3.7 Degradation of lead bromochloride, PbBrCl

Pierrard [57] has advanced the opinion that lead bromochloride in motor vehicle emissions decomposes photochemically with the rate of loss of bromine exceeding that of chlorine. The photochemical nature of the process, however, has been challenged by Robbins and Snitz [58]. Analysis of the Cl/Pb and Br/Pb ratios for a lead aerosol contained within a polyethylene covered cage in the dark also showed a rapid loss of halogen. The Cl/Pb ratio, for example, was halved in about 4 h. Initially, bromine decreases relative to chlorine at a much faster rate, but after about 4 h the loss rate becomes comparable. This result can conveniently be expressed by equation (5.113):

$$\frac{\mathrm{Br}}{\mathrm{Cl}} = a\,\mathrm{e}^{-\lambda t} + K \qquad (5.113)$$

where $a = 0{\cdot}8 \pm 0{\cdot}09$; $K = 0{\cdot}31 \pm 0{\cdot}03$; $\lambda = 0{\cdot}056 \pm 0{\cdot}007$; and t is the time in minutes.

The experiment was repeated in sunlight and there was no difference between the Br/Cl ratios between night and day.

In order to account for this rapid initial loss of bromine a diffusion model has been proposed in which the limiting factor is the mobility of Br^- ions in the interior of the crystal lattice of PbBrCl to migrate to the surface of the particle and then vaporize. The picture that emerges from this model indicates that bromine loss originates from very small lead particles. For a particle size of 0·1 μm, the model predicts that 50% of the bromine will disappear after 2 min.

Later evidence by Boyer and Laitinen [59] shows that the extent of halogen loss from lead bromochloride during photolysis is dependent upon the stoichiometry of the compound. After about 23 min irradiation the percentage halogen lost from $PbBr_2$, $PbBr_{1\cdot78}Cl_{0\cdot35}$, $PbBr_{1\cdot58}Cl_{0\cdot47}$ and $PbBr_{1\cdot25}Cl_{0\cdot74}$ is about 30, 11, 5 and 1%, respectively. These various compositions have a different morphology when viewed under an electron microscope. The composition $PbBr_{1\cdot25}Cl_{0\cdot74}$ (Br/Cl = 1·7) consists of well defined spheres with a particle diameter of about 3 μm, whilst the compound $PbBr_{1\cdot78}Cl_{0\cdot35}$ (Br/Cl = 5·0) has a crystalline needle-like appearance with a cross-sectional diameter of 0·5 μm or less.

It seems that more studies are needed before these apparently conflicting findings are understood. The stereoscan photograph of the composition $PbBr_{1\cdot25}Cl_{0\cdot74}$, however, is very similar in appearance to that found by Butler and co-workers and mentioned in section 4.2.2 and reference 42 of Chapter 4. In the latter case the sample was several days old and in the former case the composition was associated with a higher vaporization temperature of 510°C. The common factor seems to be that the samples showing spherical morphology were sufficiently mature to show the least degradation through irradiation, i.e. low Br/Cl ratio.

5.4 Stratospheric chemical reactions

As explained in section (5.2) the preservation of the ozone shield around the earth is fundamental to the continuation of life on the surface of the planet. Any man-made pollutant which upsets this balance of nature in the upper atmosphere must be carefully scrutinized and every attempt made to judge the extent of the danger. In recent years it has been suggested that nitric oxide produced by the explosion of nuclear devices in the atmosphere, nitric oxide produced by supersonic aircraft and halogens from freons could very well have this destructive capability.

Before examining these possibilities in more detail it is necessary to discuss, in general terms, some of the chemical properties of the stratosphere. These are relevant, not only from the standpoint of the compounds just mentioned, but also as sinks in the natural carbon and nitrogen cycles that were considered in Chapter 2.

5.4.1 The hydrogen–oxygen system

Photolysis of ozone through reactions (5.9) and (5.10) of Table 5.2 generates excited atomic oxygen $O(^1D)$ which can interact with water vapour, methane and molecular hydrogen to give the OH radical. Reaction (5.29) provides the main source:

$$O(^1D)+H_2O \rightarrow 2OH \tag{5.29}$$

with smaller contributions from

$$O(^1D)+CH_4 \rightarrow CH_3+OH \tag{5.114}$$

and

$$O(^1D)+H_2 \rightarrow H+OH \tag{5.115}$$

According to Nicolet [60], reaction (5.29) will produce at least 10^4 OH radicals $cm^{-3}\,s^{-1}$ above an altitude of 20 km for an overhead sun increasing to $7\times10^4\,cm^{-3}\,s^{-1}$ in the upper stratosphere. At the stratopause the HO_2 radical is formed by a three-body reaction indicated in equation (5.28), viz.

$$H+O_2+M \rightarrow HO_2+M \tag{5.28}$$

In the region of the stratosphere associated with maximum ozone concentrations reaction (5.116) can also be significant:

$$H+O_3 \rightarrow OH+O_2 \tag{5.116}$$

Nicolet maintains that the hydroxyl radical–ozone reaction:

$$OH+O_3 \rightarrow HO_2+O_2 \tag{5.117}$$

and the hydroperoxyl–atomic oxygen reaction:

$$HO_2+O \rightarrow OH+O_2 \tag{5.118}$$

are unimportant, but the photolysis of hydrogen peroxide cannot be overlooked:

$$H_2O_2+h\nu \rightarrow 2OH \tag{5.119}$$

Finally, a number of reactions contribute to the removal or exchange of OH and HO_2 radicals, several of them form water vapour:

$$OH+H_2O_2 \rightarrow H_2O+HO_2 \tag{5.34}$$

$$OH+H_2 \rightarrow H_2O+H \tag{5.33}$$

$$OH+OH \rightarrow H_2O+O \tag{5.120}$$

$$OH+HO_2 \rightarrow H_2O+O_2 \tag{5.121}$$

$$OH + OH + M \rightarrow H_2O_2 + M \quad (5.35)$$
(in the lower stratosphere)

$$HO_2 + HO_2 \rightarrow H_2O_2 + O_2 \quad (5.36)$$

These reactions are summarized schematically in Fig. 5.15.

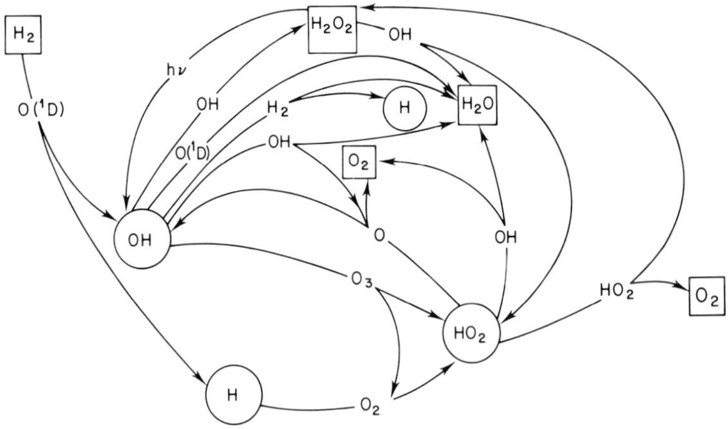

FIG. 5.15 Stratospheric chemistry of the hydrogen–oxygen system

5.4.2 The nitrogen–oxygen–hydrogen system

In the stratosphere the main source of nitric oxide is nitrous oxide which diffuses through the troposphere. When the stratospheric ozone band is reached ozone destruction initiated by u.v. solar radiation, according to reactions (5.9) and (5.10) shown in Table 5.2, produces atomic oxygen $O(^1D)$. The interaction between $O(^1D)$ and N_2O through equation (5.122) generates nitric oxide:

$$N_2O + O(^1D) \rightarrow 2NO \quad (5.122)$$

The contribution from ammonia is uncertain [61] but stratospheric ammonia is thought to be removed mainly by reactions such as:

$$NH_3 + OH \rightarrow H_2O + NH_2 \quad (5.123)$$

and

$$NH_3 + h\nu \rightarrow NH_2 + H \quad (5.124)$$

The subsequent fate of these radicals is not known but the possibility exists that NH_x radical species may operate as a sink for nitric oxide through reactions (5.125) and (5.126):

$$NH_2 + NO \rightarrow N_2 + H_2O \quad (5.125)$$

$$NH + NO \rightarrow N_2 + OH \quad (5.126)$$

but further laboratory data are needed to clarify this point.

Absorption of cosmic radiation introduces a new complexity into the NO_x system through processes of the type [62]:

$$N_2^+ + e \rightarrow N + N \quad (5.127)$$

$$NO^+ + e \rightarrow N + O \quad (5.128)$$

$$N_2 + e \rightarrow N + N \quad (5.129)$$

The product atoms are released in excited states and it is necessary to have some knowledge of the relative amounts of $N(^4S)$ and $N(^2D)$ because they behave differently. $N(^4S)$ reactions will tend to give a net destruction of NO_x through reactions (5.130) to (5.133):

$$N(^4S) + NO \rightarrow N_2 + O \quad (5.130)$$

$$N(^4S) + NO_2 \rightarrow N_2O + O \quad (5.131)$$

$$N(^4S) + NO_2 \rightarrow N_2 + O_2 \quad (5.132)$$

$$N(^4S) + NO_2 \rightarrow N_2 + O + O \quad (5.133)$$

whereas, $N(^2D)$ will be quenched by oxygen:

$$N(^2D) + O_2 \rightarrow NO + O \quad (5.134)$$

Reaction (5.134) is the major source of NO in the stratosphere above 100 km.

5.4.3 Destructive ozone processes

Anxiety expressed about the destruction of the earth's ozone layer by anthropogenic influences is well founded. Some of the details may be rather speculative owing to our lack of experience in working with chemical systems under the conditions that obtain at these altitudes. Nevertheless, this concern is based on current scientific appraisal, the logic of which is briefly outlined in this section.

5.4.3.1 The freons—chlorofluoromethanes

Dichlorodifluoromethane, CF_2Cl_2, and trichlorofluoromethane, $CFCl_3$, have found widespread application as aerosol propellants in spray preparations. Their selection for this purpose in the first instance was based on their chemical inertness and hence their harmlessness from the human health aspect. These properties, coupled with the fact that they are not soluble in water, ensure that they are not removed by rainfall in the troposphere. Accordingly, it seems unlikely that the oceans of the world can act as a sink for these compounds. All these features will permit their survival in the atmosphere for long periods so that exchange between the troposphere and the stratosphere can take place. Current tropospheric concentrations of CF_2Cl_2 and $CFCl_3$ are now about 10 and 6 parts 10^{-11} by volume, respectively [63].

Molina and Rowland [64] recognized the implications of the presence of these compounds in the stratosphere and have described the relevant chemistry. Between altitudes of 20–30 km u.v. radiation of $\lambda < 230$ nm can cause photodissociation by reactions (5.135) and (5.136):

$$CF_2Cl_2 + h\nu \rightarrow CF_2Cl + Cl \quad (5.135)$$

$$CFCl_3 + h\nu \rightarrow CFCl_2 + Cl \quad (5.136)$$

Turco and Whitten [65] have calculated the lifetimes of CF_2Cl_2 and $CFCl_3$ based on these events, they are about 300 and 30 years at 20 km and 1 and 0·1 years at 30 km, respectively. The fate of the CF_2Cl and $CFCl_2$ radicals produced by these processes is not known for sure, although the fluorine atom will end up bonded to hydrogen in the very stable HF molecule.

The release of chlorine into the stratospheric ozone band by either reactions (5.135) or (5.136) or photodissociaton of other halocarbons such as carbon tetrachloride [66] and chloroform [67] adds a new complexity to the chemistry of the stratosphere. These complications are based on reactions between chlorine atoms and ozone. In the first place reactions (5.137) and (5.138) occur to cause a net conversion of ozone into molecular oxygen [68]:

$$Cl + O_3 \rightarrow ClO + O_2 \quad (5.137)$$

$$ClO + O \rightarrow Cl + O_2 \quad (5.138)$$

This first catalytic cycle involving atomic chlorine leads to another catalytic cycle initiated by the ClO radical through reactions (5.139) and (5.140), the net effect of which as before is the conversion of ozone into molecular oxygen:

$$ClO + O_3 \rightarrow OClO + O_2 \quad (5.139)$$

$$OClO + O \rightarrow ClO + O_2 \quad (5.140)$$

The radical OClO is regarded as chlorine dioxide rather than a peroxy radical ClOO. The overall effect of this second catalytic cycle is to lower slightly the ozone destructive process. This follows since some of the ClO radicals that have been converted to OClO radicals by reaction (5.139) take part in reaction (5.140). Reaction (5.140), however, is slower than reaction (5.138) by some two orders of magnitude. Eventually, the ClO radical is removed by photodissociation,

$$ClO + h\nu \rightarrow Cl + O \qquad \lambda < 265\ \text{nm} \tag{5.141}$$

or by reaction with NO,

$$ClO + NO \rightarrow Cl + NO_2 \tag{5.142}$$

Remembering that OH, $O(^1D)$, H_2, CH_4 and u.v. radiation prevail at these altitudes, chlorine atoms can form HCl:

$$Cl + CH_4 \rightarrow CH_3 + HCl \tag{5.143}$$

$$Cl + H_2 \rightarrow H + HCl \tag{5.144}$$

and if ammonia is present as well:

$$Cl + NH_3 \rightarrow NH_2 + HCl \tag{5.145}$$

The HCl represents a temporary refuge for the chlorine atom before HCl itself undergoes further transformations that regenerate atomic chlorine:

$$HCl + OH \rightarrow H_2O + Cl \tag{5.146}$$

$$HCl + O(^1D) \rightarrow OH + Cl \tag{5.147}$$

$$HCl + h\nu \rightarrow H + Cl \tag{5.148}$$

At altitudes greater than 35 km the Cl and ClO concentrations are controlled by reactions (5.137) and (5.138), but at lower altitudes reaction (5.142) participates in the scheme. When this happens the important reactions are:

$$Cl + O_3 \rightarrow ClO + O_2 \tag{5.137}$$

$$ClO + NO \rightarrow Cl + NO_2 \tag{5.142}$$

and the ClO_x and the NO_x cycles become interlinked.

The interlinking of these cycles raises the question of the formation of chlorine nitrate, $ClONO_2$, and the influence that this compound may have on chemical reactions at these altitudes. The ability of chlorine nitrate to act as a sink for ClO_x and NO_x will have the effect of reducing the impact of the injection of NO_x and halogen species into the ozone band. The effectiveness of chlorine nitrate for preserving atmospheric ozone will depend upon its stability. Davis and co-workers [69] have considered the three most likely destructive processes to be:

(1) interaction with $O(^3P)$;
(2) interaction with OH; and
(3) direct photodissociation.

They show experimentally that direct photodissociation is the main route and that degradation of chlorine nitrate by radical reactions contributes less than 10% to the total.

The main features of these halocarbon reactions involving the freons are summarized in Fig. 5.16.

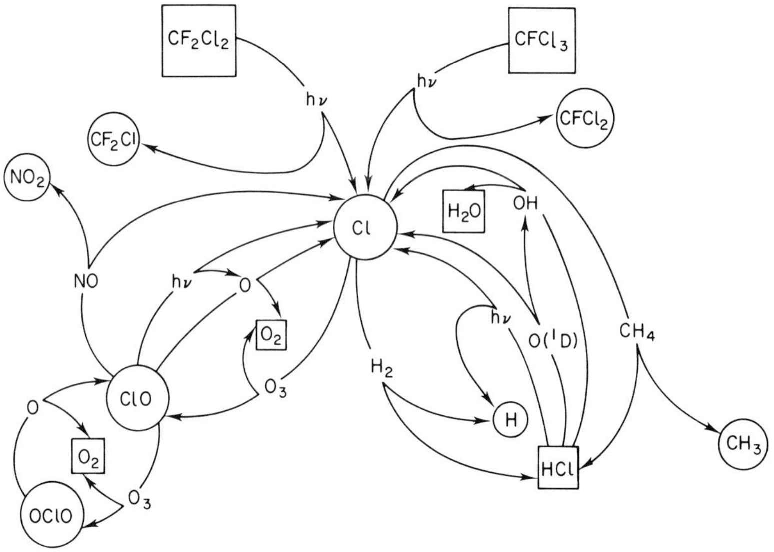

FIG. 5.16 Atmospheric reactions of the freons

Several long-term predictions on the effect on the earth's ozone layer of the discharge of chlorofluoromethanes to the atmosphere have been made. When exercises of this nature are carried out the following fundamental data are required for computer simulation:

(1) The emission rate to atmosphere.
(2) The eddy-diffusivity parameter ($cm^2 \, s^{-1}$) as a function of altitude.
(3) Concentration profiles as a function of altitude of the reacting species, i.e. mixing ratios.
(4) A list of all the relevant chemical reactions.
(5) The reaction rate constants for these reactions.

Obviously, these contributing factors are not known with equal certainty, so that usually upper and lower limits are applied to the parameters used.

This in turn means, of course, that the predictions will vary over quite large ranges.

As an illustration of source emission variations Turco and Whitten's [65] calculations may be quoted. Their model I assumes that fluorocarbons are released to atmosphere at constant rates equivalent to the 1972 production level. Model II assumes a 1972 production level which increases by 8·7% $year^{-1}$ for 20 years and then remains constant. Model III again assumes a 1972 production level which increases by 8·7% $year^{-1}$ for 5 years but after this time production ceases. In addition, to cover uncertainties in the stability of these compounds in the troposphere, calculations were performed assuming first complete stability, denoted by a suffix ∞. Second, tropospheric lifetimes of CF_2Cl_2 and $CFCl_3$ of 10 or 30 years, these are indicated by suffixes 10 and 30.

The eddy diffusivity profile used for the predictions is shown in Fig. 5.17. This parameter must be included in the model because it takes into account the variation in transport of material with altitude through the troposphere into the stratosphere.

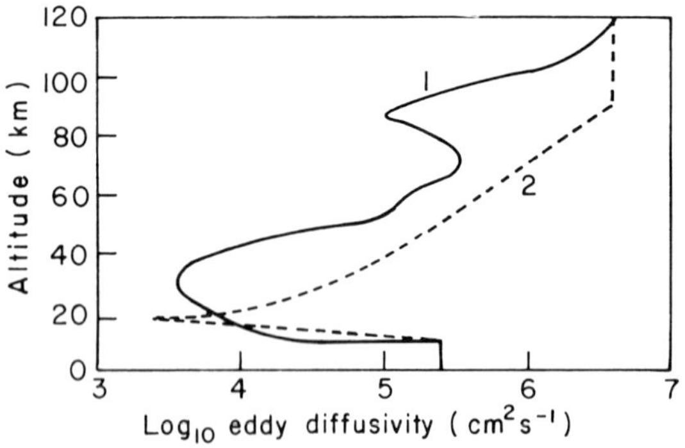

FIG. 5.17 Eddy diffusion coefficients used by Turco and Whitten [65]

The results of this analysis for the reduction in global ozone beyond 1972 are shown in Fig. 5.18. The prediction of an eventual reduction in ozone by about 10% in the year 2062 for model I_∞ is in agreement with a similar computer-simulation study by Crutzen [70]. Inspection of the lower curve in Fig. 5.18, marked III_∞, shows that if the production of fluorocarbons ceased in 1977 (which it did not), then the peak ozone deficiency would be about 2% and this would happen around 1992. On the other hand, by the year 2022 ozone depletion could amount to as much as 20% under the conditions postulated by model II_∞.

Another consequence of the depletion of the earth's ozone shield will be a decrease in temperature in the stratosphere. It will be recalled that reactions (5.11) and (5.12) are exothermic and that they are fundamental for maintaining the natural ozone balance. Less ozone at these altitudes will restrict these reactions and this will be reflected by a fall in temperature. Calculations by Reck [71] based on the computed ozone–altitude profiles of Wofsy *et al.* [72] suggest that temperatures could be reduced by 8°C at 40 km by the end of the century. The effect, if any, of these changes on climate has not been established.

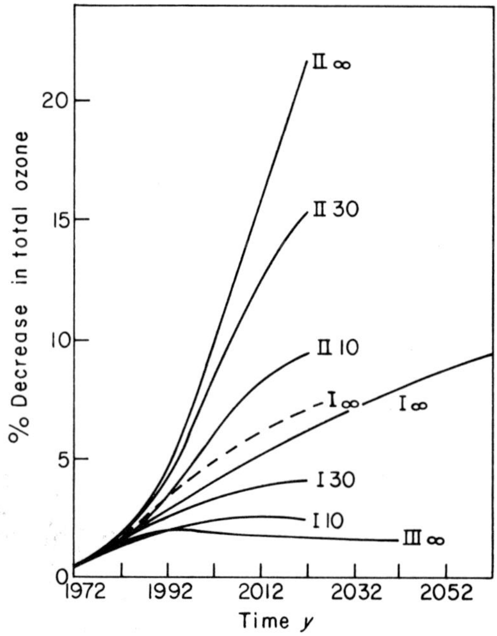

FIG. 5.18 Per cent reduction in global ozone beyond 1972 for different chlorofluoromethane production histories and tropospheric lifetimes. See the text for an explanation of the notation describing each case. The dashed curve was calculated using eddy diffusion coefficient 2 in Fig. 5.17 [65]

5.4.3.2 Ozone destruction by atmospheric nuclear explosions and supersonic transport aircraft

Calculations have been performed to estimate the formation and eventual circulation of nitric oxide in the northern hemisphere arising from 1 and 5 Mton atmospheric nuclear explosions. On the basis that each megaton of nuclear yield produces 10^{32} molecules of NO which is distributed uniformly, Whitten *et al.* [73] estimate that for a large explosion ozone depletion could be as high as 70%. Their calculations also indicate that the recovery time needed to replenish this ozone could be three years. Calculations of this type are only estimates because of

(1) the inhomogeneous distribution of nitric oxide in nuclear explosions [74];
(2) the difficulty in assessing the total amount of nitric oxide generated;
(3) changes in stratospheric circulation brought about by the depletion of the ozone layer; and
(4) changes in thermally driven winds at high altitudes.

Despite these uncertainties the calculations are not likely to be grossly in error and we may conclude that the explosion of nuclear devices in the atmosphere may be condemned on these grounds alone.

Concern about the long-term effects of supersonic high-altitude flights on the ozone layer were first expressed by Johnston [75]. Emission of nitric oxide from Concorde's Olympus Mark 602 engines, for example, is 13·8 g NO kg^{-1} of fuel consumed. At the operational altitudes of 17 km the fuel consumption is about 4550 kg h^{-1} for each engine. Assuming a trans-Atlantic flight time of 2·66 h in the stratosphere the NO produced in one crossing will be about 668 kg. One Concorde crossing the Atlantic four times per day on every day of the year will inject $9{\cdot}8 \times 10^5$ kg $year^{-1}$ of NO into the stratosphere [76].

It is instructive to compare these estimates with those predicted for nuclear weapon testing during 1961 and 1962 when this activity reached its peak. The calculated production of NO derived from nuclear tests greater than 20 kton based on the calculated yield of 5×10^9 g NO $Mton^{-1}$ is shown in Table 5.6 [76]. The table also compares the number of Concordes required to produce an equivalent amount of NO.

Calculations and comparisons of this type must be speculative because differences in emission factors have not been taken into consideration. For instance, aircraft emissions will be slower and confined to limited regions

TABLE 5.6 *Comparison of the capacity of nuclear explosions and supersonic aircraft to produce nitric oxide* [76]

Year	Mton TNT equivalent	NO production ($kg \times 10^7$)	Equivalent number of operational Concordes
1954	48·5	24·2	237
1956	26	13	127
1958	61·9	30·9	303
1961	120·6	60·3	591
1962	213·5	106·7	1047

near recognized air routes in the lower stratosphere. Nuclear injections cover a much wider altitude depending on the magnitude of the explosion. The cloud debris from a 1 Mton explosion will stabilize at around 22 km, whereas that of a 10 Mton event will reach about 32 km.

Most attention has been concentrated on injection of NO into the atmosphere but it is conceivable that SO_2 emitted at these altitudes could have the same effect. At these levels photolysis of SO_2 to $SO + O$ becomes feasible and reaction between SO and O_3 occurs at a comparable rate to that between NO and O_3 [77].

References

1. Foote, C. S. (1968). Mechanism of photosensitized oxidation. *Science* **162**, 963–969.
2. Volman, D. H. (1963). Photochemical gas phase reactions. *In* "Advances in Photochemistry", p. 48. Wiley-Interscience.
3. Cadle, R. D. and Allen, E. R. (1970). Atmospheric photochemistry. *Science* **167**, 243–249.
4. Demerjian, K. L., Kerr, J. A. and Calvert, J. G. (1974). The mechanism of photochemical smog formation. *In* "Advances in Environmental Science and Technology", Vol. 4, pp. 1–264. John Wiley.
5. Kerr, J. A., Calvert, J. G. and Demerjian, K. L. (1972). The mechanism of photochemical smog formation. *Chem. Brit.* **8**, 252–257.
6. Hampson, R. F., ed. (1972). "Chemical kinetic data survey II. Photochemical and rate data for fifteen gas phase reactions of interest in stratospheric chemistry." National Bureau of Standards, Report 10828.
7. Seiler, W. and Junge, C. E. (1970). Carbon monoxide in the atmosphere. *J. Geophys. Res.* **75**, 2217–2226.
8. Greiner, N. R. (1967). Hydroxylradical kinetics by kinetic spectroscopy. I. Reaction with H_2, CO and CH_4 at 300K. *J. Chem. Phys.* **46**, 2795–2799.
9. Ackerman, M. (1971). *In* "Mesopheric models and related experiments" D. Reidel, Ed. p 149, Dordrecht.
10. Calvert, J. G., Kerr, J. A., Demerjian, K. L. and McQuigg, R. D. (1972). Photolysis of formaldehyde as a hydrogen atom source in the lower atmosphere. *Science* **175**, 751.
11. Morris, E. D. and Niki, H. (1971). Mass spectrometric study of the reaction of hydroxyl radical with formaldehyde. *J. Chem. Phys.* **55**, 1991–1992.
12. Westenberg, A. A. and de Haas, N. (1973). Rates of $CO + OH$ and $H_2 + OH$ over an extended temperature range. *J. Chem. Phys.* **58**, 4061–4065.
13. Greiner, N. R. (1969). Hydroxyl radical kinetics by kinetic spectroscopy. V. Reaction with H_2 and CO in the range 300–500 K. *J. Chem. Phys.* **51**, 5049–5051.
14. Altshuller, A. P. and Bufalini, J. J. (1971). Photochemical aspects of air pollution: A review. *Environ. Sci. and Technol.* **5**, 39–64.
15. Agnew, W. G. (1968). Automotive air pollution research. *Proc. Roy. Soc.* **A307**, 153–181.

16. Darnall, K. R., Lloyd, A. C., Winer, A. M. and Pitts, J. N. (1976). Reactivity scale for atmospheric hydrocarbons based on reaction with hydroxyl radicals. *Environ. Sci. and Technol.* **10**, 692–696.
17. Niki, H., Daby, E. E. and Weinstock, B. (1972). Mechanisms of smog reactions. *Adv. Chem. Ser.* **113**, 16–57.
18. Yeung, C. K. K. and Phillips, C. R. (1974). Estimation of NO photooxidation reactivities of olefins from molecular structure. *Atmos. Environ.* **8**, 493–506.
19. Glasson, W. A. and Tuesday, C. S. (1970). Hydrocarbon reactivity and the kinetics of the atmospheric photoxidation of nitric oxide. *J. Air Pollut. Contr. Assoc.* **20**, 239–243.
20. Janz, J. G. (1967). "Thermodynamic properties of Organic Compounds: Estimation Methods, Principles and Practice." Academic Press.
21. Bufalini, J. J. (1974). Estimation of NO photoxidation reactivities of olefins from molecular structure. *Atmos. Environ.* **8**, 1205–1207.
22. Levy, H. (1971). Normal atmosphere: Large radical and formaldehyde concentrations predicted. *Science* **173**, 141–142.
23. Story, P. R., Alford, J. A., Ray, W. C. and Burgess, J. R. (1971). Mechanism of ozonolysis. A new unifying concept. *J. Amer. Chem. Soc.* **93**, 3044–3046.
24. Winer, A. M. and Bayes, K. D. (1966). The decay of $O_2(a^1\Delta)$ in flow systems. *J. Phys. Chem.* **70**, 302–304.
25. Pitts, J. N., Khan, A. U., Smith, E. B. and Wayne, R. P. (1969). Singlet oxygen in the environmental sciences: Singlet molecular oxygen and photochemical air pollution. *Environ. Sci. and Technol.* **3**, 241–247.
26. Leighton, P. A. (1961). "Photochemistry of Air Pollution." Academic Press.
27. Kummler, R. H., Bortner, M. H. and Baurer, T. (1969). The Hartley photolysis of ozone as a source of singlet oxygen in polluted atmospheres. *Environ. Sci. and Technol.* **3**, 248–250.
28. Khan, A. U., Pitts, J. N. and Smith, E. B. (1967). Singlet oxygen in the environmental sciences. *Environ. Sci. and Technol.* **1**, 656–657.
29. Izod, T. P. J. and Wayne, R. P. (1968). Formation of $O_2(^1\Delta g)$ in a photochemical system involving ozone. *Nature* **217**, 947–948.
30. Kummler, R. H. and Bortner, M. H. (1969). Production of $O_2(^1\Delta g)$ by energy transfer from excited benzaldehyde. *Environ. Sci. and Technol.* **3**, 944–946.
31. Steer, R. P., Sprung, J. L. and Pitts, J. N. (1969). Singlet oxygen in the environmental sciences. Evidence for the production of $O_2(^1\Delta g)$ by energy transfer in the gas phase. *Environ. Sci. and Technol.* **3**, 946-947.
32. Graedel, T. E., Farrow, L. A. and Weber, T. A. (1976). Kinetic study of the photochemistry of the urban atmosphere: *Atmos. Environ.* **10**, 1095–1116.
33. Sander, S. P. and Seinfeld, J. H. (1976). Chemical kinetics of homogeneous atmospheric oxidation of sulfur dioxide. *Environ. Sci. and Technol.* **10**, 1114–1123.
34. Roberts, P. T. and Friedlander, S. K. (1976). Photochemical aerosol formation. *Environ. Sci. and Technol.* **10**, 573–580.
35. Foster, P. M. (1969). The oxidation of sulphur dioxide in power station plumes. *Atmos. Environ.* **3**, 157–175.
36. Novakov, T., Chang, S. G. and Harker, A. B. (1974). Sulfates as pollution particulates: Catalytic formation on carbon (soot) particles. *Science* **186**, 259–261.

37. Calvert, J. G. and McQuigg, R. D. (1975). Computer simulation of the rates and mechanisms of photochemical smog reactions. *Int. J. Chem. Symp.* No. 1, 113–154.
38. Cox, R. A. and Penkett, S. A. (1972). Aerosol formation from sulphur dioxide in the presence of ozone and olefinic hydrocarbons. *J. Chem. Soc. Faraday Trans. I* **68**, 1735–1753.
39. Gartrell, F. E., Thomas, F. W. and Carpenter, S. B. (1963). Atmospheric oxidation of sulfur dioxide in coal-burning power plant plumes. *Amer. Ind. Hyg. Assoc. J.* **24**, 113–120.
40. Barrie, L. A. and Georgii, H. W. (1976). An experimental investigation of the absorption of sulphur dioxide by water drops containing heavy metals. *Atmos. Environ.* **10**, 743–749.
41. Sidebottom, H. W., Badcock, C. C., Jackson, G. E., Calvert, J. G., (1972) Photooxidation of sulfur dioxide. *Environ. Sci. and Technol.*, **6**, 72–79.
42. Okuda, S., Rao, T. N., Slater, D. H. and Calvert, J. G. (1969). Identification of the photochemically active species in sulfur dioxide photolysis within the first allowed absorption band. *J. Phys. Chem.* **73**, 4412–4415.
43. Davis, D. D., Prusazcyk, J., Dwyer, M. and Kim, P. (1974). A stop-flow time of flight mass-spectrometry kinetics study. Reaction of ozone with nitrogen dioxide and sulfur dioxide. *J. Phys. Chem.* **78**, 1775–1779.
44. Schuck, E. A., Ford, H. W. and Stephens, E. R. (1958). Air pollution effects of irradiated automobile exhausts as related to fuel composition. Report No. 26. Air Pollution Foundation, San Marino, California.
45. Smith, J. P. and Urone, P. (1974). Static studies of sulfur dioxide reactions. Effects of NO_2, C_3H_6, and H_2O. *Environ. Sci. and Technol.* **8**, 742–746.
46. Payne, W. A., Stief, L. J. and Davis, D. D. (1973). A kinetic study of HO_2 with SO_2 and NO. *J. Amer. Chem. Soc.* **95**, 7614–7619.
47. Levy, H. (1974). Photochemistry of the troposphere. "Advances in Photochemistry", p. 424, Vol. 9.
48. Badcock, C. C., Sidebottom, H. W., Calvert, J. G., Reinhardt, G. W. and Damon, E. K. (1971). Mechanism of the photolysis of sulfur dioxide–paraffin hydrocarbon mixtures. *J. Amer. Chem. Soc.* **93**, 3115–3121.
49. Sidebottom, H. W., Badcock, C. C., Calvert, J. G., Rabe, B. R. and Damon, E. K. (1971). Mechanism of the photolysis of mixtures of sulfur dioxide with olefin and aromatic hydrocarbons. *J. Amer. Chem. Soc.* **93**, 3121–3128.
50. Wei, Y. K. and Cvetanovic, R. J. (1963). A study of the vapour phase reaction of ozone with olefins in the presence and absence of molecular oxygen. *Canad. J. Chem.* **41**, 913–925.
51. Gay, B. W., Hanst, P. L., Bufalini, J. J. and Noonan, R. C. (1976). Atmospheric oxidation of chlorinated ethylenes *Environ. Sci. and Technol.* **10**, 58–67.
52. Gitchell, A., Simionaitis, R. and Heicklen, J. (1974). Inhibition of photochemical smog. I. Inhibition by phenol, benzaldehyde, and aniline. *J. Air Pollut. Contr. Assoc.* **24**, 357–361.
53. Spicer, C. W., Miller, D. F. and Levy, A. (1974). Inhibition of photochemical smog reaction by free radical scavengers. *Environ. Sci. and Technol.* **8**, 1028–1030.
54. Stockburger, L., Sie, B. K. T. and Heicklen, J. (1976). The inhibition of photochemical smog. V. Products of diethylhydroxylamine inhibited reaction. *Sci. Total Environ.* **5**, 201–202.

55. Olszyna, K. and Heicklen, J. (1976) The inhibition of photochemical smog. VI. The reaction of ozone with diethylhydroxylamine. *Sci. Total Environ.* **5**, 223–230.
56. Maugh, T. H. (1976). Photochemical smog: Is it safe to treat the air? *Science* **193**, 871–873.
57. Pierrard, J. M. (1969). Photodecomposition of lead halides from automobile exhaust. *Environ. Sci. and Technol.* **3**, 48–51.
58. Robbins, J. A. and Snitz, F. L. (1972). Bromine and chlorine loss from lead halide automobile exhaust particulates. *Environ. Sci. and Technol.* **6**, 164–169.
59. Boyer, K. W. and Laitinen, H. A. (1974). Lead halide aerosols: Some properties of environmental significance. *Environ. Sci. and Technol.* **8**, 1093–1096.
60. Nicolet, M. (1972). Aeronomic chemistry of the stratosphere. *Planet Space Sci.* **20**, 1671–1702.
61. Wofsy, S. C. and McElroy, M. B. (1974). HO_x, NO_x and ClO_x: Their role in atmospheric photochemistry. *Canad. J. Chem.* **52**, 1582–1591.
62. Warneck, P. (1972). Cosmic radiation as a source of odd nitrogen in the stratosphere. *J. Geophys. Res.* **77**, 6589–6591.
63. Lovelock, J. E., Maggs, R. J. and Wade, R. J. (1973). Halogenated hydrocarbons in and over the Atlantic. *Nature* **241**, 194–196.
64. Molina, M. J. and Rowland, F. S. (1974). Stratospheric sink for chlorofluoromethanes: Chlorine atom-catalyzed destruction of ozone. *Nature* **249**, 810–812.
65. Turco, R. P. and Whitten, R. C. (1975). Chlorofluoromethanes in the stratosphere and some possible consequences for ozone. *Atmos. Environ.* **9**, 1045–1061.
66. Molina, M. J. and Rowland, F. S. (1974). Predicted present stratospheric abundances of chlorine species from photodissociation of carbon tetrachloride. *Geophys. Res. Lett.* **1**, 309–312.
67. Yung, Y. L., McElroy, M. B. and Wofsy, S. C. (1975). Atmospheric halocarbons: A discussion with emphasis on chloroform. *Geophys. Res. Lett.* **2**, 397–399.
68. Stolarski, R. S. and Cicerone, R. J. (1974). Stratospheric chlorine: A possible sink for ozone. *Canad. J. Chem.* **52**, 1610–1615.
69. Ravishankara, A. R., Davis, D. D., Smith, G., Testi, G. and Spencer, J. (1977). A study of the chemical degradation of $ClONO_2$ in the stratosphere. *Geophys. Res. Lett.* **4**, 7–9.
70. Crutzen, P. J. (1974). Estimates of possible future ozone reductions from continued use of fluorochloromethanes (CF_2Cl_2, $CFCl_3$). *Geophys. Res. Lett.* **1**, 205–208.
71. Reck, R. A. (1976). Atmospheric temperature calculated for ozone depletions. *Nature* **263**, 116–117.
72. Wofsy, S. C., McElroy, M. B. and Sze, N. D. (1975). Freon consumption: Implications for atmospheric ozone. *Science* **187**, 535–537.
73. Whitten, R. C., Borucki, W. J. and Turco, R. P. (1975). Possible ozone depletion following nuclear explosions. *Nature*, **257**, 38–39.
74. Gilmore, F. R. (1975). The production of nitric oxide by low altitude nuclear explosions *J. Geophys. Res.* **80**, 4553–4554.
75. Johnston, H. S. (1974). Catalytic reduction of stratospheric ozone by nitrogen oxides. *In* "Advances in Environmental Science and Technology", p. 357, Vol. 4. Wiley-Interscience.

76. Goldsmith, P., Tuck, A. F., Foot, J. S., Simmons, E. L. and Newson, R. L. (1973). Nitrogen oxides, nuclear weapon testing, Concorde and stratospheric ozone. *Nature* **244**, 545–551.
77. Clyne, M. A. A. (1974). Destruction of atmospheric ozone? *Nature* **249**, 796–797.

6

Meteorological Aspects of Pollutant Dispersions

Conventional studies of chemical reactions are conducted in closed systems. Traditional chemical operations of synthesis and analysis are carried out within the confines of a reaction chamber. Conditions, thereby, can be rigorously controlled and relevant influencing parameters isolated and independently investigated. In the real world of atmospheric chemistry such ideality is not possible. Once chemical pollutants escape into the atmosphere dilution and distribution as well as chemical reactions in the atmosphere will result. Common sense indicates that meterological factors will at least affect the dilution and distribution of atmospheric emissions. These influences are considered in this chapter and equations discussed which attempt to provide quantitative predictions of the spread of pollutants from emission sources. These predictions will be made on the assumption that the aerosol has reasonable stability. In other words, if the aerosol contains particles, then these are sufficiently small to make their behaviour indistinguishable from that of a gas, and furthermore, chemical reaction within the aerosol does not take place.

6.1 Thermodynamic properties of the atmosphere

Before proceeding with the main theme of this chapter it is necessary to introduce some ideas and definitions of meteorological origin that are not usually familiar to chemists. The first of these concepts concerns the manner in which temperature changes with altitude. Figure 5.1 of the last chapter showed how the temperature profile changes with altitude within an air mass. In the troposphere up to about 10 km the temperature decreases with altitude. The rate of decrease is about $10\ \mathrm{K\ km^{-1}}$ and is called the temperature lapse rate. This fall in temperature is explained by the fact that pressure decreases with altitude. For an adiabatic process, therefore, a unit volume of air moving upwards will undergo an expansion. Since the gas itself provides the energy for overcoming the attractive forces between the molecules during the expansion, there must be an attendant fall in temperature. Air in such a condition is termed unstable.

When the surface temperature of the earth cools, as usually happens after sundown, then the temperature–altitude profile changes and the temperature increases with altitude (Fig. 6.1). A similar situation arises when a cold mass of air moves over a warm one which is lying close to the surface. This condition defines a temperature inversion in which warm air

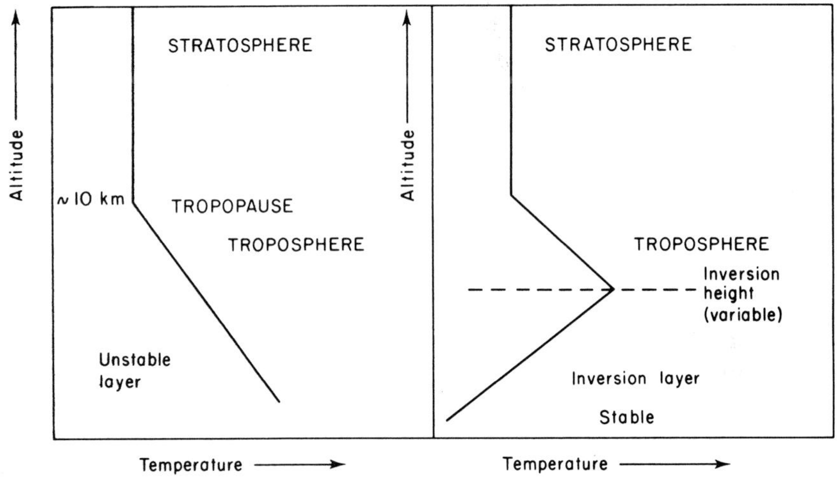

FIG. 6.1 Altitude–temperature profiles showing a positive lapse rate with unstable air conditions and inversion conditions associated with stable air masses

is trapped below cooler air. The layer of air in this inversion layer has restricted vertical movement and is stable. It will be apparent that pollutants trapped in a stable inversion layer cannot escape to higher altitudes and be dispersed. An understanding of the atmospheric dispersion of pollutants obviously depends upon atmospheric stability. This chapter is devoted to a limited non-rigorous treatment of the theoretical background which is necessary to appreciate these meteorological factors. Applications of the model for predicting pollution dispersions are provided by way of specific examples.

6.2 Plume dispersion theory

It is difficult to estimate precisely the relative importance of turbulent and molecular diffusion in a spreading plume of pollutants in air. Monin and Yaglom [1] suggest that eddy diffusion is 10^5–10^6 times greater than molecular diffusivity and that for practical purposes molecular diffusion can be neglected relative to turbulent diffusion. This assumption is implicit in many air pollution models. By analogy with diffusion from a region of high to low concentration, one can define an eddy diffusivity K so that the flux S is proportional to the gradient of the mean concentration C of the material in the X direction. Fick's law then gives the equation:

$$S = -K\frac{\partial C}{\partial X} \tag{6.1}$$

Based on this concept a general equation may be derived for the net transport of material into and out of a small element by turbulent diffusion and advection by a windspeed u [2]. This semi-empirical equation:

$$\frac{\partial C}{\partial t} + u\frac{\partial C}{\partial X} = \frac{\partial}{\partial X}\left(K_x\frac{\partial C}{\partial X}\right) + \frac{\partial}{\partial Y}\left(K_y\frac{\partial C}{\partial Y}\right) + \frac{\partial}{\partial Z}\left(K_z\frac{\partial C}{\partial Z}\right) \tag{6.2}$$

may be solved for a number of specific cases (see Table 6.1).

Equation (6.4) given in the table is the well known point source formula of Pasquill [3]. It is customary [4] to replace the diffusivities K_y and K_z in terms of σ_y and σ_z, where σ_y and σ_z are the standard deviations of the plume concentration distributions in the Y and Z directions (Fig. 6.2). In the figure a plume from a chimney stack of *effective* height H is being blown by a wind of speed u ($\mathrm{m\,s^{-1}}$) parallel to the X-axis. The origin of the coordinate system is the base of the stack such that the Y-axis is in the same plane but perpendicular to the X-axis, and the Z-axis is vertical.

Replacing the diffusivities K by σ values using the relationship

$$\sigma = (2Kt)^{1/2} \tag{6.7}$$

where $t = x/u$, in equation (6.4), x being the downwind distance from the source and u the windspeed, gives

$$C_{(x,y,z,H)} = \frac{Q}{2\pi\sigma_y\sigma_z u}\exp\left[-\frac{1}{2}\left(\frac{y}{\sigma_y}\right)^2\right] \times\left\{\exp\left[-\frac{1}{2}\left(\frac{z-H}{\sigma_z}\right)^2\right] + \exp\left[-\frac{1}{2}\left(\frac{z+H}{\sigma_z}\right)^2\right]\right\} \tag{6.8}$$

In equation (6.8), when σ_y, σ_z, H, x, y, and z are expressed in metres, the windspeed u in $\mathrm{m\,s^{-1}}$ and the source strength Q in $\mathrm{g\,s^{-1}}$, the concentration C is in $\mathrm{g\,m^{-3}}$. Occasionally, when the dispersion of radioactive

TABLE 6.1 *Solutions of equation (6.2) for the conditions specified assuming that there is no precipitation from the plume or chemical reaction within the plume released at an effective height* H *into a windspeed* u($m\,s^{-1}$) *with source strength* Q ($g\,s^{-1}$)

	Equation number	Reference
Point source: instantaneous release.		
$C_{(x,y,z,t)} = \frac{Q}{(4\pi\,\Delta t)^{3/2}(K_xK_yK_z)^{1/2}} \exp-\left[\frac{(x-u\,\Delta t)^2}{4K_x\,\Delta t}\right] \exp-\left[\frac{y^2}{4K_y\,\Delta t}\right]\left\{\exp-\left[\frac{(z-H)^2}{4K_z\,\Delta t}\right]+\exp-\left[\frac{(z+H)^2}{4K_z\,\Delta t}\right]\right\}$	(6.3)	1
Point source: continuous release.		
$C_{(x,y,z,H)} = \frac{Q}{4\pi x(K_yK_z)^{1/2}} \exp-\left[\frac{y^2u}{4K_yx}\right]\left\{\exp-\left[\frac{z-H)^2u}{4K_zx}\right]+\exp-\left[\frac{(z+H)^2u}{4K_zx}\right]\right\}$	(6.4)	1
Line source: instantaneous release. $K_z = K_x = K$		
$C_{(x,0,z,t,H)} = \frac{Q}{2\pi K\,\Delta t} \exp-\left[\frac{(x-u\Delta t)^2}{4K\Delta t}\right]\left\{\exp-\left[\frac{(z-H)^2}{4K\Delta t}\right]+\exp-\left[\frac{(z+H)^2}{4K\,\Delta t}\right]\right\}$	(6.5)	7
Line source: continuous release.		
$C_{(x,0,z,H)} = \frac{Q}{2(\pi K_zux)^{1/2}}\left\{\exp-\left[\frac{(z-H)^2u}{4K_xx}\right]+\exp-\left[\frac{(z+H)^2u}{4K_zx}\right]\right\}$	(6.6)	1

material is being considered, Q is given in Ci s^{-1} and C then becomes Ci m^{-3}.

Three fairly common situations enable equation (6.8) to be simplified. These are when

(1) concentrations are required at ground level, i.e. $z = 0$, so that

$$C_{(x,y,O,H)} = \frac{Q}{\pi\sigma_y\sigma_z u}\exp\left[-\frac{1}{2}\left(\frac{y}{\sigma_y}\right)^2\right]\exp\left[-\frac{1}{2}\left(\frac{H}{\sigma_z}\right)^2\right] \qquad (6.9)$$

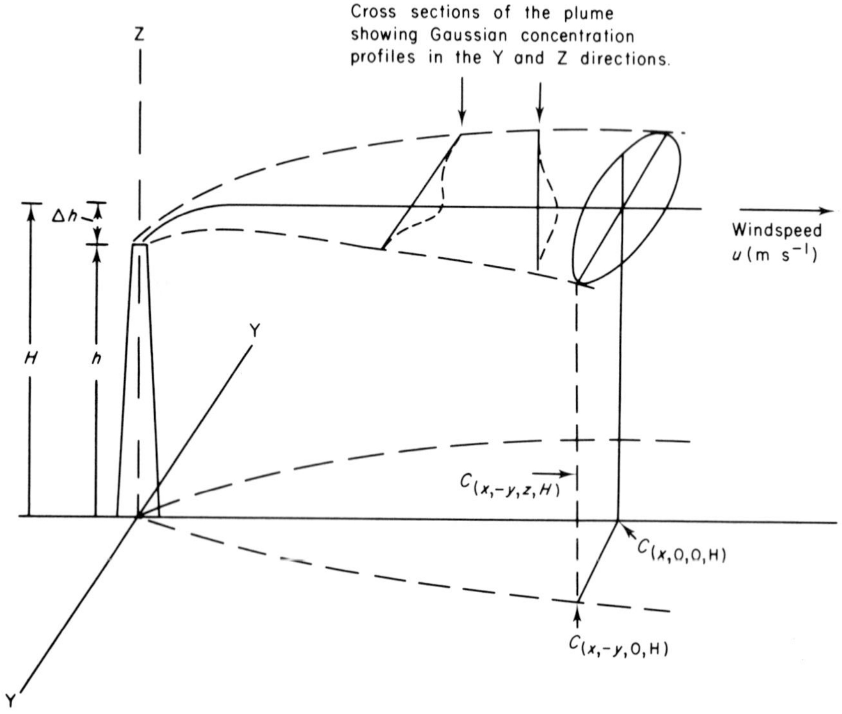

FIG. 6.2 Coordinate system of a pollution plume issuing from a point source of height, h

(2) concentrations are required along the centre line of the plume, i.e. when $y = 0$ at ground level where $z = 0$, then

$$C_{(x,O,O,H)} = \frac{Q}{\pi\sigma_y\sigma_z u}\exp\left[-\frac{1}{2}\left(\frac{H}{\sigma_z}\right)^2\right] \qquad (6.10)$$

(3) if, in addition to the conditions indicated in (1) and (2), the source is also at ground level, i.e. $H = 0$, so the equation becomes,

$$C_{(x,0,0,0,)} = \frac{Q}{\pi \sigma_y \sigma_z u} \tag{6.11}$$

6.3 Estimation of plume parameters: standard deviations σ_y and σ_z as a function of downwind distance from source

Pasquill [3] has devised a method for calculating the standard deviations σ_y and σ_z of a spreading plume from a knowledge of the atmospheric stability category based on windspeed and solar radiation. These stability categories are shown in Table 6.2. Incoming solar radiation varies with season and

TABLE 6.2 *Pasquill stability categories*

Wind speed ($m\,s^{-1}$)	Daytime (excluding 1 h after sunrise, 1 h before sunset) Incoming solar radiation ($mW\,cm^{-2}$) Strong ≥60	Moderate 30–60	Slight ≤30	Overcast	Within 1 h of sunset or sunrise	Night time[a] Cloud amount (oktas) 0–3	4–7	8
<2	A	A–B	B	C	D	F or G	F	D
2–3	A–B	B	C	C	D	F	E	D
3–5	B	B–C	C	C	D	E	D	D
5–6	C	C–D	D	D	D	D	D	D
>6	C	D	D	D	D	D	D	D

[a] Night was originally defined to include periods of one hour before sunset and after sunrise. These two hours are always categorized here as D.

time of day and the appropriate choice between strong, moderate and light insolation indicated in the table must therefore be made. This can be done by consulting Fig. 6.3, which shows the radiation in $mW\,cm^{-2}$ reaching the ground for particular months of the year and times of day under cloudless skies. This figure is valid for latitudes between 48°N and 60°N and only applies to cloudless skies. Cloud cover is normally divided into eighths and whenever clouds are present solar radiation is reduced. The extent of this reduction is given in Table 6.3.

In the stability categories shown in Table 6.2, class A is the most unstable and class F the most stable. Again, referring to Table 6.2, night refers to a period from 1 h before sunset to 1 h after sunrise. The stability

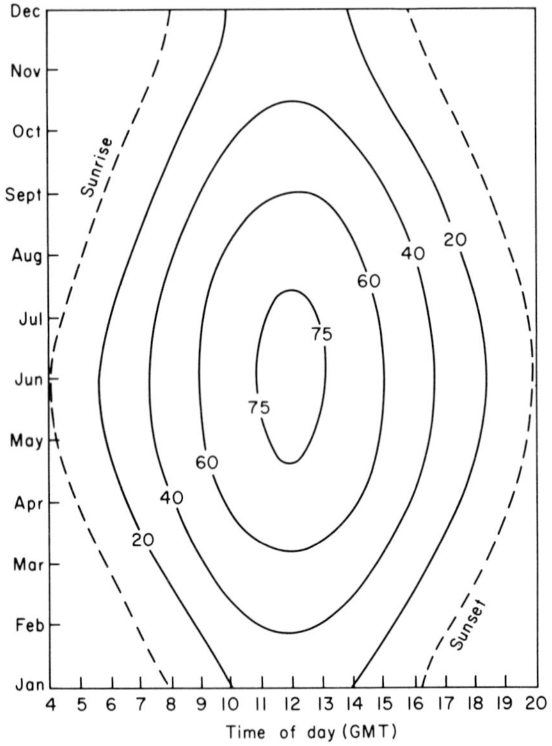

FIG. 6.3 Incoming solar radiation in mW cm^{-2} reaching the ground on a cloudless day, as a function of time of day and month. Table 6.3 shows correction factors for cloudy conditions

TABLE 6.3 *Reduction of incoming solar radiation by cloud*

Cloud amount (oktas)	Fraction to multiply I.S.R.[a]
O	1·07
1	0·89
2	0·81
3	0·76
4	0·72
5	0·67
6	0·59
7	0·45
8	0·23

[a] I.S.R. is Incoming Solar Radiation.

From Smith, F. B. (1975). Turbulence in the atmospheric boundary layer. *Sci. Prog.* **62,** 127–151.

class D, the neutral class, may be assumed to apply for any overcast conditions day or night irrespective of windspeed.

Once the appropriate weather conditions have been selected, σ_y and σ_z values for any downwind distance x can be calculated from the equations:

$$\sigma_y = ax^{0.903} \tag{6.12}$$

$$\sigma_z = bx^c \tag{6.13}$$

where values of a, b and c are given in Table 6.4 [5]. The values of the parameters a and b given in Table 6.4 have been chosen to give a close fit of the well known Pasquill–Gifford diffusion curves, shown in Figs. 6.4 and

TABLE 6.4 *Fitted constants for the Pasquill diffusion parameters* [5]

		Constants for vertical diffusion parameter σ_z^2								
	Cross wind[1]	$x \leqslant x_1$		x_1	$x_1 \leqslant x \leqslant x_2$		x_2	$x_2 \leqslant x$		
Class	a	b	c	m	b	c	m	b	c
A	0·40	0·125	1·03	250	0·00883	1·51	500	0·000226	2·10
B	0·295	0·119	0·986	1000	0·0579	1·09	10 000	0·0579	1·09
C	0·20	0·111	0·911	1000	0·111	0·911	10 000	0·111	0·911
D	0·13	0·105	0·827	1000	0·392	0·636	10 000	0·948	0·540
E	0·098	0·100	0·778	1000	0·373	0·587	10 000	2·85	0·366

[1] $\sigma_y = ax^{0.903}$ where x is the downwind distance from source, σ_y,σ_x in metres.
[2] $\sigma_z = bx^c$ where x is the downwind distance from source σ_y,σ_x in metres.
m = distance in metres.

6.5. It should be understood that these curves strictly apply to open level country and probably underestimate the plume dispersion in an urban area. When using these equations the following further limitations in their validity must be recognized [6]:

(1) Diffusion of material in the direction of transport is negligible. In other words, the source emissions must last at least as long as it takes the plume to arrive at the measurement point downwind.
(2) The material in the plume is not reacting chemically or coagulating to larger particle size material that could precipitate.
(3) A plume which comes into contact with the ground is completely reflected without deposition of material.
(4) The plume centre follows the downwind X-axis and the mean windspeed employed in the calculation is representative of the stability category.
(5) Uncertainties in the values σ_y and σ_z will increase with distance from the source.

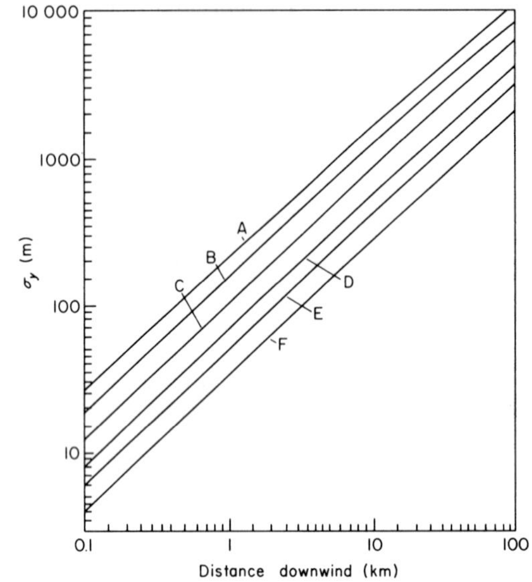

FIG. 6.4 Horizontal dispersion coefficient as a function of downwind distance from the source [6]

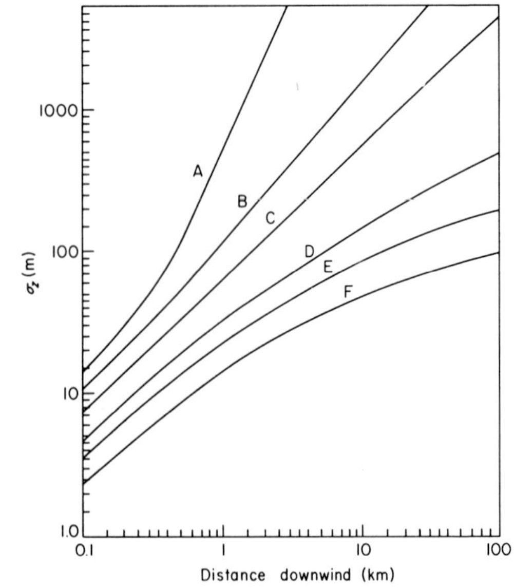

FIG. 6.5 Vertical dispersion coefficient as a function of downwind distance from the source [6]

(6) There are uncertainties in the estimation of the effective height of the plume.

6.4 Applications to specific dispersion problems

6.4.1 Emission from a point source

Example 1. A burning dump in the open is emitting 3 g s^{-1} of nitric oxide. What will be the average concentration measured over a 10 min period from this source at a distance of 1·5 km downwind, when the windspeed is 5·5 m s^{-1} on a dull overcast afternoon in November?

In this problem the concentration directly downwind at ground level is required so that $y = 0$ and $z = 0$. Since the source is a dump which is out in the open, we will assume that the source height is negligible and hence $H = 0$. Thus, equation (6.11) will apply:

$$C_{(x,0,0,0)} = \frac{Q}{\pi \sigma_y \sigma_z u}$$

The appropriate values of σ_y and σ_z must now be estimated. Inspection of Fig. 6.3 for insolation at 15·00 for a November afternoon indicates a value of about 10 mW cm^{-2} and since it is overcast, Table 6.3 suggests that this insolation value of 10 mW cm^{-2} should be multiplied by a factor of between 0·23 and 0·45. Quite clearly the insolation is "slight" and Table 6.2 shows that for a windspeed of 5·5 m s^{-1} class D stability is appropriate. The values of σ_y and σ_z can be calculated from the information given in Table 6.4. At a distance of 1500 m

$$\sigma_y = 0{\cdot}13 \times 1500^{0{\cdot}903} \quad \text{using } a = 0{\cdot}13 \text{ in equation (6.12)}$$
$$= 96 \text{ m}$$

and

$$\sigma_z = 0{\cdot}392 \times 1500^{0{\cdot}636} \quad \text{using } b = 0{\cdot}392 \text{ and } c = 0{\cdot}636 \text{ in equation (6.13)}$$
$$= 41 \text{ m}$$

Substituting these values into equation (6.11) with $Q = 3$ g s^{-1} and $u =$ 5·5 m s^{-1} gives

$$C_{(1500,0,0,0)} = \frac{3}{\pi 96 \times 41 \times 5{\cdot}5} = 4{\cdot}4 \times 10^{-5} \text{ g m}^{-3}$$

If we assume that the temperature was 10°C then this concentration of nitric oxide expressed in parts 10^{-6} comes to

$$\frac{44 \times 22{\cdot}4 \times 283}{30 \times 10^3 \times 273} = 0{\cdot}034 \text{ (parts } 10^{-6})$$

Example 2. A chimney with an effective stack height of 62 m is emitting sulphur dioxide at a rate of 50 g s^{-1} on a cloudless morning in May. Estimate the concentration of sulphur dioxide at ground level 600 m away with the wind blowing at 7 m s^{-1}

(1) directly down wind from the source,
(2) 50 m from the X-axis,
(3) 50 m from the X-axis and 20 m above the ground.

As before, the appropriate atmospheric stability category must be selected, so from Fig. 6.3 the insolation value will be about 60 mW cm^{-2} and for a windspeed of 7 m s^{-1} Table 6.2 indicates that category B should be used. At a distance of 600 m Table 6.4 gives

$$\sigma_y = 0{\cdot}295 \times 600^{0{\cdot}903} = 95 \text{ m} \qquad \text{using } a = 0{\cdot}295 \text{ in equation (6.12)}$$

$$\sigma_z = 0{\cdot}119 \times 600^{0{\cdot}986} = 65 \text{ m} \qquad \text{using } b = 0{\cdot}119 \text{ and } c = 0{\cdot}986 \text{ in equation (6.13)}$$

(1) Substituting into equation (6.10) with $Q = 50$ g s^{-1}; $u = 7$ m s^{-1}; $H = 62\ m$; $\sigma_y = 95$ m and $\sigma_z = 65\ m$.

$$C_{(x,0,0,H)} = \frac{Q}{\pi\sigma_y\sigma_z u} \exp\left[-\frac{1}{2}\left(\frac{H}{\sigma_z}\right)^2\right] \tag{6.10}$$

$$C_{(600,0,0,62)} = \frac{50}{\pi \times 95 \times 65 \times 7} \exp\left[-\frac{1}{2}\left(\frac{62}{65}\right)^2\right]$$

$$= \frac{50}{\pi \times 95 \times 65 \times 7} \times 0{\cdot}6345$$

$$= 234\ \mu\text{g m}^{-3}$$

(2) To find the concentration at a point 50 m from the X-axis equation (6.9) is required with $y = 50$ m and $H = 62$ m.

$$C_{(x,y,0,H)} = \frac{Q}{\pi\sigma_y\sigma_z u} \exp\left[-\frac{1}{2}\left(\frac{y}{\sigma_y}\right)^2\right] \exp\left[-\frac{1}{2}\left(\frac{H}{\sigma_z}\right)^2\right] \tag{6.9}$$

$$C_{(600,50,0,62)} = \frac{50}{\pi \times 95 \times 65 \times 7} \exp -\frac{1}{2}\left(\frac{50}{95}\right)^2 \exp -\frac{1}{2}\left(\frac{62}{65}\right)^2$$

$$= \frac{50}{\pi \times 95 \times 65 \times 7} \times 0{\cdot}8707 \times 0{\cdot}6345$$

$$= 203\ \mu\text{g m}^{-3}$$

(3) To find the concentration at a point 50 m from the X-axis and 20 m above ground, equation (6.8) is required with $y = 50$ m; $z = 20$ m and $H = 62$ m.

$$C_{(x,y,z,H)} = \frac{Q}{2\pi\sigma_y\sigma_z u} \exp\left[-\frac{1}{2}\left(\frac{y}{\sigma_y}\right)^2\right] \exp\left[-\frac{1}{2}\left(\frac{z-H}{\sigma_z}\right)^2\right] + \exp\left[-\frac{1}{2}\left(\frac{z+H}{\sigma_z}\right)^2\right] \tag{6.8}$$

$$C_{(600,50,20,62)} = \frac{50}{2\pi \times 95 \times 65 \times 7} \exp\left[-\left(\frac{50}{95}\right)^2\right] \exp\left[-\frac{1}{2}\left(\frac{20-62}{65}\right)^2\right] + \exp\left[-\frac{1}{2}\left(\frac{20+62}{65}\right)^2\right]$$

$$= \frac{50}{2\pi \times 95 \times 65 \times 7} \times 0{\cdot}8707(0{\cdot}8116 + 0{\cdot}4512)$$

$$= 0{\cdot}0001841 \times 0{\cdot}87 \times 1{\cdot}2628$$

$$= 202\ \mu\text{g m}^3$$

In other words, the concentration at the ground is practically the same as that 20 m above the ground at this point, 600 m away from the source and 50 m off the direct line drawn between the source along the line of the wind direction.

6.4.2 Procedure for estimating ground level concentrations—isopleths

An isopleth is a sort of contour line in the sense that instead of being the line drawn between all points of equal height above sea level, it is defined as the line joining all points of equal pollutant concentration at ground level. Using the coordinate notation shown in Fig. 6.2, it is easy to see that a point on an isopleth is the off-axis concentration at a place sufficiently downwind for the plume to contact the ground. Let such a position be x_1 on the X-axis where the concentration will be $C_{(x_1,0,0,H)}$, then the concentration required will be $C_{(x_1,y_1,0,H)}$. Substituting these values into equation (6.9) and dividing gives,

$$\frac{C_{(x_1,y_1,0,H)}}{C_{(x_1,0,0,H)}} = \exp\left[-\frac{1}{2}\left(\frac{y_1}{\sigma_{y_1}}\right)^2\right] \tag{6.14}$$

Taking logarithms of equation (6.14) gives

$$\ln\frac{C_{(x_1,y_1,0,H)}}{C_{(x_1,0,0,H)}} = -\frac{1}{2}\left(\frac{y_1}{\sigma_{y_1}}\right)^2 \tag{6.15}$$

and rearranging gives

$$y_1 = \left[2\ln\frac{C_{(x_1,0,0,H)}}{C_{(x_1,y_1,0,H)}}\right]^{1/2}\sigma_{y_1} \tag{6.16}$$

Equation (6.16) is convenient when the concentrations at the X-axis and at $C_{(x,y,0,H)}$ are specified but the distance y_1 is required.

Example 3. In example 2 we calculated that the concentration at a point 600 m from the source at ground level $C_{(x,0,0,H)}$ was 234 μg m^{-3}. How far from the X-axis will the concentration fall to 150 μg m^{-3} at this distance?

Using equation (6.16) and substituting $C_{(x_1,0,0,H)} = 234\ \mu\text{g m}^{-3}$ $C_{(x_1,y_1,0,H)} = 150\ \mu\text{g m}^{-3}$ and $\sigma_{y_1} = 95$ m, gives

$$y_1 = \left[2\ln\frac{234}{150}\right]^{1/2} \times 95$$

$$= 90\text{ m}$$

In practice, with pollution problems the maximum ground level concentration at different distances from a source is often more important than the construction of isopleths. In these cases, the maximum value of $C_{(x,y,0,H)}$ can be found as a solution to equation (6.9), such that

$$C_{(x,y,0,H)_{\max}} = \frac{2Q}{e\pi uH^2}\cdot\frac{\sigma_z}{\sigma_y} \tag{6.17}$$

where e is the base to natural logarithms, viz. 2.72, and $\sigma_z = H/\sqrt{2}$.

Maximum ground level concentrations of pollutants from a stack usually occur at distances located some 15–30 times the stack height. A relationship between the distance to maximum concentration and the normalized function $(uC/Q)_{max}$ for various effective stack heights is shown in Fig. 6.6. This concept of the normalizing function uC/Q introduced by Hilsmeier and Gifford [8] enables graphs to be constructed for different parameters under each of the stability categories. These graphs are shown in Figs 6.6, 6.7 and 6.8. The use of these diagrams allows the tedium to be taken out of the calculations.

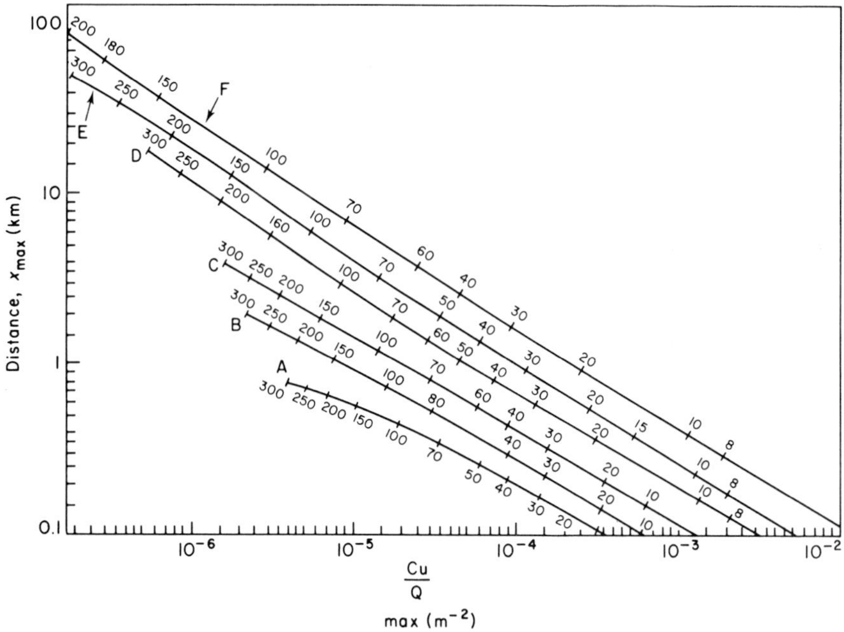

FIG. 6.6 Distance of maximum concentration and maximum Cu/Q as a function of stability (curves) and effective height (metres) of emission (numbers) [6]

6.4.3 Estimation of effective height of plume

Generally, smoke plumes from a chimney stack are emitted quite forcibly to the atmosphere at elevated temperatures. This effect of temperature and velocity propels the effluent beyond the rim of the chimney for some distance before the wind turns the plume horizontal. The effective height H of the source, therefore, is greater than the actual height of the chimney, h. Numerous investigations have developed theoretical and empirical treatments for computing the effective stack height. Some useful equations for

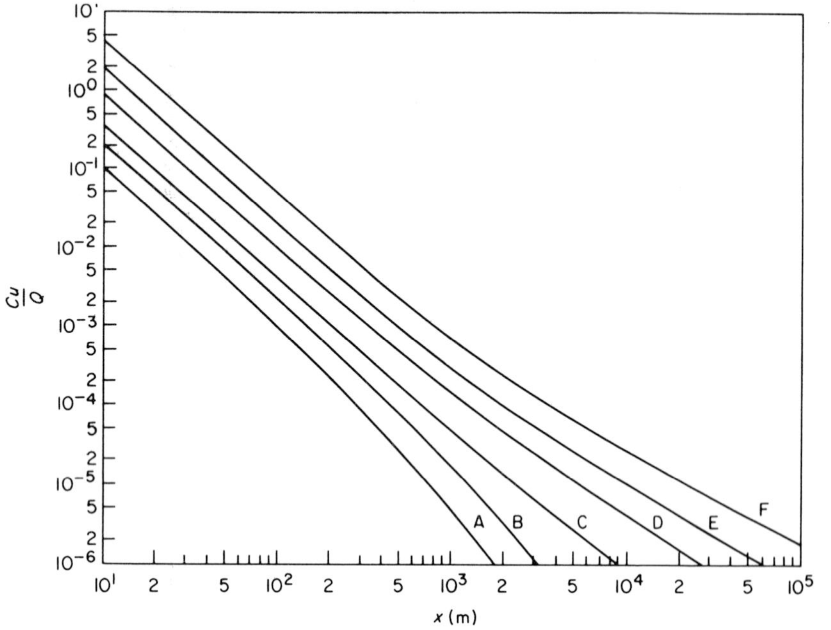

FIG. 6.7 Values of Cu/Q as a function of downwind distance x (m) for various stability categories [8]

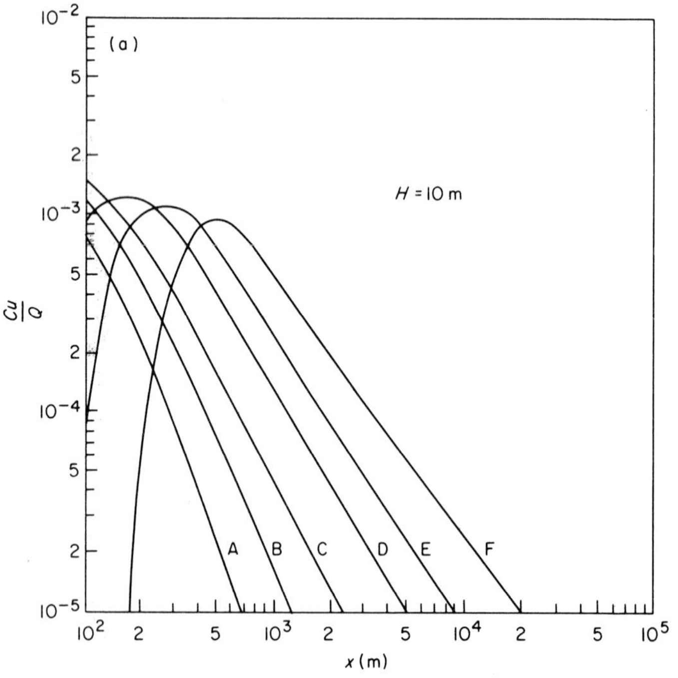

FIG. 6.8(a) Values of Cu/Q as a function of downwind distance x (m) for various stability categories for $H = 10$ m [8]

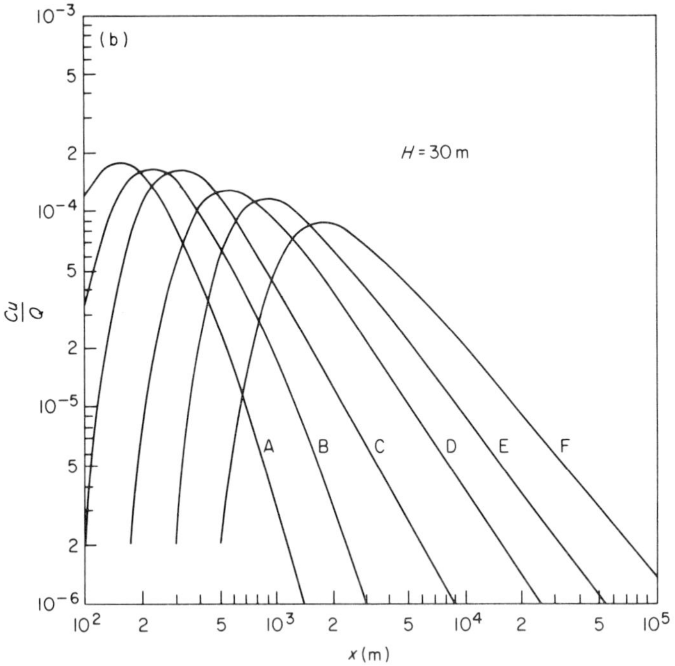

FIG. 6.8(b) Values of Cu/Q as a function of downwind distance x (m) for various stability categories for $H = 30$ m [8]

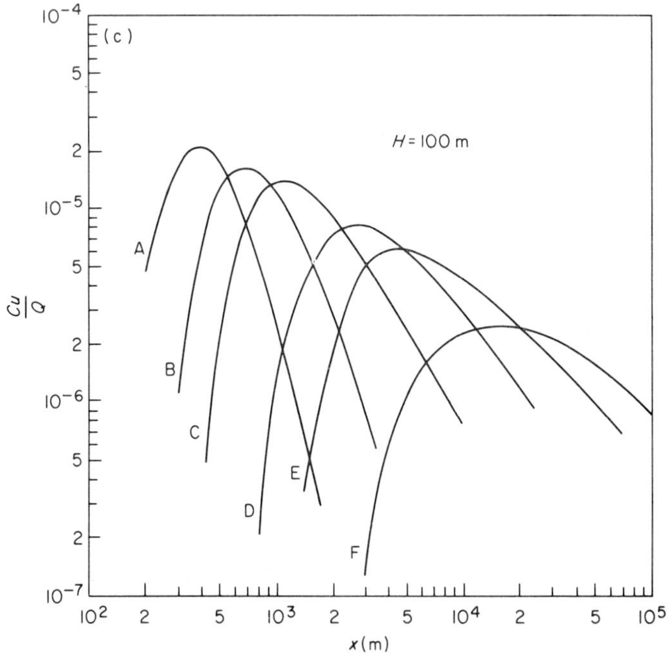

FIG. 6.8(c) Values of Cu/Q as a function of downwind distance x (m) for various stability categories for $H = 100$ m [8]

TABLE 6.5 *Summary of equations used for estimating* h *the plume rise above the height of the chimney stack, where* H = h + Δh (Fig. 6.2)

Equation	Definitions	Remarks	Reference
$\Delta h = \frac{v_s}{u} d\left[1{\cdot}5 + 2{\cdot}68 \times 10^{-3} p \frac{(T_s - T)}{T_s} d\right]$ for stack diameters 1·7 to 4·3 m	v_s = stack gas exit velocity (m s^{-1}) d = inside stack diameter (m) u = windspeed (m s^{-1}) p = pressure (mbar) T_s = stack gas temperature T = ambient temperature	$2{\cdot}68 \times 10^3$ is a constant having dimensions of $(\text{mbar})^{-1}\,\text{m}^{-1}$. For unstable conditions multiply Δh by 1·15. For stable conditions multiply Δh by 0·85	9
$\Delta h = \frac{2F^{1/3}x^{2/3}}{u}$ where $F \sim g v_s d^2 (T_s - T)/T_s$ when effluent gas has an average molecular weight and heat capacity similar to that of air	u = windspeed g = acceleration due to gravity v_s = stack gas ejection velocity d = inside diameter of stack T_s = stack gas temperature T = ambient temperature	Δh is the height of the plume axis above the top of the stack at a downwind distance of x	10
$\Delta h_{\text{max}} = \frac{\alpha Q^{1/4}}{u}$ where $\alpha = 12{\cdot}9 + 0{\cdot}02h$, for h between 50 and 150 m	Q = MW s^{-1} (1 MW = $9{\cdot}96 \times 10^3$ cal s^{-1}) Δh_{max} in metres and u in m s^{-1}	Based on plumes from moderately large power stations	11, 12
$\Delta h = 0{\cdot}047 \frac{Q^{0{\cdot}58}}{u^{0{\cdot}70}}$	Δh in metres Q = heat emission rate of stack gas	All stability categories	13

estimating Δh, the plume rise, so that $H = h + \Delta h$ can be found, are summarized in Table 6.5.

Example 4. A chimney stack 35 m high and 2 m diameter emits sulphur dioxide at a temperature of 400 K with a velocity of 10 m s^{-1}. This sulphur dioxide is derived from the burning of coal with a sulphur content of 2·6% at the rate of 4500 kg h^{-1}. Show how the maximum ground level concentration varies with windspeed for stability category D, and hence estimate the windspeed that produces the highest maximum concentration at a pressure of 1000 mbar when the ambient temperature is 294 K.

The value of Δh as a function of windspeed can be found using Holland's equation given in Table 6.5, viz.,

$$\Delta h = \frac{v_s d}{u}\left[1{\cdot}5 + 2{\cdot}68 \times 10^{-3} p \frac{(T_s - T_a)}{T_s} d\right]$$

$$= \frac{10 \times 2}{u}\left[1{\cdot}5 + 2{\cdot}68 \times 10^{-3} \times 10^{3} \frac{(400 - 294)}{400} \times 2\right]$$

$$= \frac{20}{u}[1{\cdot}5 + 1{\cdot}42]$$

$$\Delta h = \frac{58{\cdot}4}{u}$$

It is now possible to construct a table in which Δh and hence $H = h + \Delta h$ can be found as a function of windspeed. This is shown in the first three columns of the table below.

Column No.					
1	2	3	4	5	6
u (m s^{-1})	Δh (m)	$H = h + \Delta h$ $H = 35 + \Delta h$ (m)	$(Cu/Q)_{\text{maxD}}$ (m^{-2})	Q/u (g m^{-1})	C_{maxD} (g m^{-3})
1·0	58·4	93·4	$9{\cdot}4 \times 10^{-6}$	65	$6{\cdot}11 \times 10^{-4}$
2·0	29·2	64·2	$2{\cdot}3 \times 10^{-5}$	32·5	$7{\cdot}47 \times 10^{-4}$
2·5	23·4	58·4	$3{\cdot}0 \times 10^{-5}$	26·0	$7{\cdot}80 \times 10^{-4}$*
3·0	19·5	54·5	$3{\cdot}4 \times 10^{-5}$	21·7	$7{\cdot}34 \times 10^{-4}$
3·5	16·7	51·7	$3{\cdot}9 \times 10^{-5}$	18·6	$7{\cdot}25 \times 10^{-4}$

The maximum ground level concentration for these heights H in terms of the normalized function $(Cu/Q)_{\max_D}$ can now be read from Fig. 6.6 for the stability category D. These values are listed in column 4. Finally, in order to find $C_{\max_D}$ it is necessary to multiply by the factor Q/u, where Q is the source strength in $\mathrm{g\,s^{-1}}$. Q can be calculated because sulphur, mol. wt = 32, present in the coal is converted to sulphur dioxide, mol. wt = 64, during combustion. Hence, 4500 kg of coal will produce

$$4500\times 10^3\times\frac{2\cdot 6}{10^2}\ \mathrm{g\,h^{-1}}\ \text{of S or}$$

$$\frac{4500\times 10^3\times 2\cdot 6}{3600\times 10^2}\ \mathrm{g\,s^{-1}}\ \text{of S which is equivalent to}$$

$$\frac{4500\times 10^3\times 2\cdot 6}{3600\times 10^2}\times\frac{64}{32}\ \mathrm{g\,s^{-1}}\ \text{of } SO_2\text{, i.e. } Q$$

Therefore, $Q = 65\ \mathrm{g\,s^{-1}}$.

Values of Q/u shown in column 5 multiplied by the $(Cu/Q)_{\max_D}$ values indicated in column 4 of the table give the required $C_{\max_D}$ values in column 6.

Inspection of column 6 shows that maximum ground level sulphur dioxide concentration will occur for windspeed of $2\cdot 5\ \mathrm{m\,s^{-1}}$. From Fig. 6.6 when $(Cu/Q)_{\max_D}$ is $7\cdot 8\times 10^{-4}\ \mathrm{m^{-2}}$ the distance from the source will be about 1·5 km.

6.4.4 Emissions from a continuous line source

Example 5. During the rush hour a motorway running E–W carries 4000 vehicles $\mathrm{h^{-1}}$ travelling at an average speed of 40 miles $\mathrm{h^{-1}}$. Some 80% of these vehicles are petrol propelled and use fuel which contains 2·0 g of lead $\mathrm{gal^{-1}}$, that is consumed at the rate of 20 miles $\mathrm{gal^{-1}}$. Assuming that 75% of the lead is emitted in the form of a fine particulate aerosol, estimate the atmospheric lead concentrations at 250 m, 500 m and 1000 m downwind of the road for windspeeds from the N in ranges appropriate to the stability categories A to D inclusive.

The motorway can be considered a continuous infinite line source and, therefore, equation (6.6) given in Table 6.1 may be used with both z and H set equal to zero. Under these conditions equation (6.6) becomes,

$$C_{(x,0,0,0)}=\frac{Q}{2(\pi K_z u x)^{1/2}}[2]$$

$$=\frac{Q}{(\pi K_z u x)^{1/2}}$$

where by equation (6.7) $\sigma_z = (2K_z x/u)^{1/2}$. Hence,

$$C_{(x,0,0,0)} = \left(\frac{2}{\pi}\right)^{1/2} \frac{Q}{\sigma_z u}$$

is the equation that must be used, with Q expressed in $\mu g\ s^{-1}\ m^{-1}$ to give the usual units of atmospheric Pb measurement $C_{(x,0,0,0)}$ as $\mu g\ m^{-3}$. Before proceeding with the problem the source strength Q must be determined from the data. There will be $2{\cdot}0 \times 75/10^2$ g of Pb emitted in 20 miles or twice this amount in 40 miles which would be covered in 1 h or 3600 s. Therefore, the Pb emission rate for each vehicle will be

$$\frac{2 \times 2{\cdot}0 \times 75 \times 10^6}{3600 \times 10^2}\ \mu g\ s^{-1}$$

Pb emission rate = 833 $\mu g\ s^{-1}$ vehicle^{-1}. The number of vehicles m^{-1} on the motorway will be

$$\text{vehicles m}^{-1} = \frac{\text{Flow (vehicles h}^{-1})}{\text{Average speed (miles h}^{-1})\ 1609\ (\text{m mile}^{-1})}$$

where 1609 is the factor for converting miles to metres. Substituting,

$$\text{vehicles m}^{-1} = \frac{4000 \times 80/100}{40 \times 1609}$$

$$= 4{\cdot}97 \times 10^{-2} \text{ vehicles m}^{-1}$$

$$Q = (\text{vehicles m}^{-1})\ (\text{Pb emission rate in } \mu g\ s^{-1}\ \text{vehicle}^{-1})$$

$$= 4{\cdot}97 \times 10^{-2} \times 833$$

$$= 41{\cdot}4\ \mu g\ m^{-1}\ s^{-1}$$

The values of σ_z are now calculated for the distances 250 m, 500 m and 1000 m using Table 6.4, or alternatively Figs 6.4 and 6.5. The results are shown in Table (a) below.

TABLE (a) *Calculated σ_z values at 250 m, and 1000 m from a line source*

Stability category	Distance (m) 250	500	1000
A	$0{\cdot}125 \times 250^{1{\cdot}03} = 37$	$0{\cdot}00883 \times 500^{1{\cdot}51} = 105$	$0{\cdot}000226 \times 1000^{2{\cdot}1} = 451$
B	$0{\cdot}119 \times 250^{0{\cdot}986} = 27$	$0{\cdot}119 \times 500^{0{\cdot}986} = 54$	$0{\cdot}119 \times 1000^{0{\cdot}986} = 108$
C	$0{\cdot}111 \times 250^{0{\cdot}911} = 17$	$0{\cdot}111 \times 500^{0{\cdot}911} = 32$	$0{\cdot}111 \times 1000^{0{\cdot}911} = 60$
D	$0{\cdot}105 \times 250^{0{\cdot}827} = 10$	$0{\cdot}105 \times 500^{0{\cdot}827} = 18$	$0{\cdot}105 \times 1000^{0{\cdot}827} = 32$

Finally, for convenience the results of substituting the data into equation $C_{(x,0,0,0)} = (2/\pi)^{1/2}\ Q/\sigma_z u$ are given in Table (b). This table completes the calculation for the estimates of airborne Pb concentrations in $\mu g\ m^{-3}$ for windspeeds of 1, 2, 3, 4, 5·5 and 8 m s^{-1} for the various stability categories shown.

TABLE (b) *showing the calculation of* $C_{(x,0,0,0)} = (2/\pi)^{1/2}Q/\sigma_z u$

Stability category	Windspeed	σ_z m	$\sigma_z u$ $m^2\ s^{-1}$	$(2/\pi)^{1/2}Q$ $\mu g\ m^{-1}\ s^{-1}$	$(2/\pi)^{1/2}Q/\sigma_z u$ $\mu g\ m^{-3}$
			Distance 250 m		
A	2	37	74	33·0	0·45
B	4	27	108	33·0	0·30
C	5·5	17	93	33·0	0·35
D	1	10	10	33·0	3·30*
	2	10	20	33·0	1·65
	3	10	30	33·0	1·10
	4	10	40	33·0	0·82
	8	10	80	33·0	0·41
			Distance 500 m		
A	2	105	210	33·0	0·16
B	4	54	216	33·0	0·15
C	5·5	32	176	33·0	0·19
D	1	18	18	33·0	1·83*
	2	18	36	33·0	0·97
	3	18	54	33·0	0·61
	4	18	72	33·0	0·46
	8	18	144	33·0	0·23
			Distance 1000 m		
A	2	451	902	33·0	0·04
B	4	108	432	33·0	0·08
C	5·5	60	330	33·0	0·10
D	1	32	32	33·0	1·03*
	2	32	64	33·0	0·51
	3	32	96	33·0	0·34
	4	32	128	33·0	0·26
	8	32	256	33·0	0·13

It will be apparent from the last column of the table that under the conditions assumed, the neutral stability category D when associated with low windspeeds ~1 m s^{-1} gives the most lead pollution at all distances. Another point which should be appreciated is that no account has been taken of the finite plume width at the kerbside, as suggested by Calder [16]. If this is done and avalue of 27 m is added to the distances, the effect on the prediction of the D category values at the 250 m distance for windspeeds of

1, 2, 3, 4 and 8 m s^{-1} will be to reduce them from 3·3, 1·65, 1·10, 0·82, 0·41 to 3·0, 1·5, 1·0, 0·75 and 0·37 $\mu g\ m^{-3}$; respectively.

6.4.5 *Modification of σ_y and σ_z due to urban and rural diffusion or the influence of buildings*

Comparisons between urban and rural diffusion have been made by Pooler [14] and McElroy [15]. They conclude that σ_y and σ_z terms are similar to the Pasquill curves for open country provided a similar initial plume size is assumed. Pooler suggests an extra 80 m and 30 m, respectively, should be added to σ_y and σ_z values at zero distances in the case of urban diffusion. Some recommendations that have been made are summarized in Table 6.6. The modification of σ values is illustrated in the following example, where a radioactive pollutant escapes into the atmosphere from a nuclear reactor.

6.4.6 *Emissions from nuclear power stations*

Example 6. An accident at a nuclear power station occurs at 11.00 hours releasing $4{\cdot}9 \times 10^4$ curies of iodine-131, which has a half-life of 8·04 days, into the atmosphere of the containment vessel of radius 22 m. This containment vessel is hemispherically shaped and leaks activity at an estimated 0·15% day^{-1}. What will be the concentration of iodine-131, 1·5 km downwind 2 h after the accident if the windspeed is 3 m s^{-1} on a dull day?

TABLE 6.6 *Estimations of initial plume size*

Situation	Recommendation	Initial plume size	Reference
Urban diffusion	Add 80 m to σ_y Add 30 m to σ_z	$\sigma_{yo} = 80$ m and $\sigma_{zo} = 30$ m all stability categories	14
	Add 50–60 m to σ_y Add 20–30 m to σ_z	$\sigma_{yo} = 50$–60 m $\sigma_{zo} = 20$–30 m	15
	Take into account building dimensions BH = building height BL = building length	$\sigma_{zo} = BH/2{\cdot}15$ $\sigma_{yo} = BL/4{\cdot}3$	
Vehicle wake	Take into account the width of the carriage way, w, since the plume will be finite at the kerbside	$\sigma_{zo} = b(x+w)^c$ where $w = 27$ *m* *b* and *c* are the constants given in Table 6.4	16

The source strength Q (Ci s^{-1}) will be equal to the product of the leak rate and the activity corrected for decay. The decay correction factor is $\exp -0{\cdot}693t/\lambda_{1/2}$, where $\lambda_{1/2}$ is the half-life (seconds) and t is the time in seconds from the commencement of the accident. Hence,

$$Q_{131_I} = \frac{0{\cdot}15}{100\times60\times60\times24}\times4{\cdot}9\times10^4 \exp -\frac{0{\cdot}693\times60\times60\times2}{8{\cdot}04\times24\times60\times60}$$

$$= 8{\cdot}5\times10^{-4}\exp -0{\cdot}007183$$

$$= 8{\cdot}5\times10^{-4}\times0{\cdot}9928$$

$$= 8{\cdot}44\times10^{-4}\ \text{Ci s}^{-1}$$

Since the accident occurred inside a building and radiation is leaking from the containment vessel, a correction must be applied to take into account the finite size of the plume initially entering the atmosphere. From Table 6.6 and reference 15, $\sigma_{z_0} = \text{BH}/2{\cdot}15$ where BH is the building height, and since the vessel is hemispherical the same factor will be used for σ_{y_0}, so that $\sigma_{z_0} = \sigma_{y_0} = 22/2{\cdot}15 = 10{\cdot}23$ m. These values will correspond to virtual distances x_v for the Y- and Z-axes, using the data in Table 6.4 or Figs 6.4 and 6.5 for class D stability,

$$10{\cdot}23 = 0{\cdot}392x_{vz}^{0{\cdot}636} \quad \text{giving } x_{vz} = 169\ \text{m}$$

and

$$10{\cdot}23 = 0{\cdot}13x_{vy}^{0{\cdot}903} \quad \text{giving } x_{vy} = 126\ \text{m}$$

These distances must be added to $x = 1500\ m$, therefore,

$$x + x_{vz} = 1669\ \text{m}$$

and

$$x + x_{vy} = 1626\ \text{m}$$

The corrected values of σ_y and σ_z can now be calculated:

$$\sigma_y = 0{\cdot}13\times1626^{0{\cdot}903} = 103\ \text{m}$$

$$\sigma_z = 0{\cdot}392\times1669^{0{\cdot}636} = 44\ \text{m}$$

Hence, using equation (6.11),

$$C_{(x,0,0,0)} = \frac{Q}{\pi\sigma_y\sigma_z u}$$

$$= \frac{8{\cdot}44\times10^{-4}}{\pi\times103\times44\times3}$$

$$C_{(1500,0,0,0)} = 1{\cdot}97\times10^{-8}\ \text{Ci m}^{-3}$$

The MPC for iodine-131 is 3×10^{-10} μCi cm^{-3} so that it can be concluded that under these conditions 1·5 km from the accident and 2 h afterwards, atmospheric pollution due to iodine-131 is about 66 times greater than the recommended value.

The dangerous consequences of accidents culminating in the escape of radioactive material has already been mentioned in section 1.6.4. This aspect has received considerable publicity through the recognized channels of press and television coverage. Another expression of concern of a different kind that should be mentioned has recently been voiced [17]. Nuclear power stations emit argon-41, half-life 1·8 h and krypton-85, half-life 10·4 years. Unlike other gases, emission control systems are incapable of containing these radionuclides because they are chemically inert. A large power station can release into the atmosphere several hundred curies per hour of argon-41. Effective control in this instance depends upon the short half-life and a high stack policy which ensures "meteorological" dilution in the atmosphere to acceptable levels.

Unfortunately, the same is not true of krypton-85, because of the long half-life. Assuming average meterological conditions a single nuclear fuel processing plant, for example, could release as much as 10^7 Ci of krypton-85 annually without exceeding the $(MPC)_{air}$ at the boundary of the controlled area around the plant. Furthermore, the inert nature of the material will eventually ensure global dispersion. The fear, therefore, has been expressed that krypton-85 will have a modifying influence on weather, because it is capable of producing ion pairs in the atmosphere. Boeck [17] calculates that even over the oceans the additional ionization due to krypton-85 will be noticeable at concentrations well below the $(MPC)_{air}$ of 300 nCi m^{-3}. A concentration of only 3 nCi m^{-3} of krypton-85 will produce an additional $8{\cdot}55\times10^5$ ion pairs m^{-3} s^{-1}. This represents something like a 57% increase over the normal sea level ion production rate attributable to cosmic radiation. Changes of this extent in the atmospheric ionization background may well affect electrical characteristics of clouds and electrical storms. It seems feasible that this ionization could provoke detrimental climatic changes in the earth's weather patterns. Until further evidence is found to the contrary, a reasonably cautious attitude should be adopted towards this feature of nuclear power generation and processing.

References

1. Monin, A. S. and Yaglom, A. M. (1971). "Statistical Fluid Mechanics: Mechanisms of Turbulence." Vol. 1 (English translation Ed. Lumley, J. L.). MIT Press, Cambridge, Massachusetts.

2. Sutton, O. G. (1947). The theoretical distribution of airborne pollution from factory chimneys. *Quart. J. Roy. Meteorol. Soc.* **73**, 426–436.
3. Pasquill, F. (1961). The estimation of the dispersion of wind-borne material. *The Meteorol. Magazine, Met. Office, U.K.*, 33–49.
4. Gifford, F. A. (1961). Use of routine meteorological observations for estimating atmospheric dispersion. *Nuclear Safety* **2**, 47–51.
5. Koch, R. C. (1971). "Validation and sensitivity analysis of the Gaussian plume multiple-source urban diffusion model." Geomet. Inc. Rockville, M.D., Geomet EF-60 APTD 0935 CPA 70 94.351 P Nov. 1971 FLD/GP 13B.
6. Turner, D. B. (1969). "Work book of atmospheric dispersion estimates." US Department of Health, Education and Welfare, Public Health Service, Cincinnati.
7. Drivas, P. J. and Shair, F. H. (1974). Dispersion of an instantaneous cross-wind line source of tracer relased from an urban highway. *Atmos. Environ.* **8**, 475–485.
8. Hilsmeier, W. F. and Gifford, F. A. (1962). "Graphs for estimating atmospheric diffusion." ORO-545 Oak Ridge, Tenn., US Atomic Energy Commission.
9. Holland, J. Z. (1953). "Meteorological survey of the Oak Ridge Area." Report ORO-99 US Atomic Energy Commission.
10. Briggs, G. A. (1965). A plume rise model compared with observation. *J. Air Pollut. Contr. Assoc.* **15**, 433–438.
11. Lucas, D. H., Moore, D. J. and Spurr, G. (1963). The rise of hot plumes from chimneys. *Int. J. Air Water Pollut.* **7**, 473–500.
12. Lucas, D. H. (1967). Paper VI: Application and evaluation of results of the Tilbury plume rise and dispersion experiment. *Atmos. Environ.* **1**, 421–424.
13. Concawe (Conservation of Clean Air and Water, Western Europe) (1966). "The calculation of atmospheric dispersion from a stack." The Hague, Netherlands.
14. Pooler, F. (1966). A tracer study of dispersion over a city. *J. Air Pollut. Contr. Assoc.* **16**, 677–681.
15. McElroy, J. L. (1969). A comparative study of urban and rural dispersion. *J. Appl. Meteor.* **8**, 19–31.
16. Calder, K. L. (1973). On estimating air pollution concentrations from a highway in an oblique wind. *Atmos. Environ.* **7**, 863–868.
17. Boeck, W. L. (1976). Meteorological consequences of atmospheric krypton-85 *Science* **193**, 195–198.

7

Urban Atmospheres

7.1 Introduction

In this chapter we shall consider the concentrations, particle sizes and chemical composition of urban aerosols. With the exception of the photochemically initiated reactions that occur in urban aerosols, already discussed in Chapter 5, our knowledge of chemical reactions between pollutants present in city atmospheres is less well understood. The aerosol system provides the opportunity of chemical transformations between gaseous components, at gas–solid, gas–liquid and liquid–liquid interfaces. Particle growth through impaction and coagulation may or may not involve chemical reaction, but will most certainly influence particle size distributions as well as the lifetime of particulates.

Information that is available has been obtained largely on a priority basis which has some relevance to community health. For this reason, after chemical analysis to establish the type of compounds and elements present as pollutants, efforts have been mainly directed towards characterization of particle size. The sizing of particulate matter uses the Andersen impactor or some other particle-fractionating equipment which simulates the action of the human respiratory system, and thereby provides an insight into the likely extent of lung penetration and retention of airborne particles by urban populations. Particle size evaluations are also helpful for assessing the origin and lifetime of aerosol components.

7.2 Particulate aerosol characteristics in urban atmospheres

7.2.1 Total particulate concentrations and properties

Table 7.1 gives values of total particulate concentrations along with the mass median equivalent diameters (MMDs) and average geometric

TABLE 7.1 *Comparison of total particulate parameters from various cities throughout the world*

Location	Sample period	Average particulate concentration ($\mu g\ m^{-3}$)	Average MMD (μm)	Average geometric standard deviation	% Particulate $\leqslant 1\ \mu m$	% Particulate $\leqslant 2\ \mu m$	Ref.
USA:							
University of California Riverside Campus	November 1968, 15 days	82	0·9	11			3
Chicago	1970 year	86·5	0·76	8·18	55	68	1
Cincinnati	1970 year	74·3	0·70	5·49	59	74	1
Denver	1970 year	59·7	0·40	10·50	65	75	1
Philadelphia	1970 year	58·5	0·47	5·65	67	80	1
St. Louis	1970 year	73·1	0·83	6·80	54	68	1
Washington	1970 year	56·3	0·46	5·22	68	81	1
Average US data		70·06	0·64	7·55	61	74	
Ankara	April–May, 1971 10 days	115·9	1·79	5·20	37	53	4
UK:							
London	February–May, 1971	96·0	0·46	8·3	65	76	5
Birmingham	April 1977	45·5	1·10	7·0	48	62	6
Melbourne, Australia	September–November	68·8	3·44	9·4	28	40	7
Nagoya, Japan	October 1973 to October 1974	88·8	1·36				8

standard deviations that have been obtained by particle fractionation studies in various parts of the world. The USA data for the year 1970 are taken from the National Air Surveillance Cascade Impactor Network I survey [1], but since then report III in the series has appeared [2]. This last report covers data on particle size distribution parameters for the years 1970, 1971 and 1972. It shows that in general there has been a progressive increase in the MMDs of particulates in this period. During 1970 the MMDs were in the range 0·40–0·6 μm, in 1971 they were from 0·75 to 1·22 μm and in 1972 they had risen to 1·06–1·46 μm. A possible explanation is that this increase in the MMDs of particulates is mostly a consequence of emission control legislation applied to motor vehicles. Reduction of emissions of carbon monoxide and hydrocarbons, both of which are associated with finely divided soot particles, logically supports this trend.

The value of 1·79 μm for the MMD for the aerosol in Ankara given in Table 7.1 is thought to be due to the extensive use of low-grade coal. The high average particulate concentration of 115·9 $\mu g\ m^{-3}$ has been attributed to the geographical location of the city which is surrounded by hills and to the frequent temperature inversions that occur. Both of these features will lead to poor ventilation. Although the atmospheric particulate loading is high it will be noticed that the percentage particulate that falls in the respirable range, i.e. less than 3 μm in size, is comparatively low.

The UK data given in the table appear to be typical, with MMDs in the range 0·46–1·1 μm, and contrast rather sharply with the MMD values of 3·44 μm reported in Melbourne. During this Melbourne study, however, demolition and construction work was in progress two blocks away from the sampling site and these operations will almost certainly contribute significantly to the particles in the larger size range.

The Japanese study of urban particulate in Nagoya by Kadowaki [8] listed in the table shows marked bimodal characteristics. These two populations are characterized by particles having MMDs in the range 0·4–0·6 μm and MMDs in the range 3–5 μm. The smaller particle material contains mostly sulphate and ammonium ions, whereas the larger particles comprise mainly Si, Al, Fe, Ca, Mg, Na and Cl, viz. the land- or marine-derived elements. Nitrate is present in both populations but is subject to seasonal change. In summer the nitrate MMD is 3·6 μm and hence falls in the larger particle size range, but in winter this becomes 0·92 μm. Qualitatively, the reason for the variation in the nitrate distribution is thought to be the preferential formation of organic nitrates and PAN in summer through photochemical reaction, as opposed to ammonium nitrate which is the principal nitrate present in the aerosol in winter.

Much of our knowledge about the behaviour of urban aerosols, especially those involving photochemical processes comes from a concerted programme of reasearch carried out in Pasadena, California during August and September 1969. Whitby *et al.* [9] report that it is not possible to describe the properties of smog aerosols by one type of distribution function. In the particle size range 0·05–1·0 μm, however, the log-normal distribution function is adequate (section 3.2.1) and 4·7% of the total number of particles or 50% of this aerosol mass is associated with this fraction. The majority of the particles, on a number basis, are found in the submicron range below 0·05 μm. These small particles are sensitive to the proximity of combustion sources and form spontaneously during photochemical smog episodes. They have short lives, of the order of minutes, because of their rapid growth by coagulation.

The volume distribution function $\Delta V/\Delta \log$ (particle diameter) in $\mu m^3\ cm^{-3}$ when plotted against particle diameter in μm, shows a bimodal distribution (Fig. 7.1). The first mode coincides with particles of 0·30 μm MMD with a geometric standard deviation of 2·24 and fits a log-normal distribution. The second mode lies in the range 3–20 μm with the peak of the distribution curve around 12 μm.

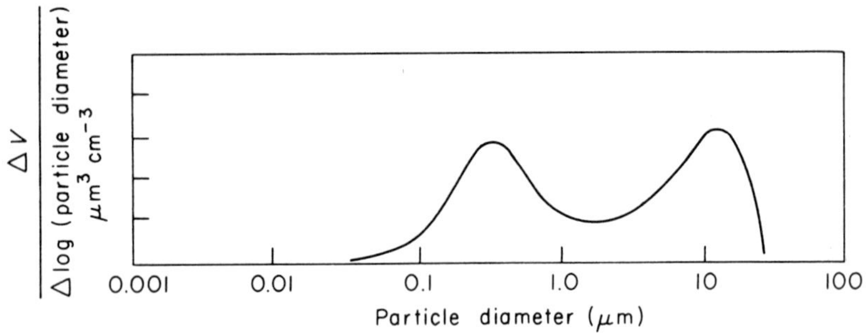

FIG. 7.1 The bimodal aspect of particle size distributions of urban aerosols [9]

Smog aerosols are subject to diurnal variations. Particle growth occurs during the morning until about midday; this is followed by particle decay during the afternoon and at night-time. This general observation is caused by the interaction of the following parameters:

(1) Source factors governed by aerosol origin, e.g. vehicle and industrial emissions or natural spontaneous production through the interaction of gases or vapours.
(2) Meteorological factors, e.g. solar radiation intensity, relative humidity, temperature inversion height, windspeed and direction.
(3) Physical and chemical rate processes, e.g. condensation, nucleation, coagulation, adsorption and absorption of gases by solids and liquids.

Results from the Pasadena study indicate that the volume distribution ($\mu m^3\ cm^{-3}$) for submicron size particles is related to solar radiation and relative humidity. Statistical analysis shows that these parameters fit the regression equation,

$$\text{volume distribution } (\mu m^3\ cm^{-3}) = -2{\cdot}17 + 1{\cdot}18[\text{RH}] + 18{\cdot}06[\text{S-R}] \tag{7.1}$$

to give a correlation coefficient of 0·71, where [RH] = relative humidity (%) and [S-R] = solar radiation (g-cal $cm^{-2}\ min^{-1}$).

Another feature of the volume distribution–relative humidity relationship discovered during the course of this work is shown in Fig. 7.2. The lower curve indicates the distribution pattern for a relative humidity of 14%, a rapid change in weather gave a 38% relative humidity within 20 min and this changed the volume distribution to that shown in the upper curve. It will be noticed that this change causes a shift in the particle diameter modes from 0·35 μm at the lower relative humidity to 0·25 μm at the higher relative humidity. On this sort of evidence it has been concluded that submicron particles in smog aerosols are hygroscopic.

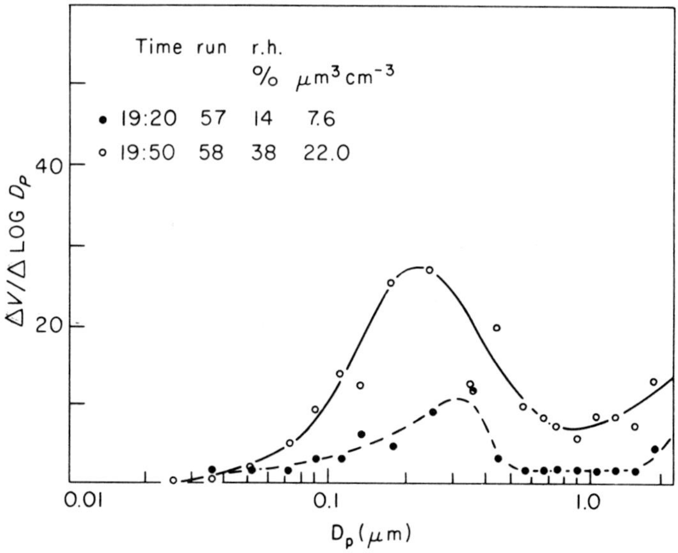

FIG. 7.2 The change of the volume distribution $\Delta V/\Delta \log D_p$ between 19·20 and 19·50 on 21 August 1971 [10]

Experiments in which a 5 m^3 polyethylene balloon was filled with air and then exposed to radiation showed that a rapid production of particles occurred. Between 10^5 and 10^7 particles cm^{-3} were produced within about

10 min and this was followed by a slow decay. The phenomenon is caused by the nucleation of photochemically produced supersaturated vapours during the growth period and this in turn is followed by coagulation to larger particles during the decay period. These experimental observations provide good evidence that nuclei formation is directly related to solar radiation. Mechanistically, the process may be visualized as a two-stage operation. In the first stage nuclei are formed and at the same time existing nuclei grow by condensation. As the number of particles increases there is a greater likelihood of coagulation, so this restricts the total number of particles allowable in the system. The slow decay of nuclei due to this cause constitutes the second stage. These changes are shown in Fig. 7.3.

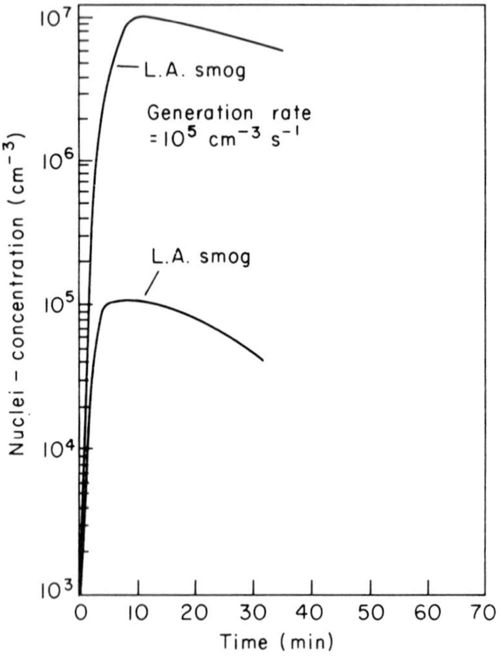

FIG. 7.3 The change of the total number concentrations in an irradiated, initially particle-free container: the initial concentration rise is due to nucleation and the subsequent decay is due to coagulation [10]

7.2.2 Sulphate, nitrate, ammonium and carbon aerosols

So far we have discussed aerosols in a rather general way from the point of view of their overall properties. In reality, of course, aerosols are conglomerates the composition of which varies from one location to another. As already mentioned, the principal influences are the source and climatic factors. Chemical analysis can be helpful for estimating the origin, and sometimes particle size can be identified with specific types of chemical compounds or radicals. Photo-electron spectral analysis of the Pasadena aerosol by Novakov *et al.* [11] has proved very valuable for the identification of different valence states of the sulphur and nitrogen compounds present therein.

Photoelectron spectroscopy is the measurement of the kinetic energies of electrons which are expelled from a sample by a flux of monoenergetic photons through the photoelectric effect. The conservation of energy equation governing the phenomenon is

$$E_k = E_h - E_b \tag{7.2}$$

where E_h is the photon energy, E_b is the binding energy and E_k is the electron kinetic energy. All electronic levels with binding energies less than the photon energy can be measured. Using Al or Mg K_α radiation as the flux, both core and valence electrons can be studied. A schematic representation of the apparatus is shown in Fig. 7.4. The sample is placed in the X-ray flux emitted by the anode, photoelectrons released from the sample are analysed in an electrostatic deflection spectrometer, by passage between two concentric spherical deflecting electrodes. The electrons within a small band are focused onto the detector where they are counted. By changing the voltage on the analyser electrodes the photoelectron spectrum can be scanned.

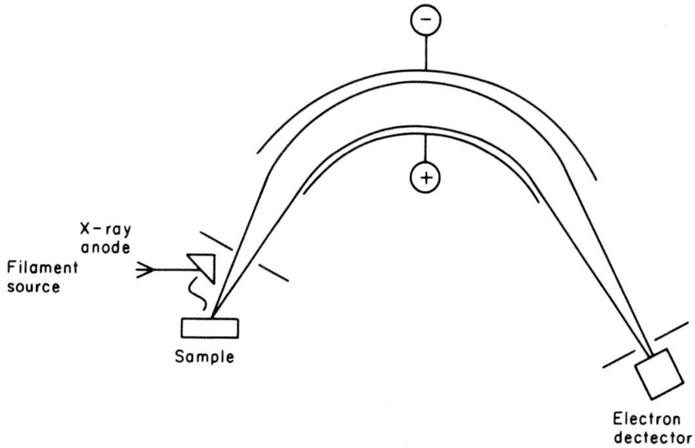

FIG. 7.4 Schematic representation of a photoelectron spectrometer. The electrons are dispersed by energy in the electric field between the two concentric spherical electrodes [11]

Application of this technique reveals different ratios of S^{6+}/S^{4+} associated with different particle size ranges. Particles between 0·6–2·0 μm have S^{6+}/S^{4+} ratios of less than unity, whereas those in the range 2–5 μm have S^{6+}/S^{4+} values between 1·5 and 1·8. Both particle size ranges exhibit a minimum in the S^{6+}/S^{4+} ratio around midday due to rapid uptake of SO_2 (Fig. 7.5). This is followed by slow oxidation to sulphate at night and will account for the overall predominance of sulphate over sulphite in the larger particle size material. Some of the sulphate in urban aerosols is in the form of ammonium sulphate, although chemical analysis indicates that there are marked departures from values expected for $(NH_4)_2SO_4$ from one region to another.

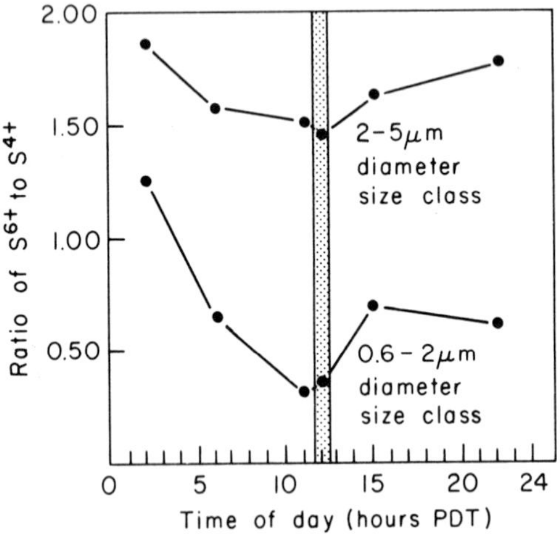

FIG. 7.5 Sulphate/sulphite sulphur ratio patterns by time of day and particle size, Pasadena, California, 3–4 September 1969 [11]

Again, using photoelectron spectroscopy different nitrogen-containing compounds can be identified in the aerosol. The photoelectron spectra of the Pasadena aerosol are shown in Figs. 7.6 and 7.7 for 0·6–2·0 μm and 2–5 μm diameter particles at different times throughout the day. These spectra reveal four charge states of nitrogen. The measured binding energies of these states are (I) 407·3 eV; (II) 402·0 eV; (III) 400·8 eV and (IV) 390·0 eV. States (I) and (II) correspond to nitrate nitrogen (NO_3^-) and ammonium ion (NH_4^+), respectively. The other two photoelectron lines (III) and (IV) are assigned to organic nitrogen. Nitrogen bonding energies of various types of amino-nitrogen are, on average, 400·3 eV and the binding energy for pyridine is 389·6 eV. On this evidence peaks (III) and

(IV) are assigned to organic amino- compounds and pyridino-compounds, respectively.

Figure 7.6 shows that nitrate is absent from the smaller particle size aerosol and that the intensities of the lines assigned to the amino- and pyridino- compounds change during the course of the day. During the morning pyridino-nitrogen is most prominent but as the day continues and on into the night amino-compounds are most pronounced. The larger

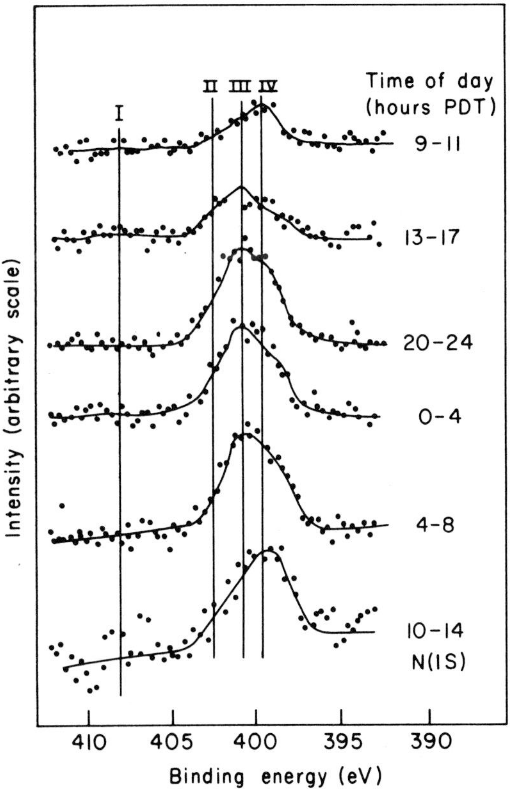

FIG. 7.6 Photoelectron spectra in N(1*s*) region for 0·6–2 μm diam. particles by time of day, Pasadena, California, 3–4 September 1969 [11]

particle size material, on the other hand, indicated in Fig. 7.7, comprises all four types of nitrogen compounds. Pyridino-nitrogen appears as the main species accompanied by nitrate and ammonium ions. This does not prove the existence of ammonium nitrate in the aerosol because the ammonium ions may be associated with other radicals such as sulphate, and the nitrate could be nitrates of metals. Nevertheless, if ammonium nitrate is present then it will exist in the larger particle size range.

Sulphate and nitrate concentrations analysed in Chicago, Cincinnati, Philadelphia and St. Louis between 1970 and 1972 show average sulphate concentrations of 15·0 $\mu g\ m^{-3}$ and nitrate concentrations of 3·5 $\mu g\ m^{-3}$ [2]

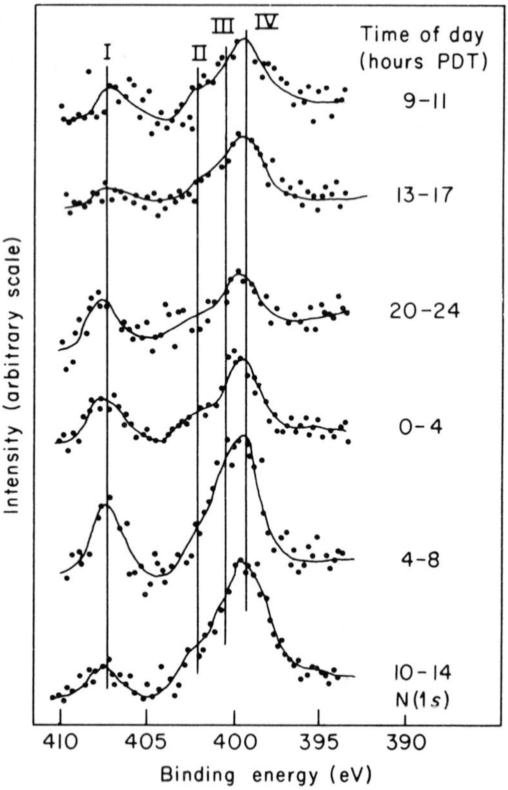

FIG. 7.7 Photoelectron spectra in N(1s) region for 2–5 μm particles by time of day, Pasadena, California, 3–4 September 1969 [11].

Ammonium concentrations for the same period are somewhat lower—around 0·6 $\mu g\ m^{-3}$.

The total carbon content of aerosols has received some attention and from the results it is generally apparent that three different forms of carbon may be distinguished. These are:

(1) elemental carbon;
(2) organic carbon present in organic compounds and biological material;
(3) inorganic carbon present in the form of carbonates.

The carbon from the first two categories is mainly of anthropogenic origin, as also are carbonates in the last class which are derived principally from cement manufacture. Carbonates also occur naturally and these are derived from soil and rock erosion. During photochemical smog episodes a typical analysis taken from the data obtained by Mueller *et al.* [12] gives as a percentage of the total particulate matter, 0·62 for carbonate and 29·0 for organic carbon. It should be mentioned that there may be errors in the estimation of carbonate in aerosols having acidic components. Inorganic carbonates, for example, collected on filter media are likely to be decomposed with release of carbon dioxide if sulphuric acid droplets are present in the aerosol.

Carbon, hydrogen and nitrogen analysis carried out on fractionated particulate by the Pregl–Dumas technique [13] gives average MMD values for carbon in Ankara, London and Chicago of 0·93 μm, 1·09 μm and 1·26 μm, respectively. When the C/H mole ratio is plotted against particle diameter, peaks occur in the C/H values which appear to be associated with specific particle sizes, but their significance is not clearly understood. A minimum in the C/H value which coincides with a particle diameter of about 0·5 μm suggests, however, that maximum saturation of organic compounds occurs at this size.

7.2.3 Metals

Published data from the National Air Surveillance Network (NASN) set up in the USA in 1970 forms the basis of much of our knowledge on this subject [1, 2, 14]. Table 7.2 summarizes information gathered by this programme for some heavy metals in the six cities mentioned in Table 7.1, viz. Chicago, Cincinnati, Denver, Philadelphia, St. Louis and Washington, along with some data from Toronto, Canada [15]. As a general rule, the soil and naturally derived elements Al, Ca, Mg, the rare earths, Na, Fe and Ti present in aerosols tend to be associated with larger aerosol particles. Those from anthropogenic sources, Sb, As, V and Pb tend to concentrate in the smaller size fractions. These elements exist principally in the submicron range and, therefore, represent the greatest inhalation threat.

Iron and lead have the highest annual average concentrations which in the US cities ranges from 1·3 to 3·2 $\mu g\ m^{-3}$ for lead and 0·6 to 1·8 $\mu g\ m^{-3}$ for iron. The metals Zn and Cu tend to fall in the 0·1–1·0 $\mu g\ m^{-3}$ range and Ni, V, Mn group have concentrations less than 0·1 $\mu g\ m^{-3}$. Metal concentrations are subject, however, to considerable variation from one location to another. The average Zn concentration in Cincinnati of 1·7 $\mu g\ m^{-3}$, for example, compares with values in the range 0·1–0·5 $\mu g\ m^{-3}$ for the other cities. Seasonal trends can also be observed for some metals. Vanadium, as we have seen in section 2.3.1.2 is present in oil and during the heating seasons in the first and last quarters of the year vanadium aerosol concentrations can be double those experienced during spring and summer.

TABLE 7.2 *Heavy metal concentrations, particle size diameters (MMDs) and percentage of particulate containing metal which is less than 1 μm in size, found in US and Canadian cities* [14, 15]

Metal	Country	Annual mean concentration ($\mu g\ m^{-3}$)	MMD (μm)	% ≤ 1 μm
Fe	USA	1·02	2·73	17
	Canada	2·20	6·0	15
Pb	USA	1·92	0·54	68
	Canada	0·97	0·70	51
Zn	USA	0·55	1·25	40
	Canada	0·32	1·20	47
Cu	USA	0·18	1·33	35
	Canada	0·33	1·50	42
Ni	USA	0·06	1·33	39
	Canada	0·02	1·20	45
Mn	USA	0·05	1·97	25
	Canada	0·07	2·40	36
V	USA	0·10	0·59	64
	Canada	0·01	0·90	54

Cadmium in air is always associated with zinc; the natural ratio of Cd/Zn in the earth's crust and soils varies between 1:100 and 1:350. Although urban air concentrations are low, about 0·01 $\mu g\ m^{-3}$ is typical, estimates of the MMD of Cd-containing airborne particulate vary. Lee *et al.* [14] give a value of 1·54 μm for a sample obtained in St. Louis in 1970 and this is in reasonable agreement with a value of about 2 μm which can be calculated from the data of Dorn *et al.* [16] for a sample site located 800 m from the base of a stack of a lead smelter.

7.2.3.1 Lead

Lead in urban aerosols is derived from motor vehicles and for the reasons discussed elsewhere (1.4.1.1) has received more attention than any other metal. Daines *et al.* [17] have shown that air lead concentrations fall off rapidly with distance from the road. A road carrying 58 050 vehicles day^{-1} had air lead concentrations of 10·1, 6·6, 3·99, 3·21, 2·65 and 2·16 $\mu g\ m^{-3}$ at distances from the road of 3, 9, 31, 46, 77 and 154 m, respectively.

Particle size studies indicate that 65% of the lead in the air from 9 to 539 m from the source consists of particles having MMDs of under 2 μm.

Table 7.3 shows some results of atmospheric lead measurements at various distances from the Midlands motorway interchange in Birmingham: the M6–A38(M) road system. The day- and night-time periods are of 12 h duration corresponding to 07.30 to 19.30 and then from 19.30 to 07.30 the next day [18]. The results indicate that maximum variation between day and night values occur closest to the traffic source.

TABLE 7.3 *Mean atmospheric lead concentrations in* $\mu g\, m^{-3}$ *in the vicinity of a complex motorway interchange carrying* 9×10^5 *vehicles week*$^{-1}$

Distance from traffic (m)	Duration of sampling	Number of determinations	Day	Night	Percentage, night/day
0	1972–73 (2 months)	104	2·50	1·24	50
5·5	1972–73 (7 months)	275	1·91	1·11	58
45	1972 (8 months), 1973–74	1814	2·05	1·25	61
510	1972–73 (16 months)	734	0·82	0·79	95
580	1972 (8 months)	510	0·83	0·83	100

As the distance from the source increases, the difference between the day and night concentrations lessens until eventually, at around 600 m from the interchange, the day and night concentrations become similar. This evidence shows that in terms of a long time average, the interchange exerts its principal influence on air quality up to distances of about 600 m. Beyond this distance the finely divided lead particles that have survived in the atmosphere will contribute to the general urban background.

This behaviour can be expressed mathematically in the form $e^{-kx} \cos px$, where the exponential term accounts for the general coalescence of the day and night concentrations with distance x from the source and the cosine function, which passes through maximum and minimum values, may be used to represent the extremes corresponding to the day and night values. Values of the constants k and p may be found empirically which give useful solutions of the function that satisfy the conditions given in Table 7.3.

In the case of the constant k, this may be chosen so that the value of the exponential term is vanishingly small at $x = 600$ m. When $k = 0{\cdot}0075$, then $e^{-0{\cdot}0075 \times 600} = 0{\cdot}01$ is an acceptable solution. At intermediate distances between 0 and 600 m the maximum and minimum solutions between $\pm 1{\cdot}0$, are given by values of $\cos px$. If p is taken as $0{\cdot}251$ and incremental distances x obtained by alternatively adding 12 and 13 m from the source to give $x=0+12=12$; $x=12+13=25$; $x=25+12=37$, etc., then the function oscillates with decreasing amplitude as x increases. This is shown in Fig. 7.8. When $x=0$, let D represent the day-time concentration and N the night-time concentration at the source where the maximum difference between day and night values is encountered. Since the night-time value never falls to zero, the zero must be represented by projecting DN to 0. At

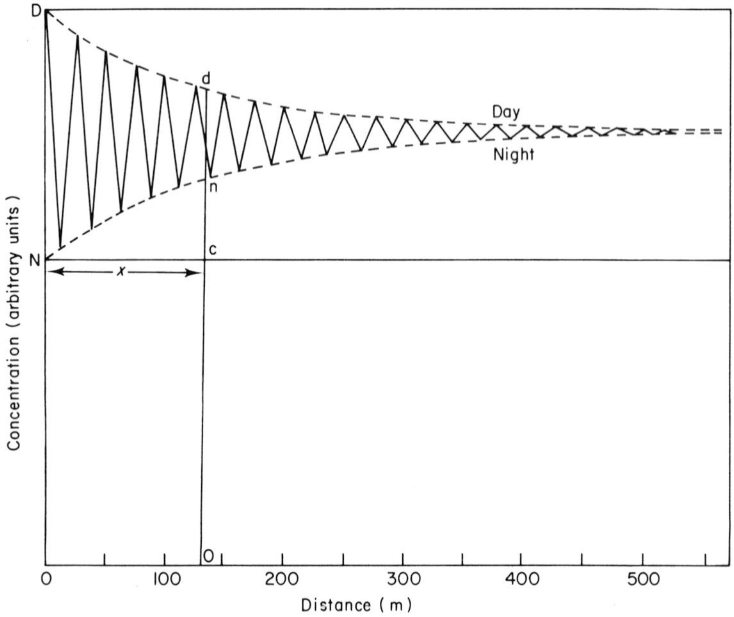

FIG. 7.8 Plot of the function $e^{-kx}\cos px$ used to represent the difference between day and night atmospheric lead concentrations with distance from a traffic source

the source the ratio of the night-time to the day-time concentration will be given by 0N/0D and at other distances x by

$$\frac{on}{od}=\frac{(oc+nc)}{(oc+dc)} \tag{7.3}$$

Since $oc=0\text{N}$, oc may always be found provided the day and night concentrations at the source are known, because by putting Pb_D = day-time lead concentration in $\mu g\ m^{-3}$, Pb_N = night-time lead concentration in

$\mu g\, m^{-3}$, and setting DN in Fig. 7.8 equal to 2·0 i.e. ±1·0, about the mid-point, gives

$$\frac{0N}{0D} = \frac{0N}{0N+ND} = \frac{Pb_N}{Pb_D} \tag{7.4}$$

or

$$0N = oc = \frac{Pb_N DN}{Pb_D - Pb_N} = \frac{2Pb_N}{Pb_D - Pb_N} \tag{7.5}$$

The values of dc and nc in equation (7.3) and Fig. 7.8 are $dc = 1{\cdot}0 + e^{-0{\cdot}0075x} \cos 0{\cdot}251x$ and $nc = 1{\cdot}0 - e^{-0{\cdot}0075x} \cos 0{\cdot}251x$, where x are the distances already defined which are the closest to those actually required. Examples of the calculation are given in Table 7.4. The results give reasonable agreement with those monitored. Provided the characteristic boundary conditions can be established in any particular case from experimental data the empirical treatment outlined should be generally applicable. In such instances the appropriate values of the constants k and p can easily be established with the aid of a small scientific hand calculator.

Colucci *et al.* [19] report good correlation between air lead and carbon monoxide concentrations. Regression equations derived by these authors under the different environmental conditions of a freeway, a residential and a commercial district are shown in equations (7.6), (7.7) and (7.8):

Freeway	$Pb = -0{\cdot}42 + 1{\cdot}42\ CO$	$R = 0{\cdot}94$	(7.6)
Residential	$Pb = -0{\cdot}85 + 0{\cdot}98\ CO$	$R = 0{\cdot}83$	(7.7)
Commercial	$Pb = 1{\cdot}49 + 0{\cdot}66\ CO$	$R = 0{\cdot}88$	(7·8)

where, Pb is the airborne lead concentration expressed in μm^{-3}, CO is given in parts 10^{-6} and R is the correlation coefficient. As pointed out in their paper, these regression equations reflect the different driving modes associated with these environments. The faster driving conditions of the freeway give the line of steepest slope of 1·42. For residential driving conditions where engine load and speed are normally between freeway and commercial conditions an intermediate slope of 0·98 obtains. Slow-running engines operating in congested commercial districts will produce more carbon monoxide and account for the line of least slope of 0·66.

Regression analysis of results, already mentioned, of an investigation into atmospheric lead concentrations at the largest motorway interchange in Europe [20, 21] the M6–A38(M), show that at Salford Circus (Fig. 7.9) lead concentrations can be represented by the equation

$$Pb = 2{\cdot}19 - 0{\cdot}2\ WSP - 0{\cdot}072\ TMP + 1{\cdot}1 \times 10^{-6}\ SCT \tag{7.9}$$

TABLE 7.4 *Calculation of the percentage night/day lead concentrations at various distances from the source*

For the conditions reported in Table 7.1, $0N = oc = \dfrac{2{\cdot}0\ Pb_N}{Pb_D - Pb_N}$. When $x = 0$,

$$oc = \frac{2{\cdot}0 \times 1{\cdot}24}{2{\cdot}50 - 1{\cdot}24} = 1{\cdot}9682 \text{ and } f(x) = e^{-0{\cdot}075x} \cos 0{\cdot}251x.$$

Distance from source (m)	Actual % night/day	$nc = 1{\cdot}0 - f(x)$	$dc = 1{\cdot}0 + f(x)$	Calculated % night/day $\frac{on}{od} = \frac{oc + nc}{oc + dc} \times 10^2$	Mean distance used (m)
0	50	0 when $x = 0$	$1{\cdot}0 + 1{\cdot}0 = 2{\cdot}0$ when $x = 0$	$\frac{1{\cdot}968 \times 10^2}{(1{\cdot}9682 + 2{\cdot}0)} = 50$	0
5·5	58	$1{\cdot}0 – 0{\cdot}9063 = 0{\cdot}0937$ when $x = 12$	$1{\cdot}0 + 1{\cdot}0 = 2{\cdot}0$ when $x = 0$	$\frac{(1{\cdot}9682 + 0{\cdot}0937)10^2}{(1{\cdot}9682 + 2{\cdot}0)} = 52$	6
45	61	$1{\cdot}0 – 0{\cdot}7505 = 0{\cdot}2495$ when $x = 37$	$1{\cdot}0 + 0{\cdot}6872 = 1{\cdot}6872$ when $x = 50$	$\frac{(1{\cdot}9682 + 0{\cdot}2495)10^2}{(1{\cdot}9682 + 1{\cdot}6872)} = 61$	43
510	95	$1{\cdot}0 – 0{\cdot}0206 = 0{\cdot}9794$ when $x = 512$	$1{\cdot}0 + 0{\cdot}0232 = 1{\cdot}0232$ when $x = 500$	$\frac{(1{\cdot}9782 + 0{\cdot}9794)10^2}{(1{\cdot}9682 + 1{\cdot}0232)} = 98$	506
580	100	$1{\cdot}0 – 0{\cdot}0116 = 0{\cdot}9884$ when $x = 587$	$1{\cdot}0 + 0{\cdot}0132 = 1{\cdot}0132$ when $x = 575$	$\frac{(1{\cdot}9682 + 0{\cdot}9884)10^2}{(1{\cdot}9682 + 1{\cdot}0132)} = 99$	581

with correlation coefficient $R = 0{\cdot}6$, where Pb is the monthly average airborne particulate lead in $\mu g\,m^{-3}$ and WSP, TMP and SCT are the monthly average windspeed ($m\,s^{-1}$), temperature (°C) and traffic entering Salford Circus, respectively. Table 7.5 shows a typical compilation of data used to predict monthly average lead values during 1975. Similar comparisons showing the actual monitored and predicted concentrations at this site for the period from 1971, before the junction was opened, up to February 1976 are shown in Fig. 7.10. This figure illustrates that with a few notable exceptions reasonably satisfactory agreement between predicted and actual values can be achieved using three comparatively easily available measurements, viz. traffic flow, windspeed and temperature. No doubt, the predictions could be significantly improved if temperature–altitude profiles were available, as these would give information on temperature inversions. The greatest discrepancies in the values given in Fig. 7.10 that occur during winter months February–March 1973 and December 1975 may be attributed to this cause. In the absence of such conditions the two meteorological parameters included in the equation—windspeed and temperature—account for aerosol dilution and vertical ventilation of the lead aerosol from the interchange. It is important to recognize that equation (7.9) will not predict maximum lead pollution at this location merely by setting WSP and TMP = 0. Under these meteorological conditions pollutants cannot be diluted and dispersed, and background concentrations therefore build up to unpredictable levels.

Another feature of Fig. 7.10 that should be noticed is that despite the increase in traffic from around 360 000 vehicles $week^{-1}$ before the interchange was opened to nearly a 10^6 vehicles $week^{-1}$ at the present time, there has not been a dramatic increase in airborne lead monitored at this site. The explanation lies in the fact that the interchange covers 13 hectares of land and has numerous elevated carriage ways. This configuration is conducive to good ventilation so that vehicle emissions, except under inversion conditions, are fairly rapidly diluted and dispersed.

Lead from motor vehicles accumulates in city dust. Table 7.6 shows some typical concentrations that have been determined in the UK and USA. Day *et al.* [24] maintain that children playing in city streets will naturally pick up significant quantities of lead, especially by eating sticky sweets contaminated with dust on fingers and hands. They suggest that although upper limits of lead in food of about 2 parts 10^{-6}, or of lead in paint for children's toys (0·5%) are recognized as acceptable, no such upper tolerable limit for lead in urban dust has ever been mentioned.

In any consideration of dust deposition upon a road there is always the possibility of particle resuspension caused by the passage of vehicles. The extent of this resuspension has been examined by Sehmel [25] using ZnS

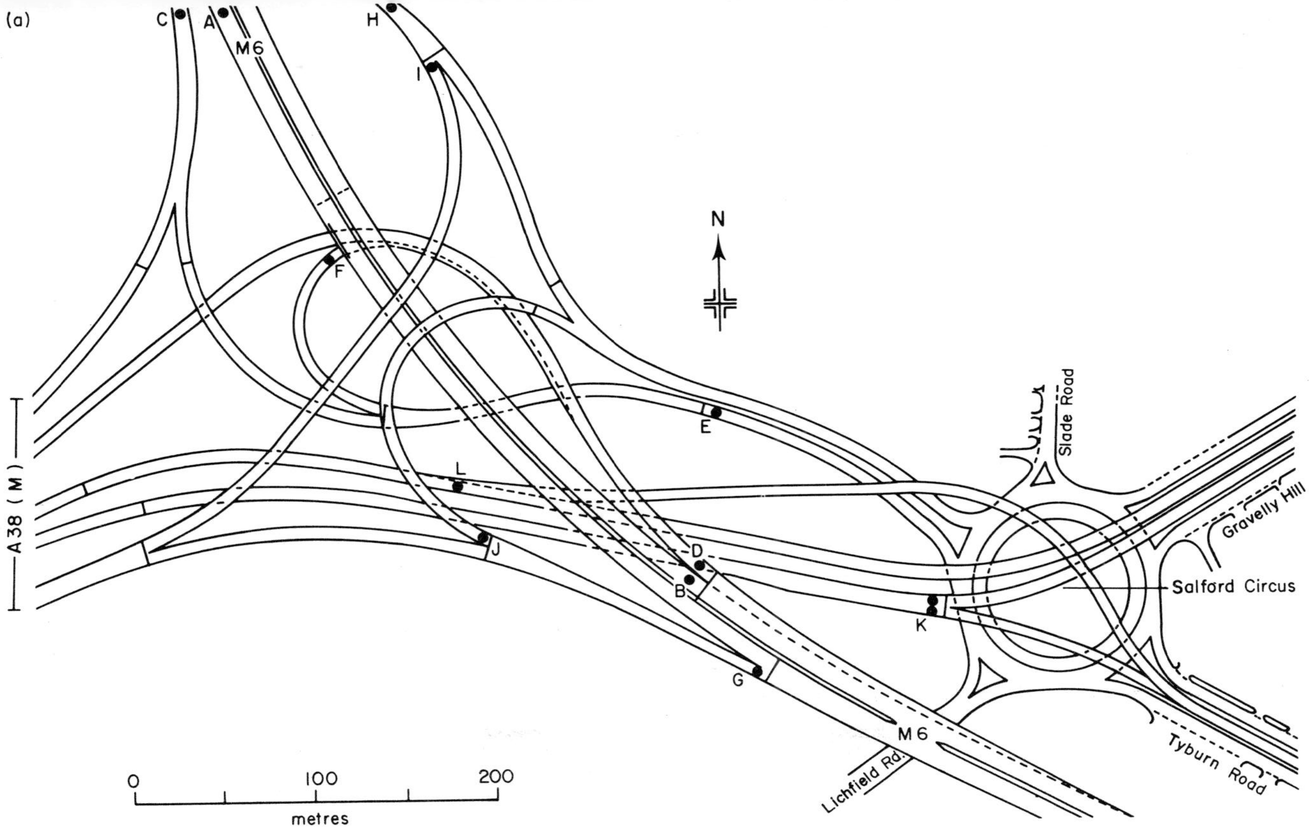

FIG. 7.9 Photograph and road plan of the Midlands motorway interchange, M6–A38(M), Birmingham, UK (the letters A to L show the locations of the sub-surface loops in the various lanes used to activate traffic counters)

TABLE 7.5 *Comparison of actual and predicted airborne lead concentrations at the M6–A38(M) interchange calculated from the equation* $Pb = 2{\cdot}19 - 0{\cdot}2\ WSP - 0{\cdot}072\ TMP + 1{\cdot}1 \times 10^{-3}\ SCT$ *for 1975*

Month	Pb_{act} ($\mu g\ m^{-3}$)	Pb_{pre}	WSP ($m\ s^{-1}$)	TMP (°C)	SCT	0·2 WSP	0·072 TMP	$1{\cdot}1 \times 10^{-6}$ SCT
Jan	1·17	1·61	6·07	6·2	974 320	1·21	0·44	1·07
Feb	1·70	2·28	3·31	3·8	925 780	0·66	0·27	1·02
Mar	1·38	1·92	5·30	4·3	1 000 451	1·06	0·31	1·10
Apr	1·50	1·67	5·10	8·1	982 080	1·02	0·58	1·08
May	1·25	1·62	4·84	9·4	984 632	0·97	0·68	1·08
Jun	1·02	1·35	5·30	12·7	1 024 513	1·06	0·91	1·13
Jul	1·50	1·19	4·13	17·3	982 319	0·83	1·25	1·08
Aug	1·70	1·36	3·16	18·2	1 011 521	0·63	1·31	1·11
Sep	1·62	1·51	4·23	13·0	1 009 852	0·85	0·94	1·11
Oct	2·23	1·91	3·42	9·7	1 002 784	0·68	0·70	1·10
Nov	2·86	2·08	4·08	5·6	1 006 333	0·82	0·40	1·11
Dec	3·56	2·16	3·85	4·8	993 817	0·77	0·35	1·09
Mean	1·79	1·72						

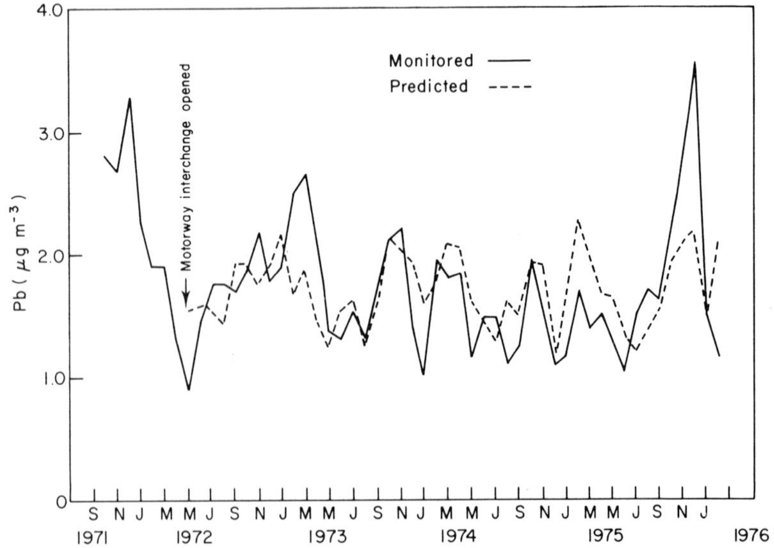

FIG. 7.10 Comparison of monitored and predicted monthly average particulate lead concentrations using equation (7.9) at the Midlands Motorway interchange, M6–A38(M), Birmingham [23]

particles of less than 25 μm in size. The results show that between 10^{-3} and 1% of the deposited material can be resuspended depending upon vehicle speed. The faster the vehicle moves the greater the resuspension.

TABLE 7.6 *Lead concentrations in city dust*

Location	Concentration (parts 10^{-6})	Reference
USA cities	1500–2400	22
Birmingham, UK	1000–2500	23
Manchester, UK	average 970	24

Most studies of airborne lead have measured particulate lead, i.e. lead retained by filters through which air has been drawn. However, organic lead in the form of tetraalkyl compounds is also present from vaporization of unburnt petrol, notably in the vicinity of petrol station forecourts. Colwill and Hickman [26] report values ranging from 28 to 59% of the total lead monitored that they attribute to volatile lead at petrol stations. Harrison *et al.* [27] found that the organic fraction of the total lead collected in street air ranged from 0·1 to 3·3%, excluding sites near petrol stations. These values are in agreement with earier work of Snyder [28].

Robinson *et al.* [29], using non-flame atomic absorption spectrophotometry, indicate that the use of high-volume sampling gives only a partial atmospheric profile because "molecular" compounds of lead are ignored. In this context, "molecular" compounds are defined as lead compounds, both inorganic and organic, which have sufficiently high vapour pressures to enable their vaporization or sublimation to atmosphere. Under atmospherically cool, dry conditions "molecular" lead compounds frequently amount to 0·1 $\mu g\ m^{-3}$. In contrast, "molecular" lead concentrations in the range 100–200 $\mu g\ m^{-3}$ were recorded during thunderstorms and a period of high humidity.

When filtration procedures are employed to separate particulates in air samples the efficiency of the filter, as discussed in section 3.4.2, always comes into question. Lamothe *et al.* [30] suggest that the existence of small particles less than 0·1 μm comprise a significant percentage of the total atmospheric mass which are not effectively collected on membrane filters. Any determination of organic lead by absorption of filtered air into iodine monochloride solution (Table 3.5) is very dependent upon filter efficiency. Failure to retain particulate lead will give a high erroneous estimate of organic lead under these circumstances.

7.2.4 Polynuclear hydrocarbons

Concentrations of polynuclear hydocarbons in city atmospheres have been studied for about the last 20 years or so. Some selected results of these investigations are summarized in Table 7.7. As we have seen from the discussion in Chapters 2 and 4 these compounds are derived from combustion processes and, as might be expected, their concentrations in the winter months are higher than in summer. These seasonal variations can be large. For instance, for the years 1958, 1959 and 1960 the average benzo(a)pyrene concentration during summer in Budapest was 25 ng m^{-3} but the corresponding average winter value during the period was 640 ng m^{-3} [40].

One of the first attempts to estimate the particle sizes associated with airborne polynuclear hydrocarbons was reported by Demaio and Corn in 1966 [41]. A two-stage fractionator was employed to separate particulate matter into two ranges, one greater than 5 μm and the other between 0·3 and 5 μm. They were able to show that on average the 0·3–5 μm size fraction contained 78% and 98% of the PNHs collected during summer and winter, respectively. With the development of the Andersen impactor operating at 566 l min^{-1} of air it now becomes possible to collect sufficient particulate per stage in a reasonable time to allow PNH analysis to be undertaken. This has been done by Pierce and Katz [42] who find that log-normal distributions are approximately obeyed and that size distributions are significantly seasonally dependent. The MMDs of chrysene, benz(a)anthracene, benzo(a)pyrene, benzo(k)fluoranthene, perylene, benzo(ghi)perylene, anthanthrene and coronene in summer were $2{\cdot}49 \pm 0{\cdot}21$ μm and in winter were $1{\cdot}45 \pm 0{\cdot}13$ μm.* Not only does this result clearly show the difference between the particle sizes between summer and winter but it also emphasizes the consistency of the MMDs of the eight different PNHs.

Butler and Crossley [43], using similar equipment to that employed by Pierce and Katz, have determined the MMD values of pyrene—1·85 μm; benzo(a)pyrene—0·39 μm; and coronene—0·56 μm; their determinations were made during April 1977, close to heavy vehicular traffic. Once more the benzo(a)pyrene and coronene MMDs are comparable with each other, although smaller than those quoted by Pierce and Katz. From all these results of particle size determinations there is mounting evidence that PNHs in urban atmospheres belong to that fraction of the aerosol which can penetrate deeply into the lung. Coupled with the fact that some members of this class of compounds are known potent carcinogens, this creates a disturbing situation. The case against the potentially deleterious

* These limits are quoted by the statistical definition described in section 4.6.

properties of the organic fraction of urban aerosols can be extended to the aza- and oxygenated derivatives [44].

7.2.5 Indoor–outdoor air pollution

The pattern of life in urban populations is governed largely by age and occupation. Old people, housewives and young children will tend to spend more time inside their homes than the adult section of the population who go out to work. On a weekly basis, however, even assuming a 45 h working week with 2–3 h per day spent travelling between home and business, this only accounts for 60 of 168 h in a week. Allowing a generous 16 h for outside activities at the week-end, this still leaves something like 90 h or 54% of the time to be spent at home. The relationship between outdoor and indoor air pollutant concentration levels and hence the quality of air within the home which is breathed by the occupants becomes a relevant feature of urban life.

The literature on the subject of outdoor–indoor air pollution relationships up to 1972 for SO_2, CO and particulates has been reviewed by Benson *et al.* [45]. Broadly, they conclude that indoor and outdoor levels are essentially the same.

7.2.5.1 Smoke and sulphur dioxide

In a study of over 60 homes in Rotterdam [46], it was found that indoor and outdoor concentrations of smoke and SO_2 when averaged over all the determinations were 141 $\mu g\, m^{-3}$ indoors and 170 $\mu g\, m^{-3}$ outdoors for smoke compared with 37 $\mu g\, m^{-3}$ and 180 $\mu g\, m^{-3}$ indoors and outdoors, respectively, for SO_2. These values give indoor concentrations of smoke of 83% and of SO_2 20% of the outdoor levels. Application of multiple regression analysis indicates a joint influence of four factors that were considered, viz.

(1) age of the property;
(2) type of heating;
(3) smoking habits of the occupants;
(4) outdoor smoke.

Taken individually, none of these factors is significant, but when combined they give a multiple correlation coefficient $R = 0{\cdot}41$ for $p < 0{\cdot}001$. Some evidence is presented which indicates that for SO_2, of the four influences considered, the age of the property is the most important. Houses built after 1960 are found to have about 6% of the outdoor concentration inside, whereas for properties constructed before 1919 this percentage increases to 30. There seems little likelihood of there being much difference in penetrability between newer and older houses so it has been concluded

TABLE 7.7 *Polynuclear hydrocarbons reported in urban atmospheres*

Compound	Urban atmospheres (ng m^{-3}) European cities		US cities		US average [34]
Pyrene	Leeds [31]	14		Trace-35 [33]	5
	Rome [32]	5·8	Los Angeles [35]	0·45	
Benz(a)anthracene	Rome	4·2	Detroit [36]	0·4–21·6	4
			Los Angeles	0·18	
Chrysene	Rome	10·2	Detroit	1·3–11·6	
			Los Angeles	0·6	
Fluoranthene	Leeds	21	Detroit	0·9–15	4
	Rome	5·8	Los Angeles	0·31	
Benzo(a)pyrene	London [37]	20–39	Detroit	0·2–17	5·7
	Leeds	42	Los Angeles	0·46	
	Hamburg [38]	134			
Benzo(e)pyrene	London	12–26	Los Angeles	0·9	5·0
	Leeds	26			
	Rome	2·3			
	Hamburg	115			
Perylene				Trace-5	0·7
			Los Angeles	0·1	
Benzo(ghi)perylene	London	12–46		2–35	8·0
	Leeds	40	Los Angeles	3·27	
Anthanthrene	London	2–6		Trace-3	0·26
	Leeds	9	Los Angeles	0·23	
Coronene	London	4–20		Trace-8	2·0
			Los Angeles	2·13	
Benzo(j)fluoranthene	Rome	9·4	Detroit	0·8–4·4	
			Los Angeles	0·17	
Benzo(k)fluoranthene				0·5–20	
			Los Angeles	0·2	

Benzo(b)fluoranthene			Los Angeles	0·54	
Indeno(1,2,3-c,d)pyrene			Detroit	1·5–8·2	
			Los Angeles	1·34	
4-Azapyrene	Rome [39]	13·1			
1-Azafluoranthene	Rome	3·0			
Benz(a)acridine					0·2
Benz(c)acridine					0·6
Dibenz(a,j)acridine					0·04
Dibenz(a,h)acridine					0·08
Benzo(f)quinoline					0·2
Benzo(h)quinoline					0·3

that the plaster of the walls and ceilings of newer houses must absorb or neutralize SO_2 more efficiently.

A similar conclusion regarding the neutralization of NO_2, the acidic component of NO_x, by alkaline dusts and surfaces which prevail in the home, was drawn by Thompson *et al.* [47]. In their investigation, indoor and outdoor concentrations of NO, NO_2, CO, PAN, total oxidants and particulate matter were monitored at hospitals, schools, a department store, a university laboratory and a private home. The work was performed during mid-to-late September when daytime temperature reached to between 29 and 34°C. During this period different establishments had different types of air-conditioning systems. For convenience the results can be expressed in terms of the effectiveness of air-handling equipment used for reducing concentrations of air pollutants in buildings (see Fig. 7.11).

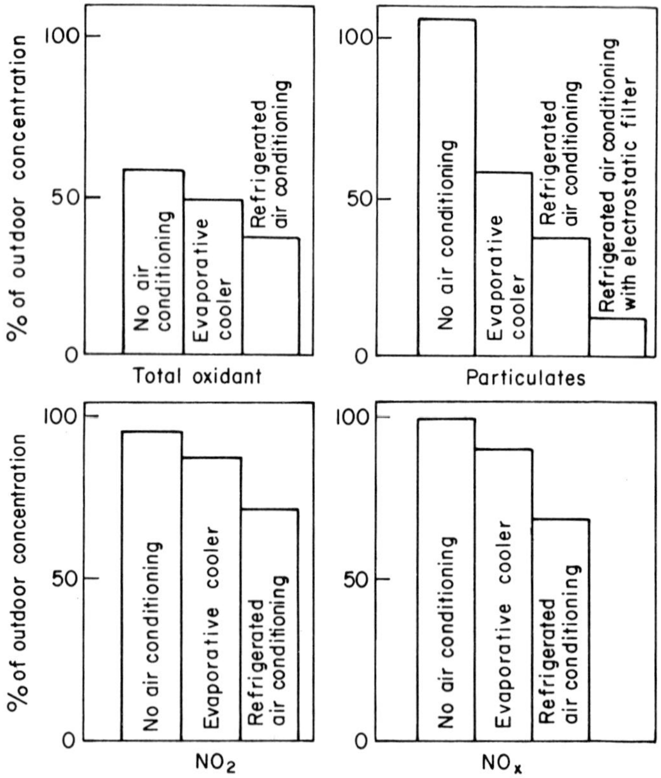

FIG. 7.11 Summary of effectiveness of air-handling equipment for reducing concentrations of air pollutants in buildings [47]

These results demonstrate, as one would expect, that the highest pollutant concentration experienced indoors occurs where no air-conditioning system is installed. In these circumstances the indoor concentrations of particulates, NO_2 and NO_x are close to those outside. Incorporation of an evaporative cooler is most beneficial for reducing indoor particulate levels, and refrigerated air-conditioning with an electrostatic filter can achieve a 90% reduction of particulate compared with outside. In the case of the other pollutants, total oxidant, NO_2 and NO_x, refrigerated air-conditioning gives only a 25% reduction in concentration of that monitored outside.

7.2.5.2 Airborne lead concentrations inside houses

Daines *et al.* [48], in their investigation of airborne lead monitored at the front porch and in the front room of houses situated 3·7 m, 38·1 m and 121·9 m from a highway carrying 33 000 cars day^{-1}, found reductions expressed as a percentage of inside/outside of 50, 38 and 30, respectively. This inquiry also noted that properties with air-conditioning generally had lower indoor lead concentrations in comparison to those without. No such reduction in inside airborne particulate lead compared with outside was found by Butler and MacMurdo [49] in an urban district monitored between 29 November 1973 and 19 December 1973. Table 7.8 shows that

TABLE 7.8 *Atmospheric lead concentration ($\mu g\ m^{-3}$) monitored inside and outside a house in an urban environment* [49]

Inside house	Outside house	house	Percentage inside/outside
Day-time (07.30 to 19.30)	1·07	1·14	94
Night-time (19.30 to 07.30)	0·66	0·62	106
Combined average	0.87	0.89	

although less lead was present at night there was little difference between inside and outside concentrations. Indeed at night, the inside concentration was marginally higher than that outside. No special precautions were taken to exclude draughts from the windows, which were kept shut, or the chimney in the house. The door to the hall, however, was kept open so that air within the house could circulate freely. The slightly higher lead value inside at night, compared with outside, probably reflects the inability of the lead-containing dust which enters the house in the day-time from escaping or settling at a rate which is comparable to the decrease in lead concentration that always occurs outside at night. Equations relating the amount of airborne lead found inside and outside this house have been constructed by regression analysis of the data. These relationships are given in Table

7.9. On this evidence it appears that the home offers little, if any, protection from lead aerosol present in city atmospheres. In view of this the two occupants of the property had venous blood samples taken as part of this investigation. Reassuringly, the values found were around 24 μg $(100\ ml)^{-1}$—well within accepted levels.

TABLE 7.9 *Equations relating inside–outside airborne lead concentration based on regression analysis of the data* [49]

Time of day	Computed equation	Validity limits	Correlation coefficient
Day-time	$(Pb)_{ins.} = 0{\cdot}66(Pb)_{outs.} + 0{\cdot}33$	$0{\cdot}27 < (Pb)_{outs.} < 2{\cdot}46$	0·74
Night-time	$(Pb)_{ins.} = 0{\cdot}82(Pb)_{outs.} + 0{\cdot}17$	$0{\cdot}27 < (Pb)_{outs.} < 1{\cdot}96$	0·94

$(Pb)_{ins.}$ denotes Pb concentration inside house ($\mu g\ m^{-3}$)
$(Pb)_{outs.}$ denotes Pb concentration outside house ($\mu g\ m^{-3}$)

7.2.5.3 Indoor airborne PNH measurement

As part of an investigation carried out on behalf of DoE/DHSS* and the EEC an indoor measurement of PNHs has recently been completed [43]. For this experiment a fully furnished detached suburban house situated 11 km north of Birmingham city centre was chosen. When originally constructed in 1935 open-hearth fireplaces were fitted, but these had not been used for the last 15 years. Instead, gas-fired central heating has been installed with supplementary heating from electric fires. None of these heating systems was in operation during PNH measurement.

A Staplex pump was positioned in the dining room with the filter 1 m above the floor. The door leading to the hall was left open, as were the doors from the hall to the kitchen and lounge. Thus, air could freely circulate through all the downstairs rooms as well as the staircase and upstairs landing. Metal-frame windows and doors leading outside were kept shut. During the period 20–24 May 1977, 4121 m^3 of air were sampled. Table 7.10 summarizes the results. These values are about one-tenth of those reported for these compounds in ambient USA city atmospheres, as given in Table 7.7. They do, however, demonstrate that measurable quantities of PNHs are present within suburban homes not using solid fuel and remote from main roads.

7.3 Gases—urban diffusion models

Historically, smoke and SO_2 are recognized as the main air pollutants of urbanization and industrial activity. As we have seen in Chapter 1, air-

* UK Department of the Environment and Department of Health and Social Security.

pollution episodes in winter are caused by elevated concentrations of these substances under atmospheric temperature inversion conditions. Legislation in some countries has been passed to prevent or at least restrict the severity of such events in the future. Although it is not possible to predict temperature inversions well in advance, it is possible to assess the amount of material likely to be discharged to the atmosphere within an urban region. Using such information, long-term trends and seasonal variations of air pollutants can be forecast by urban area modelling based on diffusion theory. These models are especially useful for predicting the future effects on air quality of policy changes which involve changes in the type of fuel consumed, e.g. from coal to natural gas.

TABLE 7.10 *Indoor concentrations of some PNHs monitored in a suburban house*

Date	TPM	PYR	CHY	BaP	BeP	COR
	($\mu g\,m^{-3}$)			($ng\,m^{-3}$)		
(1) *Expressed in weight–volume units:*						
20–24 May 1977	24	0·51	1·10	0·64	0·38	0·15
30 Nov–5 Dec 1977 (winter)	50	2·94	7·72	4·32	4·76	0·44
(2) *Expressed in weight–weight units:*						
				($\mu g\,g^{-1}$)		
20–24 May 1977 (summer)		21·6	45·2	26·4	15·6	6·2
30 Nov–5 Dec 1977 (winter)		58·5	153·6	85·4	94·9	8·6

TPM is total suspended particulate matter and PYR, CHY, BeP, BaP, COR denote pyrene, chrysene, benzo(a)pyrene, benzo(e)pyrene and coronene, respectively.

The treatment given to these matters strictly lies outside the scope of pure chemistry, but environmental chemists will almost certainly be faced with these problems sooner or later. For this reason a simple outline of the method of urban air-pollution modelling will be introduced as the final topic to be considered in this book.

7.3.1 Predictive urban atmosphere pollution model applied to SO_2

Numerous accounts of urban air-pollution models have been described since the late 1950s [50, 51, 52]. The one considered in this section, by

Goumans and Clarenburg [53] is closely similar to that proposed by Gifford and Hanna [54]. The model is based on the following concepts:

(1) A diffusion equation depending on a continuous point source model is used.
(2) An urban area source can be regarded as a large number of small point sources randomly distributed in an area. Each of these individual sources will be roughly equal.
(3) Areas downwind will receive pollutants from adjacent areas.
(4) Seasonally averaged stability and dispersion parameters may be used.
(5) Source strength concentrations can be obtained with sufficient accuracy from demographic information.

Clarenburg [55] derives an equation for the concentration of gaseous components at a point $(X, 0, 0)$ which has the form,

$$C_{(X,00)} = \left[\frac{2}{\pi}\right]^{1/2} \frac{QX^{1-\beta}}{\alpha(1-\beta)U^{1-\gamma}ab} \tag{7.10}$$

where Q is the source strength of area with dimensions ab, U is the mean windspeed, X is the length of the area source along the wind direction (Fig. 7.12), γ is a constant between 0·6 and 0·8 depending on the differences in the construction of buildings. α and β are meteorological parameters given

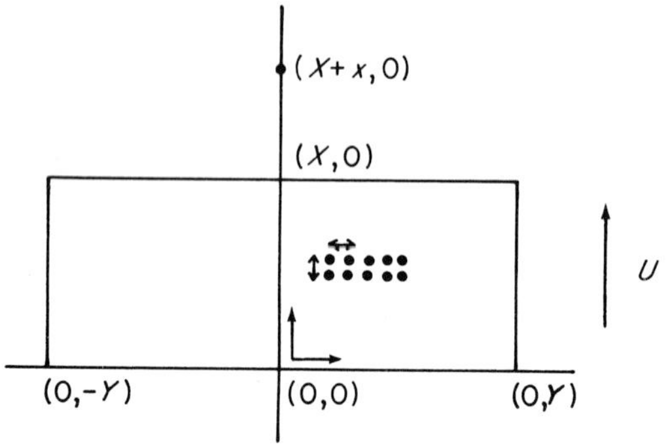

FIG. 7.12 Schematic representation of an area source showing the coordinate system used in equations (7.10) and (7.11). Adapted from reference 53.

in Table 7.11 which are dependent upon the stability category. For the point $(X+x,0,0)$ from the boundary edge shown in Fig. 7.12 the equation becomes

$$C_{(X,0,0)} = \left[\frac{2}{\pi}\right]^{1/2} \frac{Q}{\alpha(1-\beta)U^{1-\gamma}ab}[(X+x)^{1-\beta} - x^{1-\beta}] \tag{7.11}$$

The effect of wind transporting pollution from areas A and B towards monitoring point S, for the wind sector north to east is shown in Fig. 7.13.

The model has been applied to SO_2 concentrations in Holland by Goumans and Clarenburg [53]. In order to do this demographic data for different districts of an area source may be used. Let us suppose that the region of interest is shown in Fig. 7.13, where the districts are denoted by

TABLE 7.11 *Values of the meterorological constants and under various atmospheric stability conditions* [54]

Meteorological conditions	α	β
Very unstable	0·40	0·91
Unstable	0·33	0·86
Neutral	0·22	0·80
Estimated Pasquill "D"	0·15	0·75
Stable	0·06	0·71

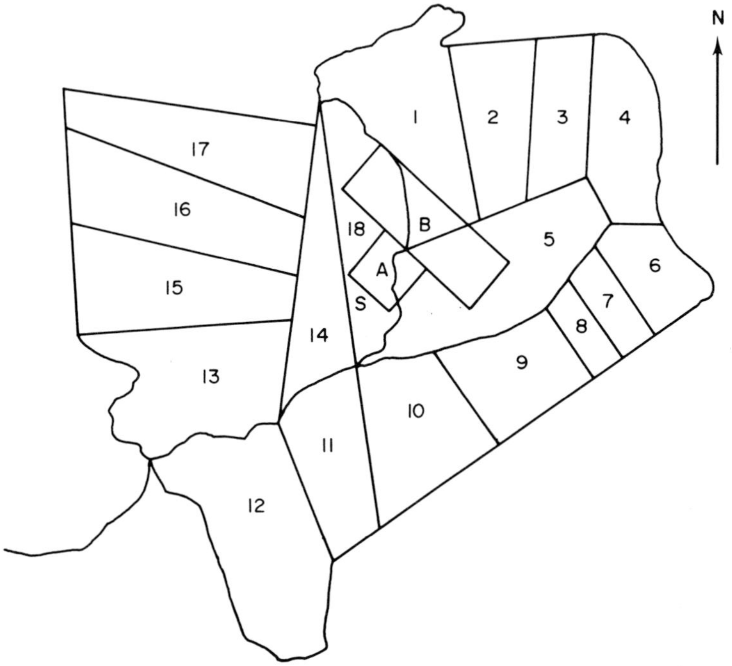

FIG. 7.13 Superimposition of area sources A and B onto different districts of a city enabling an estimate of pollution concentration to be made at S for winds from the NE sector

numbers (1–18) and the areas A and B are the closest to a sample point S on the edge of A. The district demarcations can be postal districts, electoral divisions, gas or electricity supply areas, etc., for which the population density is known. If there are n_i inhabitants using fuel containing sulphur which produces the equivalent e_i say 1·5 kg of SO_2 for each inhabitant in each winter month, then n_ie_i represents the emissions from the inhabitants in this time period. Similarly, if n_w workers come into this district to work in offices and shops during the day, then they also will consume fuel and give an estimated equivalent e_w of 0·5 kg of SO_2 for each worker during a winter month. The emission factor from this source will be n_we_w, so that the total emissions per unit area of the district will be

$$Q_D = \frac{n_ie_i + n_we_w}{S} \tag{7.12}$$

where Q_D is the emission for the districts Q_1, Q_2, Q_3, etc., of areas S_1, S_2, S_3, etc. The area A shown in Fig. 7.13 comprises approximately one-quarter of district 5 plus three-quarters of district 18, hence

$$Q_A = \tfrac{1}{4}Q_5 + \tfrac{3}{4}Q_{18} \tag{7.13}$$

where

$$Q_5 = \frac{(n_ie_i)_5 + (n_we_w)_5}{S_5} \tag{7.14}$$

The suffix 5 indicating district 5 for which n_i, n_w and S are available. Similarly Q_B can be found by considering districts 1, 5 and 18.

Once Q_A has been found, substitution into equation (7.10) enables the contribution of sources within area A to be calculated at the point S. Similarly, when the source strength for area B has been calculated, substitution into equation (7.11) allows an estimate of the SO_2 pollution from area B at S to be made. The pollutant concentration at S due to SO_2 from a mean wind direction sector north to east will be the sum from areas A+B etc. The calculation can be repeated to take into account as many upwind areas as desired but since, generally, $Q_A > Q_B > Q_C$ etc., there is a limit to the practicability of repetition and each case should be considered on merit.

Results of regression analysis of observed seasonal mean SO_2 concentration *vs* calculated values by the method outlined show correlation coefficients of 0·87 and 0·78 for The Hague and Amsterdam, respectively. In this respect the agreement is reasonable, but it should be understood that the accuracy of the predicted concentration is very much dependent upon the estimate of source strength Q used in equations (7.10) and (7.11).

References

1. Lee, R. E. and Goranson, S. S. (1972). National air surveillance cascade impactor network. I. Size distribution of suspended particulate matter in air. *Environ. Sci. and Technol.* **6**, 1019–1024.
2. Lee, R. E. and Goranson, S. S. (1976). National air surveillance cascade impactor network. III. Variation in size of airborne particulate matter over a three-year period. *Environ. Sci. and Technol.* **10**, 1022–1027.
3. Lundgren, D. A. (1970). Atmospheric aerosol compositon and concentration as a function of particle size and of time. *J. Air Pollut. Contr. Assoc.* **20**, 603–608.
4. Lee, R. E. and Smith, C. F. (1972). Size distribution of suspended particulates from lignite combustion. *Environ. Sci. and Technol.* **6**, 929–930.
5. Lee, R. E., Caldwell, J. S. and Morgan, G. B. (1972). The evaluation of methods for measuring suspended particulates in air. *Atmos. Environ.* **6**, 593–622.
6. Butler, J. D. (1977). "Urban Atmospheres". British Association meeting, University of Aston, Birmingham, Sept. 5th 1977.
7. Mainwaring, S. J. and Harsha, S. (1976). Size distribution of aerosols in Melbourne city air. *Atmos. Environ.* **10**, 57–60.
8. Kadowaki, S. (1976). Size distribution of atmospheric total aerosols, sulphate, ammonium and nitrate particulates in the Nagoya area. *Atmos. Environ.* **10**, 39–43.
9. Whitby, K. T., Liu, B. Y. H. and Husar, R. B. (1972). The aerosol size distribution of Los Angeles smog. *J. Coll. Interface Sci.* **39**, 177–210.
10. Whitby, K. T., Husar, R. B. and Liu, B. Y. H. (1972). Physical mechanisms governing the dynamics of Los Angeles smog aerosol. *J. Coll. Interface Sci.* **39**, 211–224.
11. Novakov, T., Mueller, P. K., Alcocer, A. E. and Otvos, J. W. (1972). The chemical composition of Pasadena aerosol. Chemical states of nitrogen and sulfur by photoelectron spectroscopy. *J. Col. Interface Sci.* **39**, 225–234.
12. Mueller, P. K., Mosley, R. W. and Pierce, L. B. (1972). Chemical composition of Pasadena aerosol by particle size and time of day. Carbonate and non-carbonate carbon content. *J. Coll. Interface Sci.* **39**, 235–239.
13. Lee, R. E. and Hein, J. (1974). Method for the determination of carbon, hydrogen and nitrogen in size fractionated atmospheric particulate matter. *Anal. Chem.* **46**, 931–933.
14. Lee, R. E., Goranson, S. S., Enrione, R. E. and Morgan, G. B. (1972). National air surveillance cascade impactor network. II. Size distribution of trace metal components. *Environ. Sci. and Technol.* **6**, 1025–1030.
15. Paciga, J. J. and Jervis, R. E. (1976). Multielement size characterization of urban aerosols. *Environ. Sci. and Technol.* **10**, 1124–1128.
16. Dorn, C. R., Pierce, J. O., Phillips, P. E. and Chase, G. R. (1976). Airborne Pb, Cd, Zn and Cu concentrations by particle size near a Pb smelter. *Atmos. Environ.* **10**, 443–446.
17. Daines, R. H., Motto, H. and Chilko, D. M. (1970). Atmospheric lead: Its relationship to traffic volume and proximity to highways. *Environ. Sci. and Technol.* **4**, 318–322.
18. Butler, J. D., Macmurdo, S. D. and Middleton, D. R. (1975). Motor vehicle generated pollution in urban areas. *J. Environ. Health* **83**, 24–35.

19. Colucci, J. M., Begeman, C. R. and Kumler, K. (1969). Lead concentration in Detroit, New York and Los Angeles air. *J. Air Pollut. Contr. Assoc.* **19**, 255–260.
20. Macmurdo, S. D., Studies of atmospheric pollution due to lead in the vicinity of urban motorways. Ph.D. thesis, 1975, Department of Chemistry, University of Aston, Birmingham, UK.
21. Biggins, P. D. E., "Studies in atmospheric pollution by lead in an urban environment." M. Phil. thesis, 1976, Department of Chemistry, University of Aston, Birmingham, UK.
22. Tepper, L. B. and Levine, L. S. (1972). "A survey of air and population lead levels in selected American communities: Final report." University of Cincinnati.
23. Joint Working Party on Lead Pollution around Gravelly Hill, Final Report. Central Unit of Environmental Pollution, Department of the Environment, 1978. Pollution Paper No. 14. HMSO.
24. Day, J. P., Hart, M. and Robinson, M. S. (1975). Lead in urban street dust. *Nature* **253**, 343–345.
25. Sehmel, G. A. (1973). Particle resuspension from an asphalt road caused by car and truck traffic. *Atmos. Environ.* **7**, 291–309.
26. Colwill, D. M. and Hickman, A. J. (1973). "The concentration of volatile and particulate lead compounds in the atmosphere: Measurement at four road sites." Transport and Road Research Laboratory Report LR 545.
27. Harrison, R. M., Perry, R. and Slater, D. H. (1974). An adsorption technique for the determination of organic lead in Street air. *Atmos. Environ.* **8**, 1187–1194.
28. Snyder, L. J. (1967). Determination of trace amounts of organic lead in air. *Anal. Chem.* **39**, 591–595.
29. Robinson, J. W., Rhodes, L. and Wolcott, D. K. (1975). The determination and identification of molecular lead pollutants in the atmosphere. *Anal. Chim. Acta* **78**, 474–478.
30. Lamothe, P. J., Dick, D. L., Corrin, M. L. and Skogerboe, R. K. (1974). "Evaluation of membrane filters for collection of atmospheric particulates." Environmental contamination caused by lead. Ed., Edwards, H. W., Colorado State University.
31. Stocks, P., Commins, B. T. and Aubrey, K. V. (1961). A study of polycyclic hydrocarbons and trace elements in smoke in Merseyside and other northern localities. *Int. J. Air Wat. Pollut.* **4**, 141–153.
32. Zoccolillo, L., Liberti, A. and Brocco, D. (1972). The determination of polycyclic hydrocarbons in air by gas chromatography with high efficiency columns. *Atmos. Environ.* **6**, 715–720.
33. Sawicki, E., Hauser, T. R., Elbert, W. C., Fox, F. T. and Meeker, J. E. (1962). Polynuclear aromatic hydrocarbon composition of the atmosphere in some large American cities. *Amer. Ind. Hyg. Assoc. J.* **23**, 137–143.
34. Sawicki, E., McPherson, S. P., Stanley, T. W., Meeker, J. E. and Elbert, W. C. (1965). Quantitative composition of the urban atmosphere in terms of polynuclear aza heterocyclic compounds and aliphatic and polynuclear aromatic hydrocarbons. *Int. J. Air Water Pollut.* **9**, 515–524.
35. Gordon, R. J. (1976). Distribution of airborne polycyclic aromatic hydrocarbons throughout Los Angeles. *Environ. Sci. and Technol.* **10**, 370–373.

36. Colucci, J. M. and Begeman, C. R. (1965). The automotive contribution to airborne polynuclear aromatic hydrocarbons in Detroit. *J. Air Pollut. Contr. Assoc.* **15**, 113–122.
37. Waller, R. E., Commins, B. T. and Lawther, P. J. (1965). Air pollution in a city street. *Brit. J. Ind. Med.* **22**, 128–138.
38. Hettche, H. O. (1964). Benzypyrene und spurenelemente in grosstadtluft. *Int. J. Air Water Pollut.* **8**, 185–191.
39. Brocco, D., Cimmino, A. and Possanzini, M. (1973). Determination of aza-heterocyclic compounds in atmospheric dust by a combination of thin-layer and gas chromatography. *J. Chromatog.* **84**, 371–377.
40. Kertész-Saringer, M., Mórik, J. and Morlin, Z. (1969). Benzo(a)pyrene pollution in Budapest. *Atmos. Environ.* **3**, 417–422.
41. Demaio, L. and Corn, M. (1966). Polynuclear aromatic hydrocarbons associated with particulates in Pittsburgh air. *J. Air Pollut. Contr. Assoc.* **16**, 67–71.
42. Pierce, R. C. and Katz, M. (1975). Dependency of polynuclear aromatic hydrocarbon content on size distribution of atmospheric aerosols. *Environ. Sci. and Technol.* **9**, 347–353.
43. Butler, J. D. and Crossley, P. (1979). An appraisal of relative airborne sub-urban concentrations of polycyclic aromatic hydrocarbons monitored indoors and outdoors *Sci. Total Environ.* **11**, 53–58.
44. Ciaccio, L. L., Rubino, R. L. and Flores, J. (1974). Composition of organic constituents in breathable airborne particulate matter near a highway. *Environ. Sci. and Technol.* **8**, 935–942.
45. Benson, F. B., Henderson, J. J. and Caldwell, D. E., Indoor–outdoor air pollution relationships: A literature review. US Environmental Protection Agency, Pub., AP-112, August 1972.
46. Biersteker, K., De Graaf, H. and Nass, C. A. G. (1965). Indoor air pollution in Rotterdam homes. *Int. J. Air Water Pollut.* **9**, 343–350.
47. Thompson, C. R., Hensel, E. G. and Kats, G. (1973). Outdoor–indoor levels of six air pollutants. *J. Air Pollut. Contr. Assoc.* **23**, 881–886.
48. Daines, R. H., Smith, D. W., Feliciano, A. and Trout, J. R. (1972). Air levels of lead inside and outside of homes. *Ind. Med. J.* **41**, 26–28.
49. Butler, J. D. and Macmurdo, S. D. (1974). Interior and exterior atmospheric lead concentrations of a house situated near an urban motorway. *Int. J. Environ. Stud.* **6**, 181–184.
50. Pooler, F. (1961). A prediction model of mean urban pollution for use with wind roses. *Int. J. Air Water Pollut.* **4**, 199–211.
51. Miller, M. and Holzworth, G. (1967). An atmospheric diffusion model for metropolitan areas. *J. Air Pollut. Contr. Assoc.* **17**, 46–50.
52. Clarke, J. F. (1964). A simple diffusion model for calculating point concentrations from multiple sources. *J. Air Pollut. Contr. Assoc.* **14**, 347–352.
53. Goumans, H. H. J. M. and Clarenburg, L. A. (1975). A simple model to calculate the SO_2-concentration in urban regions. *Atmos. Environ.* **9**, 1071–1077.
54. Gifford, F. A. and Hanna, S. R. (1970). "Urban air pollution modelling." 2nd International Clean Air Congress, IUAPPA, Washington, DC. Paper MS 32D p. 1146–1151.
55. Clarenburg, L. A. (1973). Penalization of the environment due to stench. *Atmos. Environ.* **7**, 333–351.

Note added in proof

When considering the dependence of one relationship with another the concept of the linear correlation coefficient R can be used, where

$$R = \frac{n \sum x_i y_i - \sum x_i \sum y_i}{[n \sum x_i^2 - (\sum x_i)^2]^{1/2} [n \sum y_i^2 - (\sum y_i)^2]^{1/2}} \quad (4.40)$$

such that x_i, y_i are the two variables and n their number. The value of $R = \pm 1$ signifies a precise correspondence, values close to ± 1 denote good correlation and those tending to zero indicate that there is no relationship between the parameters considered. As an example we can judge the data presented in Table 7.5 which attempts to predict airborne lead concentrations at a particular site. Let x_i be the actual values monitored and y_i the predicted ones, then the values $x_i y_i$, $\sum x_i y_i$, $\sum x_i$, $\sum x_i^2$, $\sum y_i$ and $\sum y_i^2$ are obtained as follows:

x_i	y_i	$x_i y_i$
1·17	1·61	1·8837
1·70	2·28	3·8760
1·38	1·92	2·6496
1·50	1·67	2·5050
1·25	1·62	2·0250
1·02	1·35	1·3770
1·50	1·19	1·7850
1·70	1·36	2·3120
1·62	1·51	2·4462
2·23	1·91	4·2593
2·86	2·08	5·9488
3·56	2·16	7·6896
$\sum x_i$	$\sum y_i$	$\sum x_i y_i$
21.49	20·66	38·7572
$\sum x_i^2$	$\sum y_i^2$	
44.6067	36.8986	
$n = 12$		

Hence, substituting into equation (4.40)

$$R = \frac{12 \times 38{\cdot}7572 - 21{\cdot}49 \times 20{\cdot}66}{[12 \times 44{\cdot}6067 - 21{\cdot}49^2]^{1/2} [12 \times 36{\cdot}8986 - 20{\cdot}66^2]^{1/2}}$$

$$R = \frac{21{\cdot}103}{34{\cdot}227} = 0{\cdot}62$$

Author Index

Numbers in italic refer to pages on which the reference is given in full

Abbasi, A. H., *44*(19), 6(19)
Abel, N., *124*(69), 105(69)
Ackerman, M., *315*(9), 275(9)
Adlard, E. R., *260*(11), 174(11)
Agnew, W. G., *315*(15), 277(15)
Alcocer, A. E., *377*(11), 351(11), 352(11), 353(11), 354(11)
Alford, J. A., *316*(23), 283(23)
Allen, E. R., *315*(3), 271(3)
Allen, H. E:, *44*(6), 3(6)
Altshuller, A. P., *166*(25)(26), *167*(40)(42), *315*(14), 139(25, 26) 145(40), 145(42), 277(14)
Amdur, M. O., *49*(115), 37(115)
Andersen, A. A., *167*(51), 155(51)
Appleby, A., *260*(28), 185(28)
Astley, R., *45*(38), 7(38)
Attala, R., *45*(35), 7(35)
Aubrey, K. V., *378*(31), 368(31)
Audric, B. N., *266*(132), 255(132)
Aue, W. A., *259*(5), 171(5)
Axelrod, H. D., *261*(31), 186(31)

Babcock, G., *260*(24) 180(24)
Badcock, C. C., *317*(41, 48, 49), 296(41), 298(48, 49)
Badger, G. M., *122*(30, 31), 80(30, 31)
Baker, C. J., *266*(127) 247(127), 248(127)
Baker, E. W., *121*(20) 72(20)
Baker, M. J., *48*(98) 26(98)
Barltrop, D., *45*(40) 7(40)
Barnes, H., *165*(17), 138(17)
Barnes, J. M., *260*(17), 178(17)
Barney, J. E., *48*(100), *260*(27), 26(100), 185(27)
Barrie, L. A., *317*(40), 295(40)
Barry, P. S. I., 17
Barry, R., *265*(118, 121), 241(118), 244(118), 245(121)
Bates, D. V., *49*(118), 119), 38(118) 39(119)
Baumgardner, R. E., *260*(22), 180(22)
Baumslag, N., *45*(26) 6(26)
Baurer, T., *121*(12), *316*(27), 58(12), 285(27)
Bayard, M. A., *122*(46), 85(46)
Bayes, K. D., 316(24), 284(24)
Beasley, W. H., *46*(49), 18(49)
Begeman, C. R., *122*(32), *378*(19), *379*(36), 81(32), 359(19), 368(36)
Belcher, R., *261*(35), 188(35)
Bender, D. F., *264*(91) 236(91)
Bennett, C. F., *122*(43), 85(43)
Benson, F. B., *379*(45), 367(45)
Berg, B., *44*(13), 6(13)
Berkeley, R. E., *265*(113), 239(113)
Berlin, M., *44*(20), 6(20)
Berry, C. M., *48*(86), 23(86)
Berry, G., *47*(78), 22(78)
Betts, P. R., *45*(38), 7(38)
Biemann, K., *265*(116), 240(116)
Biersteker, K., *379*(46), 367(46)
Biggins, P. D. E., *378*(21), 359(21),131
Birch, S. R., *166*(28), 135(28)
Blacker, J. H., *45*(22), 6(22)
Blanchard, J. W., *45*(22), 6(22)
Blessel, K., *261*(35), 188(35)
Block, C., *121*(19), 69(19)
Boeck, W. L., *344*(17), 343(17)
Boggs, W. E., *166*(22), 138(22)
Bolin, B., *120*(4), 53(4), 54(4), 55(4)

Bombaugh, K. J., *263*(77), 227(77)
Bonfield, B. A., *166*(36), 144(36)
Bonelli, E. J., *166*(32), 142(32), 143(32)
Borg, D. C., *46*(54), 18(54)
Bortner, M. H., *316*(27, 30), 285(27), 286(30)
Borucki, W. J., *318*(73), 313(73)
Boucher, L. J., *121*(24), 73(24)
Boyd, J. T., *49*(125), 43(125)
Boyer, K. W., *122*(38), 318(59), 84(38), 305(59)
Bratzel, M. P., *121*(23), 71(23)
Braverman, M. M., *165*(12), 137(12)
Brief, R. S., *45*(22), 6(22)
Briggs, G. A., *344*(10), 336(10)
Brocco, D., *264*(90), *378*(32), *379*(39), 236(90), 368(32, 39)
Brockhaus, A., *262*(52), 219(52)
Brodie, K. G., *261*(45), 203(45)
Brody, S. S., *259*(7), 171(7)
Brown, P. R., *263*(85), 229(85)
Browett, E. V., *166*(35), 144(35)
Bryce-Smith, D., *45*(30, 31), 5(30, 31)
Buchanan, W. D., *49*(116), 38(116)
Buechley, R. W., *49*(112), 37(112)
Bufalini, J. J., 299(51), *315*(14), *316*(21), *317*(51), 277(14), 280(21)
Bunch, J. E., *265*(113), 239(113)
Burg, W. R., *49*(108), *166*(28), 35(108), 135(28)
Burgess, J. R., *316*(23), 283(23)
Burland, W. L., *45*(40), 7(40)
Burton, R. M., *167*(53), 158(53)
Butler, F. F., *266*(135), 255(135)
Butler, J. D., *46*(50), *48*(96), *124*(82), *167*(41), *261*(42), *262*(63), 377(6, 18), 379(43, 49), 18(50), 24(96), 110(82), 145(41), 198(42), 199(42), 200(42), 222(63), 346(6), 357(18), 366(43), 371(49), 372(43, 49)
Butt, E. M., *44*(7), 3(7)
Byers, R. K., *45*(34), 7(34)

Cadle, R. D., *120*(1), *315*(3), 51(1), 271(3)
Cahnmann, H. J., *262*(67), 225(67)
Calder, K. L. *344*(16), 340(16), 341(16)
Caldwell, D. E., *379*(45), 367(45)
Caldwell, J. S., *377*(5), 346(5)
Calvert, J. G., *315*(5, 10), 317(37, 41, 42, 48, 49), 273(4), 274(5), 276(10), 287(5), 288(5), 289(5), 290(5), 292(37), 295(41, 42), 297(37), 298(37), 298, 48(49)
Campbell, J. M., 23
Candeli, A., *122*(33), *262*(62), 81(33), 82(33), 221(62), 236(62)
Cantuti, V. J., *166*(29), *265*(106, 107), 135(29), 237(106, 107)
Cantwell, E. N., *124*(84), 115(84)
Cardwell, T., *261*(35), 188(35)
Carnow, B. W., *48*(97), 25(97)
Carpenter, S. B., *317*(39), 293(39)
Carroll, R. E., *46*(60), 18(60), 19(60), 20(60)
Cartoni, G. P., *166*(29), *265*(106, 107), 135(29), 237(105, 106)
Castleden, C. M., *49*(109), 35(109)
Cautreels, W., *262*(59), *265*(112), 220(59), 239(112)
Cawse, P. A., *123*(62), 100(62), 101(62)
Chakrabarti, C. L., *121*(23), 71(23)
Chakraborty, B. B., *122*(36), 83(36)
Chamberlain, A. C., *46*(47), *121*(16), *124*(71), 14(47), 15(47), 16(47), 63(16), 100(16), 106(71)
Chaney, J. E., *259*(7), 171(7)
Chang, S. G., *316*(36), 292(36)
Chase, G. R., *377*(16), 356(16)
Chau, Y. K., *261*(30), 186(30)
Chesters, G., *122*(47), 85(47)
Childers, E. E., *262*(68), 225(68)
Chilko, D. M., *377*(17), 356(17)
Cholak, J., *44*(9), *46*(51, 52), 3(9), 18(51, 52)
Chow, T. J., *122*(43), *123*(68), 85(43), 105(68)
Ciaccio, L. L., *379*(44), 367(44)
Cicerone, R. J., *318*(68), 309(68)
Cimmino, A., *264*(90), *379*(39), 236(90), 368(39)
Clarenburg, L. A., *379*(53, 55), 374(53, 55), 375(53)
Clark, L. F., *121*(20), 72(20)

Clarke, T. A., *167*(53), 158(53)
Clarke, J. F., *379*(52), 373(52)
Clayton, B. E., *45*(36), 7(36)
Clayton, G. D., *44*(2), 2(2)
Cleary, G. J., *262*(69), 225(69)
Clelland, R. C., *46*(62), 20(62), 21(62)
Clemons, C. A., *167*(40), 145(40)
Clingenpeel, J. M., *165*(16), 137(16)
Clough, P. N., *260*(15), 177(15)
Clough, W. S., *46*(47), 14(47), 15(47), 16(47)
Clyne, M. A. A., *260*(14), *319*(77), 177(14), 315(77)
Cohens, I. R., *166*(26), *167*(42), 139(26), 145(42)
Cole, P. V., *49*(109, 110), 35(109, 110)
Colucci, J. M., *122*(32), 378(19), 379(36), 81(32), 359(19) 368(36)
Colwill, D. M., *378*(26), 365(26)
Commins, B. T., *49*(111), *123*(52), *166*(23), *261*(47, 48), *263*(73), *378*(31), *379*(37), 36(111), 37(111), 89(52), 138(23), 215(47, 48), 225(73), 368(31, 37)
Cooper, R. L., *263*(72), 225(72)
Corrin, M. L., *378*(30), 365(30)
Corn, M., *379*(41), 366(41)
Cotton, D. H., *122*(37), 83(37)
Cotzias, G. C., *46*(54), 18(54)
Coughlin, L. L., *46*(59), 18(59)
Coulston, F., *46*(48), 16(48)
Cox, R: A., *317*(38), 292(38)
Craig, P. P., *48*(106), *123*(61), 33(106), 98(61), 99(61)
Crawford, K. C., *124*(73), 108(73)
Crider, W. L., *122*(39), *259*(6), 84(39), 171(6)
Crossley, P., *262*(63), 379(43), 222(63), 366(43), 372(43)
Cruickshank, C. N. D., *48*(90), 23(90)
Crutzen, P. J., *318*(70), 312(70)
Cuddeback, J. E., *49*(108), *166*(28), 36(108), 135(28)
Cummins, R. L., *166*(24), 139(24)
Curwen, M. P., *47*(82), 23(82)
Cvetanovic, R. J., *317*(50), 299(50)

Daby, E. E., *316*(17), 277(17)
Daines, R. H., *377*(17), *379*(48), 356(17), 371(48)
Dahneke, B., *167*(52), 156(52), 158(52), 159(52)
Damon, E. K., *317*(48, 49), 298(48, 49)
Dams, R., *121*(19), 69(19)
Darnell, K. R., *316*(16), 277(16)
Davies, C. N., *165*(4), 132(4)
Davies, I. W., *123*(53), 91(53)
Davies, T. A. L., *44*(15), 6(15)
Davies, D. D., *317*(43, 46), *318*(69), 297(43, 46), 311(69)
Davis, D. R., *124*(82), 110(82)
Dawson, H. J., *166*(33), 142(33)
Day, J. P., *378*(24), 361(24), 365(24)
Dean, G., *47*(83), *48*(93, 94), 22(83), 23(93, 94)
De Graaf, H., *379*(46), 367(46)
de Haas, N., *315*(12), 276(12)
DeMaio, L., *379*(41), 366(41)
DeMarrais, G. A., *260*(18), 179(18)
Demerjian, K. L., *315*(4, 510), 273(4), 274(5), 276(10), 287(5), 288(5), 289(5)
Desbaume, P., *45*(32, 33), 5(32), 7(33)
Dick, D. L., *378*(30), 365(30)
Dickie, J. P., *121*(20, 24), 72(20), 73(24)
Didio, S. L., *44*(7), 3(7)
Dinman, B. D., *44*(6), 3(6)
Dolansky, V., *263*(70), 225(70)
Doll, Sir Richard, *47*(72, 73), *49*(125), 22(72, 73), 43(125)
Dolovich, M., *49*(114), 37(114)
Dong, M., *263*(84), 229(84)
Donnelly, A. J., *45*(25) 6(25)
Donnelly, J. K., *122*(30, 31), 80(30, 31)
Donovan, J. R., *49*(108), 35(108)
Doran, T., *263*(83), 229(83)
Dorn, C. R., *377*(16), 356(16)
Drake, P. F., *122*(48), 88(48)
Drew, R. T., *48*(99), 26(99)
Drivas, P. J., *344*(7), 323(7)
Dryer, F., *121*(26), 75(26)
Dubois, L., *165*(7), *263*(76), *266*(127), 136(7), 225(76), 247(127), 248(127)

Dufour, L., *123*(64), 103(64)
Dwyer, M., *317*(43), 297(43)

Earl, J. L., *122*(43), 85(43)
Eastcott, D. F., *48*(92), 23(92)
Edgar, G., 84
Edwards, R., *48*(107), 33(107)
Egorov, V. V., *122*(42), 85(42)
El-Attar, A. A., *47*(79), 22(79)
Elbert, W. C., *263*(75, 86), *264*(88, 92, 93, 94, 95), *266*(123), *378*(33, 34), 225(75), 228(86, 88), 229(86), 233(86, 88), 234(88), 236(88, 92, 93, 94, 95), 246(123), 368(33, 34)
Elmes, P. C., *47*(80), 22(80)
El-Messiri, I. A., *122*(29), 78(29), 79(29)
Enrione, R. E., *166*(36), *377*(14), 144(36), 355(14), 356(14)
Erhardt, C. L., *44*(4), 2(4)
Eshleman, A., *120*(2), 51(2)
Evans, K. P., *260*(29), 186(29)

Falk, H. L., *262*(56, 57), 220(56, 57)
Farrow, L. A., *316*(32), 292(32)
Faulds, J. S., *49*(125), 43(125)
Feliciano, A., *379*(48), 371(48)
Fenimore, C. P., *122*(34), 83(34)
Ferguson, M. B., *266*(126), 247(126)
Ferm, V. H., *44*(17), 6(17)
Ferrand, E., *263*(84), 229(84)
Field, F., *44*(4, 5), 2(4, 5)
Field, S., *49*(118), 38(118)
Filby, R. H., *121*(21, 22), 71(21, 22)
Fine, D. H., *260*(18, 19), 179(18, 19)
Fleischer, R. L., *167*(43), 146(43)
Flesch, J. P., *167*(55), 162(55), 163(55), 164(55)
Flores, J., *379*(44), 367(44)
Fontijn, A., *260*(16), 177(16)
Foote, C. S., *315*(1), 269(1), 270(1)
Foot, J. S., *319*(76), 314(76)
Ford, H. W., *317*(44), 297(44)
Foster, P. M., 316(35), 292(35), 293(35)
Fox, F. T., *263*(75), *265*(122), *378*(33), 225(75), 245(122), 368(33)
Francis, C. W., *122*(47), 85(47)
Frank, E. R., *167*(48), 149(48), 151(48)
Frank, N. R., *49*(113), 37(113)
Frei, R. W., *263*(81), 229(81)
Friedlander, S. K., *316*(34), 292(34)
Friswell, N. J., *122*(37), 83(37)
Fryčka, J., *265*(104), 237(104)

Gaeke, G. C., *165*(18, 19), 138(18, 19)
Gafefer, W. M., *44*(2), 2(2)
Galbally, I. E., *123*(55), 94(55), 95(55), 96(55)
Gandhi, H. S., *124*(74, 75, 76), 110(74, 75, 76)
Gartrell, F. E., *317*(39), 293(39)
Garvin, D., *121*(10), 57(10), 77(10)
Găvănescu, O., *45*(41), 7(41)
Gay, B. W., *317*(51), 299(51)
Gaydon, A. G., *122*(35), 83(35)
Gehrke, C. W., *259*(5), 171(5)
Gelberg, A., *165*(11), 136(11)
Georgii, H. W., *317*(40), 295(40)
Geyh, M. A., *266*(133), 255(133)
Gick, H., *46*(58), 18(58)
Gifford, F. A., *344*(4, 8), *379*(54), 322(4), 333(8), 334(8), 335(8), 374(54), 375(54)
Gillette, D. A., *122*(45), 85(45)
Gills, B. G., *123*(49), 88(49)
Gilmore, F. R., *318*(74), 314(74)
Gilmour, T. C., *44*(7), 3(7)
Gitchell, A., *317*(52), 303(52)
Giuffrida, L., *259*(4), 171(4)
Glasser, M., *44*(4, 5), 2(4, 5)
Glasson, W. A., *316*(19), 278(19)
Glassman, I., *121*(26), 75(26)
Gloyne, S. R., *47*(70), 22(70)
Golberg, L., *46*(48), 16(48)
Goldsmith, J. R., *46*(45), 14(45)
Goldsmith, P., *319*(76), 314(76)
Gonzalez, R. D., *124*(77), 110(77)
Goranson, S. S., *377*(1, 2, 14), 346(1), 347(1, 2), 355(1, 2, 14), 356(14)
Gordon, R. J., *262*(53), *378*(35), 219(53), 220(53), 368(35)
Gough, T. A., *260*(20), 180(20)
Goulden, P. D., *261*(30), 186(30)
Goumans, H. H. J. M., *379*(53), 374(53), 375(53)
Gourevitch, A., *48*(90), 23(90)

Graedel, T. E., *316*(32), 292(32)
Grant, D. W., *263*(82), 229(82)
Greenburg, L., *44*(4, 5), 2(4, 5)
Greenfield, S. M., *123*(66), 103(66)
Greiner, N. R., *120*(8), *315*(8, 13), 57(8), 275(8), 276(13)
Griffen, T., *46*(48), 16(48)
Griffiths, D. M., *47*(77), 22(77)
Grimmer, G., 223, 224
Grob, G., *265*(111), 239(111)
Grob, K., *265*(111), 239(111)
Grosjean, D., *262*(54), 219(54)
Gross, P., *47*(76), 22(76)
Grundy, R. D., *46*(42), 7(42), 13(42)
Guichon, G., *263*(79), 229(79)

Habibi, K., *122*(44), 85(44)
Haenszel, W., *48*(95), 23(95)
Hager, R. N., *260*(26), 181(26), 184(26)
Haller, W. A., *121*(21, 22), 71(21, 22)
Hammond, A. L., *44*(18), *48*(102), 6(18), 32(102)
Hammond, P. B., *45*(39), 7(39)
Hampson, R. F., *315*(6), 275(6)
Hancock, W., *46*(51), 18(51)
Hangebrauck, R. P., *123*(50, 51), 89(50, 51), 90(51), 92(51), 93(51)
Hanker, J. S., *165*(11), 136(11)
Hanna, S. R., *379*(54), 374(54), 375(54)
Hanst, P. L., *317*(51), 299(51)
Harker, A. B., *316*(36), 292(36)
Harris, C. C., *48*(98), 26(98)
Harrison, R. M., *123*(53), *166*(34), *378*(27), 91(53), 143(43), 144(34), 365(27)
Harsha, S., *377*(7), 346(7)
Hart, M., *378*(24), 361(24), 365(24)
Hartley, H. O., 257
Hartmann, H., *166*(32), 142(32), 143(32)
Hartz, S., *46*(59), 18(59)
Haskin, L. A., *122*(47), 85(47)
Hasselblad, V., *49*(112), 37(112)
Hauser, T. R., *166*(24), *263*(75), *265*(119, 120, 121), *266*(124), *378*(33), 139(24), 225(75), 242(119), 245(119, 120, 121), 246(124), 368(33)
Hayakawa, K., *44*(10), 3(10)
Haynes, C. D., *121*(27), 78(27)
Hazucha, M., *49*(118, 119), 38(118), 39(119)
Heard, M. S., *46*(47), 14(47), 15(47), 16(47)
Hecker, L. H., *44*(6), 3(6)
Heicklen, J., *317*(52, 54), *318*(55), 303(52, 54, 55)
Heimann, H., *44*(1, 2), 2(1, 2)
Hein, J., *377*(13), 355(13)
Heggestad, H. E., *49*(117), 38(117)
Helton, M. R., *48*(100), *260*(27), 26(100), 185(27)
Henderson, J. J., *379*(45), 367(45)
Henderson, S. R., *166*(37), 144(37)
Hendricks, N. V., *48*(86,), 23(86)
Henry, S. A., *48*(85, 87), 23(85, 87)
Hensel, E. G., *379*(47), 370(47)
Hettche, H. O., *379*(38), 368(38)
Hewitt, P. J., *261*(39), 195(39), 196(39)
Hexter, A. C., *46*(45), 14(45)
Hickey, R. J., *46*(62), 20(62), 21(62)
Hickman, A. J., *378*(26), 365(26)
Hildebrandt, A., 223, 224
Hill, I. D., *47*(73), 22(73)
Hilsmeier, W. F., *344*(8), 333(8), 334(8), 335(8)
Hlucháň, E., *264*(101), 237(101)
Hochheiser, S., *165*(12), *259*(1, 2), 137(12), 168(1), 169(2)
Hoffmann, D., *262*(58, 61), 20(58), 221(61), 222(58)
Holaday, D. A., *49*(124), 43(124)
Holland, J. Z., *344*(9), 336(9)
Hollis, M. G., *263*(82), 229(82)
Holmberg, R. E., *44*(17), 6(17)
Holmes, S., *47*(73), 22(73)
Holt, P. F., *47*(75), 22(75)
Holzworth, G., *379*(51), 373(51)
Horgan, D. F., *263*(77), 227(77)
Howard, J. N., *167*(53), 158(53)
Hubbard, E. H., *122*(48), 88(48)
Hudson, R. L., *166*(38), *167*(39), 144(38, 39)
Hurn, R. W., *165*(16), 137(16)
Husar, R. B., *377*(9, 10), 348(9), 349(10), 350(10)

Imbus, H. R., *44*(9), 3(9)
Ingersoll, R. B., *120*(7), 57(7)
Inman, R. E., *120*(7), 57(7)
Izod, T. P. J., *316*(29), 285(29)

Jackson, G. E., *317*(41), 296(41)
Jacobs, E. S., *124*(84), 115(84)
Jacobs, M. B., *165*(8, 9, 12, 13), 136(8, 9), 137(12, 13)
Jaeschke, W., *260*(21), 180(21)
Jalili, M. A., *44*(19), 6(19)
Janák, J., *166*(30), 135(30)
Janini, G. M., *265*(105, 109), 235(105), 237(105), 238(105,109), 239(105)
Janz, J. G., *316*(20), 280(20)
Jeffery, P. G., *165*(10), 136(10)
Jeník, M., *264*(101), 237(101)
Jenkins, D. R., *122*(37), 83(37)
Jensen, S., *48*(105) 32(105), 33(105)
Jervis, R. E., *377*(15), 355(15), 356(15)
Jick, H., *46*(58), 18(58)
Johnson, G. G., *261*(41), 197(41)
Johnson, H., *264*(87), 228(87), 230(87)
Johnson, H. A., 48(106), 123(61), 33(106), 98(61), 99(61)
Johnson, R. J., *46*(48), 16(48)
Johnston, H. S., *318*(75), 314(75)
Johnston, K., *265*(105), 235(105), 237(105), 238(105), 239(105)
Jones, D. D., *46*(49), 18(49)
Jones, S. W., *122*(34), 83(34)
Jordan, M. L., *167*(54), 161(54)
Jost, W., *262*(64), 223(64)
Junge, C. E., *120*(6), *121*(15), *123*(67), *124*(69), *315*(7), 57(6), 63(15), 103(67), 104(67), 105(69), 275(7)
Jusko, W. J., 46(59), 18(59)

Kadawaki, S., *377*(8), 347(8)
Karger, B. L., *263*(79), 229(79)
Karmen, A., *259*(4), 171(4)
Karr, C., *262*(68), 225(68)
Kats, G., *379*(47), 370(47)
Katz, M., *262*(55), *264*(89, 96, 102), *266*(126), *379*(42), 220(55), 229(89), 231(89), 236(89, 96), 237(102), 240(102), 247(126), 366(42)
Kaufman, G., *48*(98), 26(98)
Kazantis, G., *45*(27), 6(27)
Keen, P., *45*(26), 6(26)
Kehoe, R. A., *46*(46, 51, 52), 15(46), 18(51, 52)
Kennaway, E. L., *47*(82), 23(82)
Kennaway, N. M., *47*(82), 23(82)
Kerr, J. A., *315*(5, 10), 273(4), 274(5), 276(10), 287(5), 288(5), 289(5), 290(5)
Kertész-Saringer, M., *379*(40), 366(40)
Kettering, C., 84
Khan, A. U., *316*(25, 28), 285(25, 28), 286(25)
Kim, P., *317*(43), 297(43)
Kirkland, J. J., *263*(78), 227(78)
Kitamura, S., *44*(10), 3(10)
Kleinfield, M., *45*(21), 6(21)
Klepsch, I., *45*(41), 7(41)
Klimisch, R. L., *124*(78, 79), 110(78, 79)
Knelson, J. H., *46*(48), 16(48)
Knox, J. F., *47*(73), 22(73)
Koch, R. C., *344*(5), 327(5)
Kooyman, O., *45*(21), 6(21)
Kopito, L., *45*(37), 7(37)
Koppenhaal, D. W., *121*(18), 68(18)
Kopple, J. D., *46*(43), 13(43)
Kosinski, K., *264*(87), 228(87), 230(87)
Kotin, P., *262*(56, 57), 220(56, 57)
Koumides, O. P., *45*(36), 7(36)
Krost, K. J., *259*(8), *260*(18), 171(8), 179(18), 228(8)
Krstulovic, A. M., *263*(85), 229(85)
Kubota, J., *44*(8), 3(8)
Kumler, K., *378*(19), 359(19)
Kummer, W. A., *260*(25), 181(25)
Kummler, R. H., *121*(12), *316*(27, 30), 59(12), 285(27), 286(30)
Kuschner, M., *48*(99), 26(99)
Kunz, W. G., *124*(84), 115(84)

LaHue, M. D., *261*(31), 186(31)
Laitinen, H. A., *122*(38), 318(59), 84(38), 305(59)
Lamb, H. H., *124*(70), 106(70)

Lambert, P. M., *47*(84), 23(84)
Lamothe, P. J., *378*(30), 365(30)
Lancranjan, I., *45*(41), 7(41)
Lane, D. A., *264*(102), 237(102), 240(102)
Lao, R. C., *262*(50), *265*(110), 215(50), 222(50), 224(50), 239(110), 242(110) 243(110)
Larsen, R. I., *165*(3), 128(3)
Laskin, S., *48*(99), 26(99)
Laud, B. B., *122*(35), 83(35)
Laurer, G. R., *49*(123), 43(123)
Lawther, P. J., *261*(47), *379*(37), 215(47), 368(37)
Lazar, V. A., *44*(8), 3(8)
Lederer, M., *264*(100), 237(100)
Lee, R. E., *122*(39), *377*(1, 2, 4, 5, 13, 14,), 84(39), 346(1, 4, 5), 347(1, 2), 355(1, 2, 13, 14), 356(14)
Leighton, P. A., *316*(26), 285(26), 290(26), 291(26)
Leineweber, J. P., *47*(69), 21(69)
Leiper, J., *49*(125), 43(125)
Leitch, A., *48*(89), 23(89)
Leng, L. J., *166*(25), 139(25)
Levin, E., *45*(21), 6(21)
Levine, L. S., *378*(22), 365(22)
Levy, A., *317*(53), 303(53)
Levy, E. A., *120*(7), 57(7)
Levy, H., *316*(22), *317*(47), 281(22), 297(47)
Lewis, G. P., *46*(58, 59), 18(58, 59)
Liberi, V. E., *124*(84), 115(84)
Liberti, A., *265*(106, 107), *378*(32), 237(106, 107), 368(32)
Lieb, D., *260*(19), 179(19)
Lieben, J., *44*(14), 6(14)
Lillian, D., *260*(12, 28), 174(12), 185(12, 28)
Linch, A. L., *46*(44), 14(44)
Lindsay, R., *124*(73), 108(73)
Lione, J. G., *48*(86), 23(86)
Lipsky, S. R., *259*(10), 173(10)
Little, J. N., *263*(77), 227(77)
Liu, B. Y. H., *377*(9, 10), 348(9), 349(10), 350(10)
Lloyd, A. C., *316*(16), 277(16)
Lob, M., *45*(32, 33), 5(32), 7(33)
Locke, D. C., *263*(84), 229(84)
Lodge, J. P., *166*(20), *167*(46, 47, 48), *261*(31), 138(20), 148(46), 149(47, 48), 151(48), 186(31)
Lofquist, G. A., *46*(52), 18(52)
Loheac, J., *263*(79), 229(79)
Long, J. V. P., *266*(132), 255(132)
Long, R., *122*(36), 83(36)
Lonneman, W. A., *261*(32), 186(32)
Loosli, H. H., *266*(134), 255(134)
Lord, E. E., *45*(34), 7(34)
Losee, F., *44*(8), 3(8)
Lovelock, J. E., *166*(31), *259* (10), *260*(11), *318*(63), 142(31), 173(10), 174(11), 309(63)
Lucas, D. H., *344*(11, 12), 336(11, 12)
Ludmann, W. F., *259*(2), 169(2)
Luly, A. M., *264*(98), 236(98)
Lundgren, D. A., *377*(3), 346(3)
Lutkins, S. G., *46*(49), 18(49)
Lynch, K. M., *47*(71), 22(71)

Macaull, J., *46*(57), 18(57)
MacFarlane, E. M., 24
MacMurdo, S. D., *46*(50), *261*(42), *377*(18), *378*(20), *379*(49), 18(50), 198(42), 199(42), 200(42), 357(18), 359(20), 371(49), 372(49)
Magee, P. N., *260*(17), 178(17)
Maggs, R. J., *260*(11), *318*(63), 174(11), 309(63)
Mainwaring, S. J., *377*(7), 346(7)
Malakov, S. G., *122*(42), 85(42)
Malý, E., *264*(101), 237(101)
Manahan, S. E., *121*(18), 68(18)
Mancuso, T. F., *47*(7), 3(7)
Manganelli, R. M., *166*(21), 138(21)
Sister Mariano, *44*(7), 3(7)
Markham, M. C., *266*(129), 250(129)
Markey, S. P., *265*(114), 240(114)
Martin, A. E., *44*(3), 2(3)
Martin, B. E., *260*(22), 180(22)
Martin, M., *263*(79), 229(79)
Massmann, H., *261*(44), 203(44)
Mastromatteo, E., *48*(91), 23(91)
Mathias, A., *260*(29), 186(29)
Matousek, J. P., *261*(45), 203(45)
Maugh, T. H., *318*(56), 304(56)
May, G., *48*(103), 32(103)
McClenny, W. A., *260*(22), 180(22)

McConnell, J. C., *121*(9), 57(9)
McElroy, J. L., *344*(15), 341(15)
McElroy, M. B., *121*(9), *123*(57, 59), *318*(61, 67, 72), 57(9), 96(57), 97(57, 59), 307(61), 309(67), 313(72)
McGuire, T., *165*(2), 128(2), 129(2), 130(2)
McIlhinney, J. G., *46*(52), 18(52)
McLafferty, F. W., *265*(115), 240(115)
McLaughlin, M., *46*(44), 14(44)
McPherson, S., *264*(88, 97), *378*(34), 228(88), 232(97), 233(88), 234(88), 236(88, 97), 378(34)
McQuigg, R. D., *315*(10), *317*(37), 276(10), 292(37), 297(37), 298(37)
McRae, J., *265*(113), 239(113)
McTaggart, N. G., *263*(83), 229(83)
Meeker, J. E., *123*(50, 51), *263*(71, 74), *264*(88), *378*(33, 34), 89(50, 51), 90(51), 92(51), 93(51), 225(71, 74), 228(88), 233(88), 234(88), 236(88), 368(33, 34)
Megit, A., *46*(49), 18(49)
Mehler, A., *262*(56), 220(56)
Meier, P., *48*(97), 25(97)
Meiris, R. B., *263*(82), 229(82)
Mellor, N., *260*(29), 186(29)
Menser, H. A., *49*(117), 38(117)
Mercer, T. E., *167*(50), 155(50)
Merritt, W. F., *123*(63), 102(63)
Messite, J., *45*(21), 6(21)
Middleton, D. R., *261*(37), *377*(18), 190(37), 192(37), 193(37), 357(18)
Middleton, F. M., *262*(60), 220(60), 224(60)
Midgley, T., 84
Miller, D. F., *317*(53), 303(53)
Miller, L. H., *44*(9), 3(9)
Miller, M., *379*(51), 373(51)
Miller, R., *265*(120), 245(120)
Milnes, M. H., *48*(104), 32(104)
Moe, H. K., *264*(102), 237(102), 240(102)
Molina, M. J., *123*(56, 60), *318*(64, 66), 96(56), 98(60), 309(64, 66)
Moncrieff, A. A., *45*(36), 7(36)
Monin, A. S., *343*(1), 322(1), 323(1)
Monkman, J. L., *165*(7), *262*(55, 66), *263*(76), *265*(110), *266*(127), 136(7), 220(55), 225(66, 76) 226(66), 239(110), 242(110), 243(110), 247(127), 248(127)
Monafy, R., *49*(123), 43(123)
Moore, D. J., *344*(11), 336(11)
Moore, G. E., *262*(55, 66), 220(55), 225(66), 226(66)
Moore, J., *121*(28), 78(28)
Morgan, G. B., *377*(5, 14), 346(5), 355(14), 356(14)
Morgan, M. J., *263*(71, 74), *264*(94, 97), 225(71, 74), 232(97), 236(94, 97)
Mórik, J., *379*(40), 366(40)
Morlin, Z., *379*(40), 366(40)
Morozumi, M., *123*(68), 105(68)
Morozzi, G., *122*(33), *262*(62), 81(33), 82(33), 221(62), 236(62), 237(62)
Morris, E. D., *315*(11), 276(11)
Morris, J. C., *123*(58), 96(58)
Mosley, R. W., *377*(12), 355(12)
Moss, R., *166*(35), 144(35)
Mostecky, J., *263*(70), 225(70)
Motto, H., *377*(17), 356(17)
Moyes, J. L., *261*(46), 203(46), 204(46)
Mueller, P. K., *377*(11, 12), 351(11), 352(11), 353(11), 354(11), 355(12)
Mukai, M., *262*(51), 216(51), 217(51)
Mulik, J. D., *259*(8), 171(8), 228(8)
Murray, J. J., *261*(49), *262*(50), 215(49, 50), 222(50), 224(50)
Muschik, G. M., *265*(109), 238(109)

Naegeli, D., *121*(26), 75(26)
Nandi, M., *46*(58), 18(58)
Nass, C. A. G., *379*(46), 367(46)
Nauman, R. V., *165*(19), 138(19)
Nederbragt, G. W., *260*(23), 180(23)
Neel, J. V., *44*(6), 3(6)
Newhall, H. K., *122*(29), 78(29), 79(29)
Newhouse, M. L., *47*(78), 22(78)
Newhouse, M. T., *49*(114), 37(114)
Newson, R. L., *319*(76), 314(76)
Newton, D., *46*(47), 14(47), 15(47), 16(47)
Nicolet, M., *318*(60), 306(60)
Nielson, J. M., *49*(120), 39(120)
Niki, H., *120*(5), *315*(11), *316*(17), 56(5), 58(5), 276(11), 277(17)

Noe, J. L., *265*(122), 245(122)
Noll, K. E., *165*(2), 128(2), 129(2), 130(2)
Noonan, R. C., *317*(51), 299(51)
Nordberg, G. F., *44*(20), 6(20)
Novák, J., *166*(30), 135(30)
Novakov, T., *316*(36), *377*(11), 292(36), 351(11), 352(11), 353(11), 354(11)
Nusbaum, R. E., 44(7), 3(7)

Oeschger, H., *266*(134), 255(134)
Oyanguren, H., *44*(16), 6(16)
O'Keeffe, A. E., *259*(8), *260*(22), *261*(36), 171(8), 180(22), 188(36), 189(36), 228(8)
Okuda, S., *317*(42), 296(42)
Olszyna, K., *318*(55), 303(55)
Ortman, G. C., *261*(36), 188(36), 189(36)
Otvos, J. W., *377*(11), 351(11), 352(11), 353(11), 354(11)

Paciga, J. J., *377*(15), 355(15), 356(15)
Palmer, T. Y., *123*(54), 91(54), 94(54)
Paolacci, A., *122*(33), *262*(62), 81(33), 82(33), 221(62), 236(62), 237(62)
Parent, C., *49*(118), 38(118)
Parizek, J., *46*(61), 20(61)
Parker, W. W., *166*(38), *167*(39), 144(38, 39)
Pasquill, F., *344*(3), 322(3), 325(3)
Pate, J. B., *166*(20), 138(20)
Patrick, A. D., *45*(36), 7(36)
Patterson, C., *45*(29), *122*(41), *123*(68), 5(29), 84(41), 105(68)
Patterson, R. K., *122*(39), 84(39)
Payne, W. A., *317*(46), 297(46)
Pearson, E. S., 257
Pellizzari, E. D., *265*(113), 239(113)
Penkett, S. A., *317*(38), 292(38)
Penley, R. L., *167*(53), 158(53)
Pérez, E., *44*(16), 6(16)
Perkins, R. W., *49*(120), 39(120)
Perlstein, M. A., *45*(35), 7(35)
Perry, R., *123*(53), *166*(34), *378*(27), 91(53), 143(34), 144(34), 365(27)
Pfaff, J. D., *264*(99), 235(99), 237(99)
Phillips, C. R., *316*(18), 278(18), 281(18)
Phillips, P. E., *377*(16), 356(16)
Pich, J., *167*(44, 45), 147(44, 45)
Pierce, J. O., *377*(16), 356(16)
Pierce, L. B., *377*(12), 355(12)
Pierce, R. C., *264*(89), 96), *379*(42), 229(89), 231(89), 366(42)
Pierrard, J. M., *318*(57), 304(57)
Pitts, J. N., *260*(25), *316*(16, 25, 28, 31), 181(25), 277(16), 285(25, 28), 286(25, 31)
Pooler, F., *344*(14), *379*(50), 341(14), 373(50)
Pooley, F. D., *47*(74, 77), 22(74, 77)
Popescu, H. I., *45*(41), 7(41)
Popl, M., *263*(70), 225(70)
Porter, K., *165*(6), *259*(3), 130(6), 171(3)
Porterfield, J. D., 24
Possanzini, M., *264*(90), *379*(39), 236(90), 368(39)
Pottie, R. F., *261*(49), *262*(50), 215(49, 50), 222(50), 224(50)
Prager, M. J., *259*(9), 173(9)
Pratt, J. W., *266*(136), 259(136)
Pravica, M., *261*(35), 188(35)
Price, P. B., *167*(43), 146(43)
Prusazcyk, J., *317*(43), 297(43)
Pueschel, S. M., *45*(37), 7(37)
Pupp, C., *261*(49), *262*(50), 215(49, 50), 222(50), 224(50)
Purdue, L. J., *166*(36), 144(36)

Rabe, B. R., *317*(49), 298(49)
Rabinowitz, M. B., *46*(43), 13(43)
Raine, D. N., *45*(38), 7(38)
Ramsay, P. A., *167*(53), 158(53)
Ranweiler, L. E., *261*(46), 203(46), 204(46)
Ranz, W. E., *167*(49), 153(49)
Rao, T. N., *317*(42), 296(42)
Ratnayaka, D., *123*(53), 91(53)
Ravishankara, A. R., *318*(69), 311(69)
Ray, S., *263*(81), 229(81)
Ray, W. C., *316*(23), 283(23)
Reck, R. A., *318*(71), 313(71)
Reed, J. I., *44*(4), 2(4)
Reid, D. D., *47*(84), 23(84)
Reinhardt, G. W., *317*(48), 298(48)

Renard, K. G. St. C., *49*(126), 43(126)
Renwick, A. G. C., *45*(36), 7(36)
Rhodes, L., *378*(29), 365(29)
Rhodes, R. E., *121*(20), 72(20)
Ricardo, Sir Henry, 84
Riggan, W. B., *49*(112), 37(112)
Ripley, D. L., *165*(16), 137(16)
Roach, S. A., *165*(1), 126(1), 127(1)
Robbins, J. A., *318*(58), 304(58)
Robbins, R. C., *120*(3), *121*(11, 13, 14), 51(3), 52(3), 58(11), 59(13), 69(13, 14), 60(13), 61(13), 62(14), 84(3), 104(3)
Roberts, G. E., *45*(36), 7(36)
Roberts, P. T., *316*(34), 292(34)
Robinson, E., *120*(3), *121*(11, 13, 14), 51(3), 52(3), 58(11), 58(13), 59(13, 14), 60(13), 61(13), 62(14), 84(3), 104(3)
Robinson, J. W., 378(29), 365(29)
Robinson, M. S., *378*(24), 361(24), 365(24)
Roch, G., *264*(100), 237(100)
Ronco, R. J., *260*(16), 177(16)
Roschin, A. V., *44*(12), 5(12)
Rosen, A. A., *262*(60), 220(60), 224(60)
Rosie, D. M., *263*(85), 229(85)
Ross, W. D., *261*(33, 34), 187(33, 34)
Rossman, C. M., *49*(114), 37(114)
Rounbehler, D. P., *260*(18, 19), 179(18, 19)
Rowland, F. S., *123*(56, 60), *318*(64), (66), 96(56), 98(60), 309(64, 66)
Rubino, R. L., *379*(44), 367(44)
Rufeh, F., *260*(19), 179(19)
Russell, J. W., *166*(27), 135(27)
Ruyle, C. D., *259*(5), 171(5)
Ryan, T., *121*(15), 63(15)

Sabadell, A. J., *260*(16), 177(16)
Saffioti, U., *48*(98), 26(98)
Sakabe, H., *262*(65), 224(65)
Sakodynsky, K., *264*(98), 236(98)
Saltzman, B. E., *165*(14, 15), 137(14, 15)
Sander, S. P., *316*(33), 292(33), 297(33)
Sanderson, H. P., *266*(126), 247(126)
Santner, J., *259*(2), 169(2)
Sawicki, E., *260*(18), *263*(71, 74, 75, 86), *264*(87, 88, 91, 92, 93, 94, 95, 97, 99), *265*(118, 119, 120, 121, 122, 123, 124, 125), *378*(33, 34), *179*(18), 225(71, 74, 75), 228(86, 87, 88), 229(86), 230(87), 232(97), 233(86, 88), 234(88), 235(99), 236(88, 91, 92, 93, 94, 95, 97), 237(99), 241(118), 242(119), 244(118), 245(119, 120, 121, 122), 246(123, 124, 125), 368(33, 34)
Scala, R. A., *45*(22), 6(22)
Schaefer, H. J., *49*(122), 42(122)
Schoff, E. P., *46*(62), 20(62), 21(62)
Schrenk, H. H., *44*(2), 2(2)
Schroeder, H. A., *46*(55), 18(55)
Schuck, E. A., *317*(44), 297(44)
Schultze, H. F., *266*(130), 252(130)
Schwachman, H., *45*(37), 7(37)
Scott, A. N. B., *46*(47), 14(47), 15(47), 16(47)
Sehmel, G. A., *378*(25), 361(25)
Seiler, W., *120*(6), *315*(7), 57(6), 275(7)
Seinfeld, J. H., *316*(33), 292(33), 297(33)
Seitz, W. R., *259*(9), 173(9)
Selleck, B., *46*(54), 18(54)
Serbanescu, M., *45*(41), 7(41)
Serenius, F., *44*(20), 6(20)
Shah, K. R., *121*(21, 22), 71(21, 22)
Shair, F. H., *344*(7), 323(7)
Shapiro, J., *45*(21), 6(21)
Shapiro, S., *46*(58), 18(58)
Sheesley, D. C., *167*(48), 149(48), 151(48)
Shelef, M., *124*(74, 75, 76), 110(74, 75, 76)
Shepherd, M., *165*(5), 136(5)
Shibata, T., *44*(10), 3(10)
Sidebottom, H. W., *317*(41, 48, 49), 296(41), 298(48, 49)
Sie, B. K. T., *317*(54), 303(54)
Siegel, S. M., *120*(2), 51(2)
Sievers, R. E., *261*(33, 34,), 187(33, 34)
Siezel, B. Z., *120*(2), 51(2)
Silverman, F., *49*(118), 38(118)

Silvester, R., *260*(29), 186(29)
Simionaitis, R., *317*(52), 303(52)
Simmons, E. L., *319*(76), 314(76)
Simpson, M. J. C., *47*(80), 22(80)
Singh, B. H., *260*(12, 28), 174(12), 185(12, 28)
Skoberboe, R. K., *378*(30), 365(30)
Slater, D. H., *166*(34), *317*(42), *378*(27), 143(34), 144(34), 296(42), 365(27)
Slavin, W., *261*(43), 201(43)
Sloane, D., *46*(58), 18(58)
Smith, C. F., *377*(4), 346(4)
Smith, D. W., *379*(48), 371(48)
Smith, E. B., *316*(25, 28), 285(25, 28), 286(25)
Smith, F. B., 326
Smith, G., *318*(69), 311(69)
Smith, G. Z., *166*(38), 144(38)
Smith, J. E., 167(54), 161(54)
Smith, J. M., *48*(98), 26(98)
Smith, J. P., *317*(45), 297(45)
Smith, W. A., *47*(71), 22(71)
Snees, R. D., *46*(44), 14(44)
Snitz, F. L., *318*(58), 304(58)
Snyder, L. J., *166*(37), *378*(28), 144(37), 365(28)
Southam, A. H., *48*(88), 23(88)
Spence, J. A., *46*(51), 18(51)
Spencer, J., *318*(69), 311(69)
Speil, S., *47*(69), 21(69)
Spicer, C. W., *317*(53), 303(53)
Sporn, M. P., *48*(98), 26(98)
Spotswood, T. M., *122*(30, 31), 80(30, 31)
Sprung, J. L., *316*(31), 286(31)
Spurny, K., *167*(45, 46, 47, 48), 147(45), 148(46), 149(47, 48), 151(48)
Spurr, G., *344*(11), 336(11)
Stalling, D. L., *259*(5), 171(5)
Stanley, C. W., *48*(100), *260*(27), 26(100), 185(27)
Stanley, T. W., *263*(75, 86), *264*(88, 92, 93, 94, 95, 97), *265*(119, 120), *266*(124, 125), *378*(34), 225(75), 228(86, 88), 229(86), 232(97), 233(86, 88), 234(88), 236(88, 92, 93, 94, 95, 97), 242(119), 245(119, 120, 121), 246(124, 125), 368(34)
Stauff, J., *260*(21), 180(21)
Stedman, D. H., *260*(13), 176(13)
Steer, R. P., *260*(25), *316*(31), 181(25), 286(31)
Steffenson, D. M., *260*(13), 176(13)
Stephen, W. I., *261*(35), 188(35)
Stephens, E. R., *261*(38), *317*(44), 192(38), 297(44)
Sterling, T. D., *44*(9), *46*(52), 3(9), 18(52)
Stevens, R. K., *259*(8), *260*(22), 171(8), 180(22), 228(8)
Stewart, C. J., *261*(42), 198(42), 199(42), 200(42)
Stief, L. J., *317*(46), 297(46)
Stockburger, L., *317*(54), 303(54)
Stocks, P., *47*(81), *378*(31), 22(81), 23, 23(81), 368(31)
Stolarski, R. S., *318*(68), 309(68)
Story, P. R., *316*(23), 283(23)
Sullivan, R. J., *45*(24), 6(24)
Sumino, K., *44*(10), 3(10)
Sunderman, F. W., *45*(25), 6(25)
Sutton, O. G., *344*(2), 322(2), 338(2)
Sze, N. D., *318*(72), 313(72)

Tanaka, S., *44*(14), 6(14)
Taylor, K. C., *124*(78, 79), 110(78, 79)
Tebbens, B. D., *262*(51), 216(51), 217(51)
Ter Haar, G. L., *122*(46), 85(46)
Terraglio, F. P., *166*(21), 138(21)
Tepper, L. B., *378*(22), 365(22)
Testi, G., *318*(69), 311(69)
Thind, G. S., *46*(56), 18(56)
Thomas, F. W., *317*(39), 293(39)
Thomas, J. F., *262*(51), 216(51), 217(51)
Thomas, R. S., *262*(66), *265*(110), 224(66), 226(66), 239(110), 242(110), 243(110)
Thompson, C. R., *379*(47), 370(47)
Thompson, R. J., *166*(36), 144(36)
Thorne, D. M., *167*(41), 145(41)
Thorpe, J. J., *48*(86), 23(86)
Thrush, B. A., *260*(14, 15), 177(14, 15)
Timbrell, V., *47*(77), 22(77)
Timofeevskaya, L. A., *44*(12), 5(12)

Tindle, R. C., *259*(5), 171(5)
Torri, A. G., *265*(107), 236(107)
Tron, F., *165*(19), 138(19)
Trout, J. R., *379*(48), 371(48)
Tuck, A. F., *319*(76), 314(76)
Tuesday, C. S., *316*(19), 278(19)
Turco, R. P., *318*(65, 73), 309(65), 312(65), 313(65, 73)
Turner, D. B., 344(6), 327(6), 328(6), 333(6)
Turok, M. E., *47*(78), 22(78)
Tynan, E. C., *121*(24), 73(24)

Uden, P. C., *261*(35), 188(35)
Unland, M. L. *124*(80, 81), 110(80, 81)
Urone, P. F., *166*(22), *317*(45), 138(22), 297(45)

Van Bruggen, J. B., *49*(112), 37(112)
van Cauwenberghe, K., *262*(59), *265*(112), 220(59), 239(112)
Van der Horst, A., *260*(23), 180(23)
Van Duijn, J., *260*(23), 180(23)
Vartuli, J. C., *124*(77), 110(77)
Vašák, V., *166*(30), 135(30)
Vaughan, C. G., *263*(80), 228(80), 229(80)
Vaughan, G. V., *121*(24), 73(24)
Volman, D. H. *165*(6), 259(3), 315(2), 136(6), 171(3), 270(2)
von Lehmden, D. J., *123*(50, 51), 89(50, 51), 90(51), 92(51), 93(51)

Wade, N., *48*(101), 26(101)
Wade, R. J., *318*(63), 309(63)
Wagman, J., *122*(39), 84(39)
Wagner, J. C., *47*(78), 22(78)
Waldron, H. A., *45*(31), 5(31), 6
Wahlen, M., *266*(134), 255(134)
Walker, R. M., *167*(43), 146(43)
Waller, R. E., *49*(111), *379*(37), 36(111), 37(111), 368(37)
Warneck, P., *120*(6), *318*(62), 57(6), 308(62)
Warner, W. C., *262*(68), 225(68)
Warren, G. J., *260*(24), 180(24)
Wartburg, A. F., *165*(14), *166*(20), 137(14), 138(20)

Wayne, R. P., *260*(14), *316*(25, 29), 177(14), 285(25, 29), 286(25)
Weaving, J. W., *121*(27), 78(27)
Webb, K. S., *260*(20), 180(20)
Weber, T. A., *316*(32), 292(32)
Wei, Y. K., *317*(50), 299(50)
Weinstock, B., *120*(5), *316*(17), 56(5), 58(5), 277(17)
Wellings, R. A., *123*(53), 91(53)
Wells, A. C., *46*(47), 14(47), 15(47), 16(47)
Went, F. W., *123*(65), 103(65)
West, P. W., *165*(18, 19), 138(18, 19)
Westeberg, A. A., *315*(12), 276(12)
Wetherill, G. W., *46*(43), 13(43)
Wexler, H., *44*(2), 2(2)
Wheals, B. B., *263*(80), 228(80), 229(80)
Whitakker, E. J. W., *47*(63, 64, 65, 66, 67, 68), 21(63, 64, 65, 66, 67, 68)
Whitby, K. T., *377*(9, 10), 348(9), 349(10), 350(10)
White, E. W., *261*(41), 197(41)
Whitehouse, M. J., *263*(80), 228(80), 229(80)
Whitten, R. C., *318*(65, 73), 309(65), 312(65), 313(65, 73)
Williams, A. E., *260*(29), 186(29)
Williams, D., *165*(10), 136(10)
Wilmshurst, J. R., 265(108), 237(108)
Wilson, R. M., *264*(91), 236(91)
Wilson, S. R., *48*(88), 23(88)
Winell, M., *44*(11), 5(11), 12(11)
Winer, A. M., *316*(16, 24), 277(16), 284(24)
Witten, B., *165*(11), 136(11)
Wofsy, S. C., *121*(9), *123*(57, 59), *318*(61, 67, 72), 57(9), 96(57), 97(57, 59), 307(61), 309(67), 313(72)
Woldseth, R., *261*(40), 196(40), 197(40)
Wolcott, D. K., *378*(29), 365(29)
Wolff, R. K., *49*(114), 37(114)
Wong, J. B., *167*(49), 153(49)
Wong, P. T. S., *261*(30), 186(30)
Woodwell, G. W., *48*(106), *123*(61), 33(106), 98(61), 99(61)
Wosczyna, K., *266*(129), 250(129)

Wrenn, M. E., *49*(123), 43(123)
Wynder, E. L., *262*(58, 61), 220(58), 221(61), 222(58)

Yaglom, A. M., *343*(1), 322(1), 323(1)
Yen, T. F., *121*(20, 24, 25), 72(20, 25), 73(24, 25)
Yeung, C. K. K., *316*(18), 278(18), 279(18), 281(18)
Yobs, A. R., *48*(100), *260*(27), 26(100), 185(27)
Young, D. K., *47*(75), 22(75)
Yung, Y. L., *123*(57, 59), *318*(67), 96(57), 97(57, 59), 309(67)

Zahor, A., *46*(61), 20(61)
Zane, A., *264*(103), 237(103)
Zdrojewski, A., *165*(7), *266*(127), 136(7), 247(127), 248(127)
Zenz, C., *44*(13), 6(13)
Zhigalovskaya, T. N., *122*(42), 85(42)
Zielinski, W. L., *265*(105, 109), 235(105), 237(105), 238(105, 109), 239(105)
Zlatkis, A., *166*(31), 142(31)
Zoccolillo, L., *122*(33), *262*(62), *378*(32), 81(33), 82(33), 221(62), 236(62), 237(62), 368(32)

Subject Index

Aberdeen, 42
Absorption cross-section, 268
Aceanthrylene, 207
Acenaphthylene, 233
analysis, 244
Acetaldehyde, 8, 277, 287–291, 303
Acetic acid, 8
Acetonitrile, 8, 220, 229
Acetylene, 80, 81, 83, 179, 189, 278
analysis, 175, 176
Acridine, 208, 227
analysis, 240, 243
Acrolein, 8
analysis, 139
Aerodynamic diameter, 152
Air-fuel ratio, 75, 78, 87, 88
Air travel, 65, 66
supersonic, 40, 42, 314
Aitken particles, 103, 105
Akron, Ohio, 19
Albuquerque, 19
Aldehydes, 79, 274, 282, 283, 298, 299
analysis, 139, 184
Aldrin, 8, 28
analysis, 185
Alizarin, 136
Alkali-flame ionization detector, 171
Alkaline earth oxides, 69
Alkenes, 274, 277, 280, 282, 284, 287, 288, 298
Alkoxyl radicals RO, 281, 283, 287, 288, 290–292
Allen Town, Pa., 19
Allyl alcohol, 8
Aluminium, 3, 4, 64, 68, 69, 101, 197
analysis, 203, 204
in air, 102, 347, 355
Alumina, 69, 114, 224–227, 236
Alumina-cellulose acetate, 232–234, 236
Alveoli, 4, 5, 7
Aminco-Bowman spectrophotofluorometer, 231
Amino radical, NH_2, 307, 308, 310
p-Aminodimethylaniline, 137
Ammonia, 8, 58, 59, 61, 63, 109, 110, 113, 138, 169, 178, 188, 310
analysis, 137, 184
Ammonium, 51, 61, 69, 347, 350, 352–354
Ammonium sulphamate, 8
Amosite, 20
Amphiboles, 20, 22
Amsterdam, 376
Amyl alcohol, 8
Andersen cascade impactor, 152, 153, 155, 160, 164, 345, 366
Angiosarcoma, 180
Aniline, 8, 245, 303
p-Anisidine, 8
Ankara, 346, 347, 355
Ann Arbor, Mich., 3
Antarctic, 105
Anthanthrene, 90, 92, 93, 210
analysis, 226, 227, 231, 233, 243–245
in air, 366, 368
Anthophyllite, 20
Anthracene, 82, 92, 207, 217
analysis, 227, 228, 233, 235, 239, 240, 243
9-Anthraldehyde, 242, 244
α- and β-Anthramine, analysis, 245

Antimony, 3, 8, 68, 71, 72
analysis, 194, 198
in air, 101, 196, 355
Antimony-125, 251
Argentina, 5
Argon-37, -39, 255
Argon-41, 343
Arithmetic mean diameter, 160
Arsenic, 3, 4, 6, 8, 67, 68, 71, 72, 194
analysis, 198
in air, 101, 102, 355
Asbestos, 4, 20, 21
definition of fibre, 249
analysis, 250
Atmospheric turbidity, 55
Atomic absorption, 143, 144, 184, 186, 194, 201–204, 365
Atomic emission, 194, 200–202
Australia, 23, 346
Austria, 117
Averaging times, 128, 129
4-Azabenzenediazonium fluoborate, 246
1-Azafluoranthene, 369
4-Azapyrene, 369

Baltimore, 179
BBBT, *N,N'*-bis(*p*-butoxybenzylidene)-α, α'-bi-*p*-toluidine, 237, 238
Barium, 3, 4, 69, 83
analysis, 198
Barium-140, 43
Belgium, 2, 24, 117
Benzaldehyde, 286, 303
analysis, 184
Bentone-34, dimethyldioctadecylammonium montmorillonite, 229
Benz(j)aceanthrylene, 213
Benz(a)acridine, analysis, 227
in air, 369
Benz(c)acridine,
analysis, 227, 235
in air, 369
Benz(a)anthracene, 82, 90, 93, 208, 215, 217, 222, 224
analysis, 226–228, 233, 239, 240, 243
in air, 366, 368
Benzindene, analysis, 243
Benzene, 8, 20, 71, 80, 82, 188, 219, 220, 223–225, 236, 255, 277, 278
analysis, 183, 184
Benzo(a)carbazole, 213
Benzo(b)chrysene, 233
analysis, 243
Benzofluoranthenes, 82
Benzo(b)fluoranthene, 211, 222
analysis, 231
in air, 369
Benzo(f)fluoranthene,
analysis, 243
Benzo(j)fluoranthene, 211, 222
in air, 368
Benzo(k)fluoranthene, 211, 215, 222
analysis, 225, 226, 230–232, 235, 240, 243, 247, 248
in air, 366, 368
Benzo(ghi)fluoranthene, 212,
analysis, 243
Benzo(mno)fluoranthene,
analysis, 239
Benzofluorenes, 210, 211
analysis, 243
11-H-Benzo(a)fluorene, 233,
analysis, 240
11-H-Benzo(b)fluorene, 233,
analysis, 240
7-H-Benzo(c)fluorene, 233
Benzo(rst)pentaphene,
analysis, 231
Benzo(ghi)perylene, 82, 89–93, 212, 215, 222, 224, 233
analysis, 225–227, 231, 233, 238–240, 243–245
in air, 366
Benzo(c)phenanthrene, 209, 233
Benzo(lmn)phenanthridine,
analysis, 227
Benzo(a)pyrene, 25, 26, 80–82, 89–93, 209, 215–219, 224
analysis, 225–236, 238–241, 243–245, 247, 248
in air, 366, 368, 373
Benzo(e)pyrene, 82, 89, 90, 92, 93, 209, 215, 217, 222, 224, 233
analysis, 226, 227, 231, 235, 238–240, 243–245
in air, 368, 373

Benzo(f)quinoline, analysis, 227
in air, 369
Benzo(h)quinoline, analysis, 227
in air, 369
Benzoyl peroxide, 8
Benzyl chloride, 8
Beryllium, 8, 67, 68, 197
analysis, 187, 203, 204
Beryllium-7, 39
Bethlehem, Pa., 19
BHC, 29, 32
analysis, 185
Bimodal distribution, 161, 347, 348
Biological half-life time, 7, 13, 15, 43, 125–127
Biosphere, 53, 58, 61, 98
Birmingham, UK, 131, 160, 346, 357, 362–365
Bis-acetylacetoneethylenedi-imine, 188
Bis(chloromethyl)ether, analysis, 184, 186
p-Bis-2-(5-phenyloxazolyl)benzene, 255
Bismuth, analysis, 203, 204
Blackness index, 247, 248
BMBT, *N,N'*-bis(p-methoxybenzylidene)-α,α'-bi-*p*-toluidine, 237, 238
Bondapak, 229
Boron, 3, 68
Boron oxide, 8
Boron trifluoride, 8
Bradford, UK, 198
Bromide, 68, 69, 97, 100, 305
analysis, 198
in air, 101
9-Bromoanthracene, 233
Bromoform, 8
1-Bromopyrene, 233
Brownian diffusion, 103, 148
Budapest, 366
Buildings, 341
1, 3-Butadiene, 8, 80, 278
Butane, 71, 278
2-Butanone, 8
Butene, 80, 181, 278, 279, 287–290, 298
analysis, 181
Butyl acetate, 8
Butyl alcohol, 8
Butylamine, 8

Caesium, 69, 198
analysis, 204
in air, 101, 203
Caesium-137, 16, 42, 43
analysis, 251, 253, 254
Cadmium, 3, 4, 6, 8, 18–20, 67–69, 74, 100, 137
analysis, 194, 198, 203
in air, 19, 101, 356
Calcium, 3, 4, 68, 69, 173
analysis, 197, 203
in air, 101, 204, 347, 355
Calcium sulphate, 69, 119, 186
Calcium sulphite, 119
Calcium sulphide, 119
Cambridge, Mass., 196
Camphor, 8
Canada, 355, 356
Canton, Ohio, 19
Carbazole, 139, 208,
analysis, 236, 240
Carbon, 68, 69, 237, 350, 354
charcoal, 134, 135, 180, 255
analysis, 355
Carbon-14, 39, 41, 43, 56
analysis, 255
Carbon cycle, 52, 305
Carbon dioxide, 15, 50, 52–55, 58, 74, 75, 77, 86, 97, 109–112, 119, 169, 171, 175, 176, 178, 273, 275, 287–290, 303
analysis, 175, 176, 255
Carbon disulphide, 8, 135
Carbon monoxide, 4, 8, 15, 33–36, 55–58, 63–67, 74, 75, 77, 83, 86–89, 106–114, 130, 133, 170, 171, 178, 273–276, 281–284, 287–290, 300, 302, 303, 359, 370
analysis, 136, 171, 175, 176
in air, 35, 129, 130
Carbon tetrachloride, 8, 94–96, 220
analysis, 174, 185
Carboxyhaemoglobin, 33, 35
Carboxylic acids, 71, 79
Carcinogenicity, 186, 207–214
Cerium, in air, 101

Cerium -144, 43
analysis, 251
Cerium(IV)sulphate, 180
Charleston, W. Va., 19
Chattanooga, Tenn., 19
Chemiluminescence, 171, 176–181, 191
Chicago, 19, 346, 354, 355
Chile, 17
Chloracne, 32
Chlordane, 29
analysis, 185
Chlorin, 72
Chlorine/chloride, 9, 68, 69, 94, 97, 102, 126, 137, 138, 180, 188, 347
analysis, 136, 146, 187
in air, 101
Chlorine(atomic), 96, 309–311
Chlorine-36, -38, -39, 39
Chlorine dioxide, OClO, 9, 309–311
Chlorine mono-oxide, ClO, 309–311
Chlorine nitrate, 310
Chloroacetylchloride, 301
Chlorobenzene, 9
Chlorodiphenyl, 9
Chloroform, 96, 97, 219, 236
Chlorophyll, 52, 73
Chloroprene, 9
Chromium (chromates), 3, 4, 6, 9, 68, 69, 71, 102, 137
analysis, 194, 195, 197, 203, 204
in air, 20, 101, 196
Chrysene, 82, 207, 222, 233
analysis, 226, 227, 239, 240, 243, 245
in air, 366, 368, 373
Chrysotile, 20–22
analysis, 250
Cincinnati, 3, 19, 164, 346, 354, 355
Cloud cover, 326
Coal, 59, 63, 65, 68, 69, 88–90, 118
Cobalt, 3, 9, 20, 68, 69, 71, 74
analysis, 194, 197, 203, 204
in air, 101, 102, 196
Coefficient of restitution, 157
Columbus, Ohio, 19
Column chromatography, 224–227
Combustion, 1, 59, 63, 65, 69, 70, 74–81, 83, 84, 86–88, 94, 107, 118, 119, 129, 273, 274
Copper, 3, 4, 6, 59, 64, 68, 69, 71, 74, 120
analysis, 188, 194, 197, 203, 204
in air, 101, 102, 104, 355, 356
Corasil, 227–229
Coronene, 89–93, 213, 215
analysis, 225–227
in air, 366, 368, 373
Cosmic radiation, 41, 56
Counter efficiency, 253
Crocidolite (blue asbestos), 20
analysis, 249, 250
Crotonaldehyde, 9
Cumene, 9, 278
Curie, 40
Cyanides, 69
Cyclohexane, 9, 219–221, 223, 225, 226, 233
Cyclohexanone, 9
Cyclopentadiene, 9
4-H-Cyclopenta(def)phenanthrene, 233
Cyclopentane, 71
Czechoslovakia, 8–12, 17

2,4-D, 9, 29
analysis, 185
DDT, 9, 26, 27, 32, 98–100
analysis, 185
DDE, 28, 32
analysis, 185
Denmark, 24, 117
Denver, 3, 346, 355
Detroit, 368, 369
Derivative spectroscopy, 181–184
Dexsil-300,(meta-carborane units[B_{16} H_{10} C_2]connected to siloxane)
Dibenz(a,h)acridine, 214
analysis 227
in air, 369
Dibenz(a,j)acridine, 214
analysis, 227
in air, 369
Dibenz(c,h)acridine, 214
Dibenz(a,c)anthracene, 209, 233
analysis, 243
Dibenz(a,h)anthracene, 209, 217, 222, 233
analysis, 238–240, 243

Dibenz(a,j)anthracene, 208, 233
analysis, 243
Dibenzo(a,g)carbazole, 213
Dibenzo(a,i)carbazole, 214
7-H-Dibenzo(c,g)carbazole, 213
analysis, 240
Dibenzo(b,def)chrysene, 212
analysis, 231
Dibenzo(def,p)chrysene, 212
analysis, 231
Dibenzo(g,p)chrysene, 233
Dibenzo(a,h)fluorene, 210
Dibenzo(a,g)fluorene, 211
Dibenzo(a,c)fluorene, 211
Dibenzofuran, 32
analysis, 245
Dibenzo(brst)pentaphene, 233
Dibenzo(a,e)pyrene, 233
Dibenzo(a,h)pyrene, 210, 217
Dibenzo(a,i)pyrene, 210, 217
Dibenzo(a,1)pyrene, 209
Dibromotetrafluorethylene, 174
Dichloroacetaldehyde, 302
Dichloroacetylchloride, 300, 302
1,1-Dichloroethylene, 299–301
1,2-Dichloroethylene, 299, 300, 302
Dichloromethane (methylene chloride), 180, 220, 233, 234
1,2-Dichloropropane, 11
Dieldrin, 9, 29
analysis, 185
Diesel engine, 74, 83, 109, 113, 114
Diesel fuel, 70, 82
Diethylamine, 9
Diethylamino ethanol, 9
Diethyl ether, 9, 220, 221, 225, 226, 236
Diethylhydroxylamine, DEHA, 303, 304
Diethylnitrosamine, 304
Dihydroanthracene, analysis, 243
Dihydrobenz(a)anthracene, analysis, 243
Dihydrobenzofluorenes, analysis, 243
Dihydrochrysene, analysis, 243
Dihydrofluoranthene, analysis, 243
Dihydromethylbenz(a)anthracene, analysis, 243
Dihydrophenanthrene, analysis, 243
Dihydropyrene, 243
Dihydrotriphenylene, analysis, 243
Diisopropylamine, 9
Dimethylamine, 9
N,N'-Dimethylaniline, 9, 246
7,10-Dimethylbenz(c)acridine, analysis, 235
7,12-Dimethylbenz(a)anthracene, 208
analysis, 243
2,3-Dimethylbutadiene, 181
2,3-Dimethylbut-2-ene (tetramethylethylene, TME), 278, 285, 286
Dimethylformamide, 9, 236, 237, 246, 247
2,5-Dimethylfuran, 181
2,3-Dimethyl-2-hydroxybut-2-ene, 285, 286
Dimethylpentane, 80
N,N'-Dimethyl-*p*-phenylenediamine, 137
Dimethyl selenide, analysis, 184, 186
Dimethyl sulphide, analysis, 171, 181
Dimethylsuphoxide, 222, 223
Dinitrobenzene, 9
Dinitro-*o*-cresol, 9
Dinitrotoluene, 9
Dioxane, 9
Dithizone, 144, 174
DPEP (deoxophylloerythroetioporphyrin), 72
Driving cycles, 107, 108
Dry deposition, 61, 63, 102, 105
definition, 100
Dundee, 42

Eastern Germany, 8–12
Edinburgh, 42
Effective cut-off diameter, 156, 160, 161, 163
definition, 155
Effective height, 322, 333, 336
Egypt, 17
Eire, 24, 117
Electron affinity, 173
Electron capture detector, 142, 143, 173, 174, 185, 186, 239
Electrostatic precipitation, 118, 163–165

Endrin, 29,
 analysis, 185
England, 24, 32
Enrichment factor, 101,
 definition, 100
Enthalpy of sublimitation of PNHs, 215
Entropy of sublimitation of PNHs, 215
Epichlorhydrin, 9
Equilibrium vapour concentration (EVC), 215, 216, 222, 223
Ethanol, 9, 220, 234, 236
Ethene(ethylene), 58, 80, 83, 178–180, 255, 277–279, 299, 303
Ethyl acetate, 9
Ethylamine, 9
Ethylanthracene, 243
Ethylbenzene, 278
Ethylbromide, 9
Ethylchloride, 9
Ethylene chlorohydrin, 9
Ethylene diamine, 10
Ethylene dibromide, 84, 97, 142
Ethylene dichloride, 84, 97, 142
Ethylene imine, 10
Ethylene oxide, 10
Ethyl mercaptan, 9
Ethylphenanthrene, analysis, 243
o-, *m*-, *p*-Ethyltoluene, 278
Etioporphyrin, 72
Eugene, Ore., 19
Europe (European), 24, 107, 108, 117
Europium, in air, 101
Exponential dilution, 190–192

Fick's law, 322
Filter collection efficiency, 147–150, 155
Finland, 17, 117
Flame ionization detector, 142, 169, 170, 239
Flame photometric detector, 171, 172, 180, 185
Fluoranthene, 82, 90, 92, 93, 207, 219, 222
 analysis, 226–228, 239, 240, 243, 245
 in air, 268
Fluorene, 207,
 analysis, 239, 240, 243, 245, 246
9-Fluorenone, 240
Fluorescence, 221, 225, 231, 233, 235
Fluoride, 10, 67, 68
 analysis, 136
Florisil, 185
Forest fires, 51, 94
Formaldehyde, 10, 57, 276, 281, 283, 287, 288, 300, 303
 analysis, 184
Formic acid, 288, 300
Formyl chloride, 300
Formyl radical (CHO), 57, 276, 282, 283
France, 24, 117
Franck–Condon diagram for oxygen, 268
Freons (CCl_3F and CCl_2F_2), 299, 305, 308, 309, 311–313
 analysis, 173, 174, 184, 185
 atmospheric life-times, 98, 312
Fuchsin-formaldehyde reagent, 13
Furfural, 9, 242, 244

Gallium, 68, 69
 analysis, 198
Gary, Ind., 19
Gas chromatography/gas-liquid chromatography, 183–187, 235, 237–240, 242
Gas-turbine engine, 86, 88
Geometric standard deviation, σ_g, 159, 161, 345–347
 definition, 160
Germanium, 68
 analysis, 198
Gibbs–Duhem equation, 140, 141
Glyceryltrioctanoate, 219
Gold, 69
 analysis, 198
 in air, 101
Greece, 117
"Greenhouse effect", 55
Greenland, 105
Griess-Ilosvay analysis of NO_2, 137

Haemoglobin, 7

Hafnium, 69
Half-life, radioactive, 16, 43, 194, 253
biological, 43, 125–127
Halocarbons, 4, 32, 173, 299
Hamburg, 368
Helium, 206
Heptachlor, 10, 30
analysis, 185
Heptachlor epoxide, 30
n-Heptane, 229
Hexachlorophene, 26, 31
n-Hexane, 220, 229, 236, 278
2,2′, 4,4′, 6,6′-Hexanitrodiphenylamine, 180
Hex-3-ene, 279
4-Hexylresorcinol, 139
High pressure liquid chromatography, 227–229
"Hi-Vol" air sampler, 138, 153, 158
Holland, 17, 24, 117
Hydrazine, 10
Hydrocarbons, 52, 63–67, 70, 71, 74, 75, 78, 79, 81–84, 86–88, 94, 103, 106, 107, 109, 110, 114, 133, 170, 273, 274, 291, 292
reactivities, 277, 278
Hydrogen (atomic), 69, 75–77, 83, 95, 271, 273, 275, 276, 283, 295, 303, 306, 307, 310, 311
Hydrogen, 36, 76, 83, 109, 110, 113, 120, 136, 169, 173, 206, 271, 276, 283, 306, 307, 310, 311
Hydrogen bromide, 98
Hydrogen chloride/hydrochloric acid, 10, 51, 133, 138, 204, 236, 237, 300, 302, 310, 311
analysis, 136
Hydrogen cyanide, 10
analysis, 136
Hydrogen fluoride/hydrofluoric acid, 10, 51, 133, 204, 309
Hydrogen peroxide, 168, 180, 276, 277, 306, 307
analysis, 137
Hydroperoxyl radical (HO_2), 77, 271–278, 281–283, 287, 288, 291, 295, 297, 298, 306
Hydroxyl radical (OH), 57, 58, 75–77, 83, 95, 97, 271, 273–278, 280–283, 288, 289, 291, 292, 295, 297–299, 302, 303, 306–308, 310, 311
Hydrogen sulphide, 10, 51, 58, 59, 61, 69, 126, 133, 137, 144, 169, 189
analysis, 137, 171, 181
Hypochlorous acid, 97

Imino radical, NH, 308
Incinerators, 66, 91, 92
Indeno(1,2,3-ij)isoquinoline, analysis, 227
Indeno(1,2,3-cd)pyrene, 82, 210
in air, 369
11-H-Indeno(1,2-b)quinoline, analysis, 227
Indianapolis, 19
Indole, analysis, 245
Indole-3-aldehyde, 244
Indoor air pollution, 35, 367, 370–373
Indium, 69
analysis, 198
in air, 101
Internal combustion engine, 65, 74, 80, 84, 86, 89
Iodine, 10, 69, 97, 144
analysis, 198
in air, 101
Iodine-131, 4, 42, 43, 253, 341–343
analysis, 255
Iodine monochloride, 144, 365
Iridium, analysis, 198
Iron, 3, 4, 26, 43, 64, 68, 69, 71, 102, 136, 138, 194, 293–296
analysis, 197, 203, 204
in air, 101, 196, 347, 355, 356
Iron(111)oxide, 26, 69, 84, 126
Iron pyrites, 69
Isatin tests, 245
Isobutane, 80
Isobutene, 80
Isocyanate, 110–112
Isokinetic sampling, 132
Iso-octane, 229
Isopentane, 278
Isopleths, 331
Isopropylamine, 10
Isreal, 17
Italy, 17, 24, 32, (117

Japan (Japanese), 3, 17, 18, 26, 32, 108, 196, 346, 347

Knudsen number, definition, 149
Krypton-81, 255
Krypton-85, 43, 255, 343

Lambert–Beer law, 267
Lanthanum, in air, 101
Las Vegas, 19
LD_{50} definition, 32
Lead, 3–7, 10, 13–15, 17, 18, 64, 67, 68, 71, 74, 84, 85, 97, 100, 102, 104, 105, 109, 114–117, 126, 340, 365
 analysis, 198, 204
 in air, 15, 20, 85, 101, 131, 203, 355–365, 371, 372
Lead-203, 14–16
Lead alkyls, 14, 84, 85, 115
 analysis, 135, 142–144
 in air, 365
Lead bromochloride, 84, 97, 98, 102, 304, 305
Leeds, UK, 368
Legislation, 18, 106, 107, 113, 114, 117, 118
d-Limonene, 278
Lindane, 10
Lithium, 196
 analysis, 203, 204, 206
Log-normal distribution law, 158, 161, 348
London, 2, 36, 346, 355, 368
Los Angeles, 2, 19, 129, 130, 273, 290, 291, 350, 368, 369
Lung cancer, 22–25, 43, 254

Magnesium, 68, 69, 73, 74, 136, 173
 analysis, 197, 203, 204
 in air, 101, 347, 355
Maleic anhydride, 10
Manchester, UK, 365
Manganese, 3, 4, 6, 10, 67–69, 71, 74, 136, 293–295
 analysis, 194, 197, 203, 204
 in air, 101, 355, 356
Manganese-54, 251
Mass spectrometry, 186, 194, 205, 206, 239–241, 243
MBTH, 3-methyl-2-benzothiazolone hydrazone, 139
Malathion, analysis, 185
Melbourne, 346, 347
Medford, Ore., 19
Mercury, 3, 4, 6, 10, 67–69, 71, 72, 74, 126
 analysis, 184, 194, 198
 in air, 51, 101
Mercury alkyl, 10
Mesothelioma, 22
Methane, 36, 57, 58, 70, 80, 94, 110, 179, 189, 255, 277, 278, 280, 281, 283, 306, 310, 311
 analysis, 136, 171, 175, 176
Methanethiol, analysis, 181
Methyl acetate, 10
Methyl acrylate, 10
Methyl alcohol (methanol), 10, 220, 228, 229, 237
Methylamine, 10
9-Methylanthracene, analysis, 245
Methylbenz(a)anthracene, analysis, 243
Methylbenzo(j)fluoranthene, analysis, 243
Methylbenzo(k)fluoranthene, analysis, 243
Methylbenzo(a)pyrene, analysis, 243
Methylbenzo(e)pyrene, analysis, 243
Methyl bromide, 10, 97
2-Methylbut-2-ene, 278–280
Methyl chloride, 10, 91, 94
Methyl chloroform, 10
3-Methylcholanthrene, 213
 analysis, 244, 245
Methylchrysene, analysis, 243
Methylcyclohexane, 10
Methylene chloride (dichloromethane), 10, 220, 234
Methylethylketone, 278
Methylfluoranthenes, analysis, 243
Methylfluorenes, analysis, 243
Methyl iodide, analysis, 174, 185
Methylisobutylketone, 144, 278
Methylisocyanate, 10
1-Methylnaphthalene, analysis, 237
2-Methylnaphthalene, analysis, 237
Methyl nitrate, 287–290
2-Methylpentane, 278

3-Methylpentane, 278
2-Methyl-2,4-pentandiol, 185
2-Methylpent-2-ene, 279, 280
Methylphenanthrene, analysis, 243
1-Methylpyrene, 233
 analysis, 243
4-Methylpyrene, analysis, 226, 243
Methyl radical, CH_3, 57, 281–283, 306, 310, 311
Methylstyrene, 10
Methyl suphide, analysis, 171
Methyltriphenylene, analysis, 243
Miami, 3
Mixing ratio, 275
MMD, mass median equivalent diameter, 159, 160, 345–348, 355–357, 366
Molar extinction coefficient, 267
Molecular sieve, 186
Molozonide, 284
Molybdenum, 3, 4, 10, 68, 69, 71, 136
 analysis, 198
 in air, 101
Monchloracetaldehyde, 301
Monochloroacetylchloride, 300, 302
Morpholine, 11
MPC, maximum permissible concentration of radionuclides, 43, 343

Nagoya, 346, 347
Naphtha, 11
Naphthacene, 207, 233
 analysis, 227, 239, 240
Naphthalene, 11
 analysis, 228, 245
Naphtho(1,2,3,4-def)chrysene, analysis, 231
Naphtho(2,1,8-qra)naphthacene, 212
 analysis, 231, 233
α- and β-Naphthylamine, 245
N-(1-Naphthyl)-ethylenediaminedihydrochloride, 137
Nematic liquid crystal, 235, 237–239
Neobium, analysis, 198
Nessler's solution, 137
Neutron activation analysis, 194–196
Newark, N. J., 19
New Guinea, 17
New York, 2, 3, 17, 19, 37, 117
New Zealand, 23
Nickel, 3, 4, 6, 20, 67–69, 71, 73, 74, 79
 analysis, 188, 194, 197, 200, 203, 204
 in air, 101, 355, 356
Nickel-63, 173
Nickel carbonyl, 6, 11, 113
Nitrate (nitrate aerosol), 51, 52, 61, 104, 347
 analysis, 138, 146, 352–354
Nitric acid, 180, 204, 275, 292, 300, 303
p-Nitroaniline, 11
Nitrobenzene, 11, 229, 230, 236
p-Nitrochlorobenzene, 11
Nitroethane, 11
Nitrogen, 1, 54, 69, 71, 73, 77, 107, 109, 112, 173, 274, 308
Nitrogen-14, 56, 171
Nitrogen cycle, 60, 102, 305
Nitrogen as amino- 352–354, as pyridino-, 352–354
Nitrogen $N(^4S)$, $N(^2D)$, 308
Nitrogen oxides, $NO_x/NO/NO_2$, 4, 11, 34, 37, 38, 51, 58–61, 63–67, 74–79, 86–88, 103, 106–114, 118, 119, 126, 129, 130, 133, 136–138, 180, 188, 191, 271–275, 277, 278, 280–284, 287–292, 296–299, 302, 303, 305, 307, 308, 310, 311, 313–315, 329, 330, 370
 analysis, 137, 177, 178, 183, 184
 in air, 34, 36, 130
 NO_3, 61, 274, 275, 292, 297
 N_2O_4, 297
 N_2O_5, 275, 297
 N_2O (nitrous oxide), 58–61, 109, 111–113, 303, 307, 308
  analysis, 186
3-Nitro-4-dimethylaminobenzaldehyde, 242, 244
Nitromethane, 11, 220, 221, 229, 233
p-Nitrophenylisocyanate, 229
Nitropropane, 11, 236
N-Nitrosamines, 178–180
Nitrosyl radical, (NO), 179
Nitrous acid, 178, 275, 276, 283, 303

Northern Ireland, 24
Norway, 5, 12, 117
Nuclear weapon testing, 39, 314, 315

Octahydrofluoranthene, analysis, 243
Octahydropyrene, analysis, 243
Octane number, 84, 115
Ohio, 17, 19
Oil, 65, 66, 70, 71, 89, 90, 118–120
Oklahoma City, 19
Omaha, 19
Osaki, 196
OV-1, (dimethylsilicone gum), 185, 186
OV-7, (phenylmethyldimethylsilicone 20% phenyl), 238
Oxidants, 129, 370
 analysis, 137
 in air, 130
Oxygen, 1, 54, 69, 71, 73, 76, 77, 109, 169, 272–275, 281–283, 286, 299, 303, 306, 307, 309–311
 $O_2(^1\Delta_g)$, 219, 268–270, 275, 285, 286
 $O_2(^1\Sigma_g^+)$, 268–270, 276, 285, 286
 $O_2(^3\Sigma_g^-)$, 268–270, 285, 286
 $O_2(^3\Delta_u)$, 268
 $O_2(^3\Sigma_u^+)$, 268
 $O_2(^1\Sigma_u^-)$, 268, 270
 $O_2(^3\Sigma_u^-)$, 268, 270
Oxygen (atomic), 76, 77, 271–274, 276, 281, 282, 287, 301, 302, 306, 308–310, 315
 $O(^1D)$, 60, 95, 268, 270, 275, 276, 283, 285, 306, 307, 310, 311
 $O(^3P)$, 95, 268, 270–272, 275, 278, 285, 296, 299, 303, 311
Ozone, 11, 34, 38, 39, 129, 133, 138, 176–181, 270–278, 282–285, 287, 290–292, 297–302, 305–315
 analysis, 178, 184
 in air, 34, 36, 274

Palladium, 110
 analysis, 188, 198
PAN, peroxyacylnitrates, 184, 274, 287–291, 347, 370
 analysis, 186, 192
 chlorinated PAN, 300–303
Paper chromatography, 235–237
Parathion analysis, 185
Particle size distribution, 153, 155, 156, 158–162, 164, 345–356
Particulates, 51, 52, 63–67, 74, 79, 85, 88, 99, 102, 103, 117–120, 133, 145, 152–165, 187, 203, 204, 220, 223, 247, 251, 254, 345–367, 370
Partisil, 229
Pasadena, 3, 348–355
Pasquill, stability categories 325,
 diffusion parameters, 327
PCBs, polychlorobiphenyls, 30, 32
Pentachlorophenol, 11
n-Pentane, 222, 223, 225, 236
2-Pentanone, 11
Pent-2-ene, 278, 279
Perchloroethylene, 11
Permeation tube, 188–190, 192
Peroxyalkyl radicals, ROO, 274, 281–283, 287, 288, 291, 292, 299, 302
Peroxybenzylnitrate, 303
Peru, 5, 12, 17
Perylene, 92, 93, 208, 217, 218, 222
 analysis, 227, 231, 233, 235, 238–240, 244
 in air, 366, 368
Pesticides, 27–32, 98–100, 173, 184
 analysis, 185
Phenanthrene, 82, 207, 233
 analysis, 239, 240, 243
Phenanthridine analysis, 227
Phenol, 11, 303,
 analysis, 245
o-Phenylenepyrene, indeno(1,2,3,-cd)pyrene, 210
 analysis, 243
Pheonix, Ariz., 19
Philadelphia, 19, 346, 354, 355
Phorbin, 72
Phosgene, 11, 300–302,
 analysis, 185
Phosphine, 11
Phosphorus, 11, 68
 analysis, 171, 173, 185, 197
Phosphorus-32, -33, -35, 39
Phosphorus oxychloride, 242
Phosphorus pentachloride, 242, 245
Photochemical laws, 267
Photoelectron spectroscopy, 351–354
Photosynthesis, 52–54
Phthalic anhydride, 11

Picene, 233
β-Pinene, 278
Piperonal, 245
Platinum, 109, 110, 171, 255
 analysis, 188, 198
Plutonium, 4, 43, 254
 analysis, 255
Poland, 17
Polyethyleneglycoladipate, 186
Polynuclear hydrocarbons (PNHs), 22, 74, 80–83, 89–93, 113, 115, 156, 206–248
 analysis, 219–248
 in air, 366, 368, 369, 372, 373
Polyvinyl chloride (PVC), 94
Porapak (cross-linked polystyrene beads) Q, 186, T, 229
Porasil (porosity controlled silica base), 229
Porphyrin, 72–74
Portland, 3
Portugal, 117
Potassium, 68, 69, 74
 analysis, 197, 203, 204
 in air, 101
Potassium-40, 16, 39, 41
Potassium bis(5-sulphoxino)palladium(11), 136
Potassium borohydride, 246
Potassium permanganate, 180
Pregl-Dumas, 355
Propane, 277, 278
Propargyl alcohol, 11
Propene, 80, 189, 278, 282
n-Propyl acetate, 11
Propyl alcohol (propanol), 11, 236
n-Propylbenzene, 278
Propylene oxide, 11
Pulmonary flow resistance, 37
Pyrene, 82, 89, 90, 92, 93, 208, 215, 217, 219, 222, 224, 283
 analysis, 226, 227, 239–241, 243–247
 in air, 366, 368, 373
Pyridine, 11, 69, 71, 220, 236
Pyrosynthesis, 80–82

Quinone, 11

Racine, Wis., 19
Radiation, α-, 40, 43, 251
 β-, 40, 43, 251, 253, 254
 γ-, 40, 43, 194, 251
Radioactivity, 39, 251–253
Radionuclides, 39, 42
 half-life, 43, 194, 252
Radon, 43, 252, 255
Rain-fall, -out, -precipitation, 61, 63, 98, 102, 103, 251
 efficiency, 103, 104
Rare earths, 69, 355
Relative humidity, 85, 293, 348, 349
Reynolds number, 145
Roentgen, 40
 Roentgen-equivalent-man(rem), 40, 41
Rome, 368, 369
Rotterdam, 367
Rubidium, 69, 171, 198
 analysis, 204
 in air, 101, 203
Ruthenium-103, -106, 251

Sakai, 196
Samarium, in air, 101
Scandium, 69, 100
 analysis, 197
 in air, 101
Scanning electron microscope (SEM), 198, 199
Scintillation detector, 251, 255
Scotland, 24, 42
Scranton, Pa., 19
SE 52 (methylphenyl silicone), 187, 237
SE 30 (methyl silicone), 237
Sea spray, 51, 59, 103
Selenium, 11, 68, 69, 71
 analysis, 184, 198
 in air, 101
Seveso, 32
Shetland Isles, 100, 101
Silica/silica gel, 69, 158, 220, 224, 225, 236, 255
Silicon, 68
 analysis, 203, 204
 in air, 347
Silicon-32, 39
Silicon semiconductor, 196, 197

Silver, 3, 69, 74
analysis, 198
in air, 101
Silver nitrate, 142
Sodium, 68, 69, 74
analysis, 197, 203, 204
in air, 101, 102, 347, 355
Sodium-22, 39
Sodium aluminate, 114
Sodium hydroxide, 11
Sodium nitrate, 180
Sodium sulphite, 119
Sodium tetrachloromercurate reagent, 138
Soil, 51, 60, 98, 100, 102
Solar radiation, 326, 348
Solvent strength, 219, 220
Smelting, 59, 64
Smog, 2, 79, 129, 273, 287, 302, 304, 348, 355
Smoke, 36, 82, 87, 88, 129
Soot, 74, 79, 81, 83, 84, 87, 88, 216, 217, 247, 292
South Africa, 23
Spain, 117
Standard deviation (SD), 127, 253, 256, 259
Standardized mortality ratio (SMR), 19
Stationary combustion emissions, 64–67, 87, 118
St. Louis, 346, 354–356
Stoddard solvent, 11
Stokes' law, 105, 145
Stratified charge, 78
Strontium, 3
analysis, 198, 203, 204
Strontium-89, 43, strontium-90, 4, 42, 43
analysis, 253, 254
Styrene, 11
analysis, 184
Sulphate (aerosol), 50–52, 61, 63, 104, 146, 158, 297, 347, 350–352
Sulphinic acids, 298
Sulphur, 58, 59, 70, 71, 73, 119, 173
analysis, 171, 197, 200
Sulphur-35, -38, 39
Sulphur cycle, 61, 62, 102, 103
Sulphur hexafluoride, 174
Sulphur oxides, SO_x, 64–67, 74, 78, 79, 88, 119
SO_2, 4, 11, 34–39, 50, 51, 61–63, 117–120, 126, 129, 130, 133, 137, 178, 188, 189, 273, 292–299, 315, 367, 370, 373–376
analysis, 138, 168, 171, 172, 175, 176, 180, 184
in air, 34–36
$^1SO_2$, 296, 299
$^3SO_2$, 296, 298, 299
SO, 296, 299, 315
SO_3, 37, 88, 120, 133, 296–299
SO_3^-, 295, SO_4^-, 295, SO_5^-, 295
HSO_3, 297, 299
Sulphuric acid, 11, 293, 298, 299
analysis, 138
Sweden, 9–12, 17, 32, 106, 117
Switzerland, 117

2,4,5-T, 32
TCDD, 31–33
TDE, 27
analysis, 185
Tellurium, 11
analysis, 198
Temperature inversion, 131, 321, 347, 348
Terpenes, 51, 206
1,1′,2,2′-Tetrachloroethane, 11
Tetrachloroethylene, 299, 300, 302
analysis, 185
Tetraethylammonium hydroxide, 246
Tetraethyl lead, 11; *see also* Lead alkyls
Tetrahydrofuran, 11
Tetrahydronaphthalene, 71
Tetramethylethylene, TME, 181, 285, 286
Tetranitromethane, 11
Thallium, 11
analysis, 198
2-Thenaldehyde, 242, 244
Thin-layer chromatography, 230–236
Thiram, 12
Thorium, 39, 69
in air, 101, 102
Thoron, 252
Tin, 68
analysis, 198

Titanium, 68, 69
analysis, 197, 203, 204
in air, 101, 355
Threshold limit value (TLV),
definition, 5,
metals, 6, 8–12
gases and vapours, 8–12, 34
Tokyo, 32
Toluene, 12, 71, 80, 82, 220, 234, 236, 278
analysis, 183, 184
Toluene-2,4-diisocyanate, 12
o-Toluidine, 12, 136
Toronto, 355
Trachea, 5
Trichloroacetylchloride, 300, 302
1,1,1-Trichloroethane, analysis, 185
Trichloroethylene, 12, 299, 300, 302
1,2,3-Trichloropropane, 12
Triethylamine, 12
Trifluoroacetic acid, 229, 242
Trifluoroacetylacetone, 187
Trimethylbenzenes, 278
Trimethylethylene, 181
Trimethylpyrene, analysis, 243
2,3,4-Trimethylpentene, 80
Trimethylpyrene, analysis, 243
2,4,7-Trinitrofluorenone, 229
Trinitrotoluene, 12
Triorthocresylphosphate, 12
Triphenylene, 233
analysis, 239, 240, 243
Triplet sensitizers, 286
1,2,3-Tris(2-cyanoethoxy)propane (TCEP), 143
Tritium, 39, 43
analysis, 255
Tucson, Ariz., 19
Tungsten, 69
analysis, 198
in air, 101
Tungsten oxide, 179
Turpentine, 12
"t" values, 257

UK, 2, 5, 7, 22, 23, 84, 85, 100, 101, 117, 118, 131, 160, 196, 346, 365, 368, 369
blood lead concentration of population, 17, 18, 372
TLVs of chemicals, 8–12
Uranium, 12, 39, 43, 253
Uranium-235, 146
Urban diffusion parameters, 341, 375
USA, 2, 3, 5, 7, 22, 23, 63–67, 84, 85, 98, 100, 106, 107, 113, 117, 118, 196, 304, 346, 355
population blood concentration of metals, 3, 17, 19
TLVs of chemicals, 8–12
USSR, 5, 85
TVLs of chemicals, 8–12

Vanadium, 3, 6, 12, 20, 67–69, 71, 73, 74, 79, 120
analysis, 194, 197, 199, 200, 203, 204
in air, 101, 102, 355, 356
Volcanoes, 50, 51
Vinyl chloride (monomer), 12, 299–302
analysis, 135, 180
Vinyl toluene, 12

Wales, 24
Washington, DC, 346, 355
Washout factor, 102
definition, 100
Western Germany, 24, 106, 117, 118
TLVs of chemicals, 8–12
West-Gaeke, 138, 169, 180
Wheeling, W. Va., 232
Wilmington, Dl., 19

XE 60 (cyanomethylethyl silicone), 237
X-radiation, 40, 42, 351
X-ray fluorescence, 194, 196–200
Xylenes, 12, 80, 278
analysis, 184
2,4-Xylenol, 138
Xylidine, 12

Yugoslavia, 17
Yttrium-90, 254

Zeldovich reaction, 77, 274
Zinc, 3, 6, 12, 20, 59, 64, 68, 69, 71, 74
 analysis, 194, 198, 203, 204
 in air, 101, 102, 196, 355, 356
Zipax, 227, 229
Zirconium, 12, 68, 198
Zirconium-95, 43, 251
Zorbax, 229
Zwitterion, 282, 284, 298